Nineteenth-C
of Identities

Also by Laurel Brake

THE ENDINGS OF EPOCHS (*editor*)

INVESTIGATING VICTORIAN JOURNALISM (*editor with Aled Jones and Lionel Madden*)

PATER IN THE 1990s (*editor with Ian Small*)

SUBJUGATED KNOWLEDGES

WALTER PATER

Also by Bill Bell

THE COLLECTED LETTERS OF THOMAS AND JANE WELSH CARLYLE: Volumes 22–24 (*editor with K. J. Fielding et al.*)

ACROSS BOUNDARIES: The Book in Culture and Commerce (*editor with P. Bennett and J. Bevan*)

Also by David Finkelstein

AN INDEX TO BLACKWOOD'S MAGAZINE, 1901–1980

PHILIP MEADOWS TAYLOR (1818–1876): A Bibliography

Nineteenth-Century Media and the Construction of Identities

Edited by

Laurel Brake
Birkbeck College
London

Bill Bell
University of Edinburgh

and

David Finkelstein
Queen Margaret University College
Edinburgh

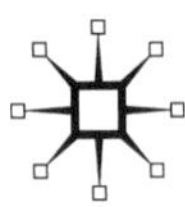

First published 2000 by
PALGRAVE
Houndmills, Basingstoke, Hampshire RG21 6XS and
175 Fifth Avenue, New York, N. Y. 10010
Companies and representatives throughout the world

PALGRAVE is the new global academic imprint of St. Martin's Press LLC Scholarly and Reference Division and Palgrave Publishers Ltd (formerly Macmillan Press Ltd).

Outside North America
ISBN 0–333–68151–7 hardback
ISBN 0–333–68152–5 paperback

In North America
ISBN 0–312–23215–2 hardback

This book is printed on paper suitable for recycling and made from fully managed and sustained forest sources.

A catalogue record for this book is available from the British Library.

Library of Congress Cataloging-in-Publication Data
Nineteenth-century media and the construction of identities / edited by Laurel Brake, Bill Bell, and David Finkelstein.
p. cm.
Includes bibliographical references and index.
ISBN 0–312–23215–2 (cloth)
1. English periodicals—History—19th century. 2. Great Britain—Social life and customs—19th century. I. Brake, Laurel, 1941– II. Bell, Bill. III. Finkelstein, David.
PN5124.P4 N56 2000
052'.09'034—dc21

00–021404

10 9 8 7 6 5 4 3 2 1
09 08 07 06 05 04 03 02 01 00

Printed in Great Britain by Antony Rowe Ltd, Chippenham, Wiltshire

Contents

Part V National and Ethnic Identity

List of Illustrations

Figures

Tables

Acknowledgements

Chapter 1 by Kate Jackson, 'George Newnes and the "loyal Tit-Bitites"', and Chapter 2 by Richard Salmon, '"A Simulacrum of Power"' are articles that first appeared in *Victorian Periodicals Review*; they are printed here with the permission of the editor. Chapter 10 is reprinted from *Nineteenth-Century Prose*, Spring 1997, with the permission of the University Press of Colorado. Chapter 15 by Mark Turner, '*Saint Pauls Magazine* and the Project of Masculinity' has appeared in part in Chapter 5 of his book, *Trollope and the Magazines*, Macmillan, 2000. Chapter 18 by Leslie Williams, 'Bad Press', is reprinted from *Representing Ireland*, UP of Florida, 1997, with the permission of the University Press of Florida. Figures 1–4 at shelfmark PP 6930, Figures 14–17 at shelfmark PP 236, and Figure 20–22 at shelf mark PP 1931.pcl are printed by the kind permission of the British Library. Figures 8 and 9 are reproduced by permission of the Department of Special Collections and Rare Books, University of Minnesota. The editors acknowledge with thanks the help of the University of Edinburgh Interdisciplinary Symposium Fund, and that of the Faculty of Continuing Education, Birkbeck College, in the preparation of this manuscript for the press.

Notes on the Contributors

Amy Beth Aronson received her PhD in American literature and culture from Columbia University in 1996. Involved in the magazine industry as a writer and editor, she has held masthead positions with the business magazine *Working Woman* and the feminist magazine *MS*. Her most recent project is a forthcoming edition of Charlotte Perkins Gilman's *Women and Economics*, and she is currently at work on a cultural history of early American women's magazines.

Margaret Beetham is Reader in the Department of English at Manchester Metropolitan University. She is the author of *A Magazine of her Own? Domesticity and Desire in the Woman's Magazine 1800–1914*, and co-author of *Women's Worlds: Ideology, Femininity and the Woman's Magazine*. She has also published on popular culture, the periodical press and feminist theory and pedagogy. She is currently researching and writing on the domestic and on feminist theory at the end of the second millennium.

Bill Bell is Senior Lecturer and Director of Studies at the University of Edinburgh. He is general editor, with Jonquil Bevan, of *A History of the Book in Scotland*, to be published in four volumes by Edinburgh University Press. From 1990–5 he was a member of the editorial team of the *Collected Letters of Thomas and Jane Welsh Carlyle* (vols 19–24). He is co-editor of a forthcoming collection of essays entitled *Across Boundaries: The Commerce and Culture of the Book*, and has published widely on nineteenth-century literary culture.

Laurel Brake lectures in literature at Birkbeck College. She is the author of *Subjugated Knowledges* (1994) and *Walter Pater* (1994), and has co-edited *Investigating Victorian Journalism* (1990) with Aled Jones and Lionel Madden, and *Pater in the 1990s* (1991) with Ian Small; *The Endings of Epochs* (1995) was edited for the English Association. She has published articles and reviews on nineteenth-century literature, publishing and cultural theory.

Kate Campbell lectures at the University of East Anglia in English and American Studies and is a supervisor of English at Cambridge University. She has edited the interdisciplinary essay collection *Critical Feminism* (1992), as well as *Journalism, Literature, and Modernity: From Hazlitt to

Modernism. She is currently completing a study on perceptions of journalism.

Dean de la Motte teaches French at Guilford College in Greensboro, North Carolina. He is co-editor of *Making the News: Modernity and the Mass Press in Nineteenth-Century France* and *Approaches to Teaching Stendhal's 'The Red and the Black'*. Current work includes a study of narratives of progress in nineteenth-century France.

Alexis Easley is Assistant Professor of English at the University of Alaska Southeast. Her work has appeared in *Women's Writing*, *Nineteenth-Century Prose*, and *Victorian Women Writers and the 'Woman Question'* (Cambridge, 1999). Easley is currently at work on a new project, entitled *Constructing the Victorian Home: Literary and Domestic Economies*.

David Finkelstein is Head of the Media and Communication Department at Queen Margaret University College, Edinburgh. He has written extensively on nineteenth-century cultural history and was editor of *SHARP News*, the quarterly bulletin of the Society for the History of Authorship, Reading and Publishing. Co-editor of the forthcoming collection of essays *Negotiating India in the Nineteenth-Century Media*, his previous publications include *An Index to Blackwood's Magazine 1901–1980* (1995) and, as part of the Victorian Fiction Research Guides Series, *Philip Meadows Taylor* (1990).

Michael Hancher teaches English at the University of Minnesota. He has published a study of John Tenniel, as well as articles on John Everett Millais and William Holman Hunt. Other published work include articles on pragmatics and literary and legal hermeneutics.

Anne Humpherys is Professor of English at Lehman College and the Graduate School, City University of New York. She is the author of *Travels into the Poor Man's Country: The Work of Henry Mayhew*, and has published extensively on G. W. M. Reynolds, Dickens and nineteenth-century popular culture and cultural history in such journals as *Dickens Studies Annual*, *Victorian Poetry* and *Victorian Studies*.

Kate Jackson has recently completed a PhD in History at the University of Sydney, producing a thesis entitled 'Culture and Profit: George Newnes and the New Journalism in Britain, 1880–1910'. She has published a number of articles on the periodicals of George Newnes and the late nineteenth- and early twentieth-century British press.

Toni Johnson-Woods is a lecturer in Contemporary Studies at the University of Queensland. Her research interests include nineteenth-century popular fiction and popular culture, and she has published articles on convicts, Australian crime fiction and feuilletons. She is currently completing an index of the serialized fiction in popular nineteenth-century Australian periodicals for Mulini Press.

Aled Jones teaches history at the University of Wales, Aberystwyth, where he is Sir John Williams Professor of Welsh History and Head of the Department of History and Welsh History. His main research interest lies in print media, particularly newspapers, and he is the author of a number of articles on newspaper history as well as of *Press Politics and Society. A history of journalism in Wales* (1993), and *Powers of the Press. Newspapers, power and the public in nineteenth-century England* (1996). He is currently researching the history of public discussion in Wales, in relation to which he is writing a book on the cultural history of water.

Andrew King has taught in universities in Italy, Poland, and Romania. His teaching and research focus on the intersections of theory, history, and cultural studies. He recently has completed his PhD at Birkbeck College, University of London, on a history of the *The London Journal*.

Margaret Linley teaches nineteenth-century literature at Simon Fraser University. She has published extensively on Christina Rossetti, Felicia Hemans, Letitia Landon and Alfred Tennyson, and is currently completing a book-length study of Christina Rossetti and the female literary tradition of the Poetess. She is also researching the poetics of travel and empire in early Victorian women's writing for a study on transcontinental representations of colonial Canadian frontiers.

Brian E. Maidment is Professor and Head of English at the University of Huddersfield. His most recent book is *Reading Popular Prints 1790–1870* (1996). He is currently completing a study of the graphic, theatrical and literary representations of dustmen in the nineteenth century.

Robert L. Patten is Lynette S. Autrey Professor in Humanities at Rice University. He edits *SEL Studies in English Literature 1500–1900* and writes about nineteenth-century British literature and graphic arts. He recently published a two-volume biography of George Cruikshank and is completing a book on copyright and constructions of authorship in the early Victorian period.

Meri-Jane Rochelson is Associate Professor of English at Florida International University. She has published numerous essays and reviews on Victorian and Anglo-Jewish literature and culture. She co-edited

Transforming Genres: New Approaches to British Fiction of the 1890s (1994), and is editor of a forthcoming edition of Israel Zangwill's *Children of the Ghetto*.

Richard Salmon is Lecturer in Victorian Literature in the School of English at the University of Leeds. He is the author of *Henry James and the Culture of Publicity* (1997), and has also published several articles on nineteenth-century journalism and other forms of mass culture.

Joanne Shattock is Professor of Victorian Literature and Director of Victorian Studies at the University of Leicester. She is editor of the nineteenth-century volume of *The Cambridge Bibliography of English Literature, 3rd edition*, and co-editor of The Nineteenth Century, a monograph series published by Scolar Press. She is the author of *The Oxford Guide to British Women Writers* and has published widely on Victorian journalism.

Mark W. Turner is Lecturer in English Literature at King's College, London. He is the author of *Trollope and the Magazines: Gendered Issues in Mid-Victorian Britain*, and of articles on serial literature and illustration. He is co-editor of *Media History*, and has co-edited a collection of essays entitled *From Author to Text: Re-Reading George Eliot's* Romola.

Lynne Warren is currently completing her PhD at Liverpool John Moores University. Her research interests include gender and identity, readerships and popular culture.

Leslie Williams is an art historian at the University of Cincinnati who specializes in the Victorian period. She received her PhD from Indiana University. She has published a chapter on the image of the Irish famine in the *Illustrated London News* in the essay collection *Representing Ireland*, as well as contributed articles on Irish subjects to the *New Hibernia Review*. She is currently completing a study of British press representations of the Irish famine.

List of Abbreviations

BL	British Library
BMC	*British Museum Catalogue of Political and Social Satires*
CR	*Contemporary Review*
ILN	*Illustrated London News*
LM	*Longman's Magazine*
PMG	*Pall Mall Gazette*
PMLA	*Publications of the Modern Language Association of America*
SDUK	Society for the Diffusion of Useful Knowledge
SEL	*Studies in English Literature*
VPR	*Victorian Periodicals Review*
WAH	*Woman at Home*

Introduction

This collection presents some of the most important new research in nineteenth-century media history and, taken together, these essays represent some of the salient developments in the field in recent years. It has been a productive time: monographs on single serial titles have been supplemented by studies of particular markets and discourses, often comparatively; likewise attempts both to measure and to theorise the media of the past have flourished. So too has archival research which, contrary to the assumptions of some anti-empiricist critiques, has increased simultaneously with the pursuit of new methods.

The 'subject' of media, like other categories of culture, has undergone increasingly detailed analysis, particularly as it has come under the redefining lens of poststructuralism. The advent of cultural studies, along with its assumed interdisciplinarity and acceptance of the 'popular', has contributed to the development of analytical methods particularly suited to both general and nuanced study of the media and their history. Theories of intertextuality, most notably emerging from the discourses of Bakhtin and Barthes, and theories of mass production and consumption such as those of Benjamin, Baudrillard, Bourdieu, and Habermas, have also been formative. The rise of Communication Studies, with its dual orientation to the market and industry has provided a plethora of methods useful for identifying distinct discourses within media of the past. 'The sociology of text', a method and field resulting, remarkably, from the intersection of cultural theory and the more traditional constituencies of historical and descriptive bibliography, has served to incorporate the practice of many past historians of print, connecting – at a stroke, and retrospectively – disparate scholarship from a variety of disciplines into something like an emergent field.

Moreover, the explosion of technologies throughout the latter half of the twentieth century – from photocopying to database access, CD ROM and electronic catalogues – has greatly facilitated archival research. The appearance of dedicated periodicals over the past three decades, with titles such as *Victorian Periodicals Review, Journal of Newspaper History* and more recently *Media History*, have created awareness while also giving impetus to this emergent field. The related printed indices of serials which were to appear shortly before general computerisation, such as the *Wellesley Index* and the *Waterloo Directory* in the 1970s and 80s, constituted a great North American fillip of definition and consolidation to an area of study which has flourished in its wake as the electronic age has progressed. One of the purposes of this volume is to take stock of such gains, while attempting to suggest ways in which advances might be made in the future.

It is perhaps appropriate that the majority of serials discussed in the pages that follow emerge from a growing scholarly interest in what is often termed 'the popular press', treated here not as a homogeneity but in terms of its various discourses and in its diversity of forms – from the daily, weekly, and monthly, to the annual. It is this carnival of everyday life, represented and produced by titles ranging from *Tit-Bits* to *The Keepsake*, that figures in a first section which explores some of the more general discursive aspects of the period's print journalism. Opening the collection is Kate Jackson's account of the growth and influence of one of the century's most popular periodicals. Demonstrating some of the ways in which editorial and audience values were closely allied in its weekly pages, Jackson attributes the phenomenal rise of *Tit-Bits* to Newnes's personal ability to exploit with consummate skill the new journalistic market of the late nineteenth century. Arguing that the intimate note struck in many articles disguised the real conditions of popular authorship within an increasingly organized newspaper press, Richard Salmon's accompanying contribution on the rhetoric of the New Journalism examines the tendency of popular writing to become more 'personal' after the 1860s. While Kate Campbell demonstrates how the terms of popular journalism were appropriated by modernism, thereby providing evidence of a discursive link between two modes of address often taken to be oppositional, Margaret Linley offers a similarly detailed account of that much neglected though highly popular form of nineteenth-century publication, the annual gift book. Overall, the net is cast deliberately wide here in an effort to register something of the wide range of discursive shifts that can be witnessed in the print journalism of the period. The result, as these chapters together show, was a press that through subtle modulation and transformation was able both to reflect and to mediate social consciousness throughout the long nineteenth century.

In the four decades that have passed since the publication of Richard Altick's *English Common Reader*, researchers have come increasingly to recognise the importance of reading audiences to the definition of Victorian culture. The nineteenth-century periodical has over the years offered a particularly fruitful line of enquiry in this respect, not least because of the way in which the producers of print were to assume, and at times to dictate, particular audience values. The regular engagement with an individual periodical from week-to-week or month-to-month, it has been argued, made the nineteenth-century reader part of a clearly definable, and defining, textual community with its own ideologies, social aspirations, and cultural assumptions. Continuing with an examination of audience formation in this period, Part II offers insight into their representation in the periodical press..

Overall, the reader in the nineteenth century was coming to inhabit an increasingly textual environment, as a massive effort for the propagandization of print consumption was undertaken by benevolent societies, church organizations, and publishers themselves. As Andrew King demonstrates, it was not only the content but the format of the periodical that was to have a profound effect on its reception. King's reading of the interaction between the text and audience of the *London Journal* is one of three chapters taking up the issue of popular reading in the light of increasingly sophisticated forms of visual representation. Despite the conservative agendas of organizations like the Society for the Diffusion of Useful Knowledge, increasing literacy was often promoted as an agent of social liberation for the reader, so that attempts by the Victorian press to 'educate' and 'improve' the individual through text and image rendered the definition of the nineteenth-century reading audience a site of intense political contestation. Exploring the representation of and response to the growth of popular literacy in mass-circulation periodicals, Brian Maidment offers a persuasive argument for the examination of graphic satire as a way of uncovering contemporary social anxieties about literacy and class. The pages of the *Penny Magazine* and other periodicals, as Maidment demonstrates, give an acute insight into the way in which the ambiguously situated artisan reader was regarded by the popular press in the first half of the century. Michael Hancher also offers a reading of the way in which the improving ethos, or the 'March of Intellect' as it was often ironically described in the satiric press, was represented within nineteenth-century print culture, from broadsides to novels, and within periodicals themselves. Finding precedent for such responses to popular aspiration in earlier forms of satire, Hancher demonstrates how, in reaction to changing class structures, such a defensive strategy reached its peak in the nineteenth century.

We would do well to remember, however, that in the struggle for social expression readers have not always responded to texts in prescriptive or even predictable ways. If we see readers as having a relative autonomy from the texts they confront, then it follows that the definition of a contemporary reading audience from the printed evidence becomes particularly problematic. Unless we are aware of such difficulties, the concept of audience, as one historian of reading reminds us, can only ever be constructed as 'an otherness of one's own discourse'. Part II concludes with Lynne Warren's examination of the tension between 'implied' and 'real' readers to be found in the correspondence columns of *Woman*. Presenting the relationship between the magazine and its readers as a struggle to impose divergent meanings on the world of the text, Warren shows how the processes by which identities are formed, as well as appropriated, can be found within the pages of such magazines. Taken together, these chapters demonstrate that the identity of the reader is not the unity it is often imagined, but rather the product of a complex set of negotiations and exchanges between historically informed discursive practices and the individuals and communities with whom they came into contact.

It is not only in regard to the category of the reader that an awareness of such complex negotiations are now necessary. Authorship has similarly become an issue that has required increasingly sophisticated treatment in critical discussions of the period. Before the pioneering work undertaken by, for example, the monumental *Wellesley Index to Victorian Periodicals*, the pages of many of the major journals of the Victorian era – *The Cornhill, Blackwood's, Bentley's*, and others – presented an anonymously authored and therefore unified front to the researcher. Nineteenth-century print media in this context tended towards the monolithic, with textual meaning mediated primarily through its identification with the specific periodical in which it appeared. The result of such a model was that each periodical was often seen as little more than the product of a singular editorial policy – the vehicle of a guiding Editor – one which correlated unproblematically with the ideological position of the 'target' audience.

One of the outcomes of recent authorial attribution is that the formerly monovocal periodical text is increasingly to be seen as a site for competing voices, contending within and even, at times, reorienting the very textual spaces they occupy. The journalism of canonical writers has been recuperated, collected, and circulated in its own right, identified through passing references in correspondence, memoirs, and biographies, as well as from official business records. The important cultural contribution made by the non-canonical writer is more apparent than it has been since the great age of journalism itself, the nineteenth century. New research is also providing insight into the manner in which the nineteenth-century writer constructed and forged cultural, professional, and authorial identities within such contexts. Robert L. Patten's contribution on Dickens presents the serialization of *Oliver Twist* in *Bentley's Miscellany* as a case study in order to suggest how an awareness of periodical form can itself alter the manner in which authorship is viewed. The productive processes that such a work undergoes create a multivalent text, constructed not only by the 'author', but by the other contributors and editors, as well as the readers of the publication in which the work appears. The complexity of the authorial position is similarly addressed by Alexis Easley and Joanne Shattock, who offer discussions of the ways in which two major figures in nineteenth-century journalism – Harriet Martineau and Margaret Oliphant, respectively – negotiated their way in the periodical as a male-oriented space, while Meri-Jane Rochelson provides an account of how one of the best known Jewish writers in the English-speaking world tailored his journalism to create separate yet interconnected media identities within the Anglo-Jewish and general British press. Such approaches remind us that the study of the seemingly ephemeral pages of the nineteenth-century media offered important opportunities not only for the negotiation of authorial identity, but also for addressing a range of pressing contemporary issues, not least the national, ethnic and sexual politics of the period which provide the topics of our final two sections.

The negotiation of gender, that deft movement among categories of address, invocation, inclusion, and marginalization, was in the nineteenth century often visible in the periodical press. Its most self-conscious form involved an address to women, variously referred to as 'ladies', 'the sex', or 'the family reader'. Consequently, the word 'woman' retained its derogatory class identity as *something other than* a 'lady' until late in the period, when it was finally embraced by the 'new woman'. Although an assumed masculinity remained the default position for most kinds of writing and reading, 'male' itself was a fissured category throughout the period, divided not only along lines of class and education but also in terms of degrees of masculinity, ranging from the 'manly' and 'muscular', to the 'effeminate', Greek, and 'invert'. While it might be argued that a late twentieth-century interest in gender politics has doubtless alerted contemporary critics to such issues, their articulation is more than simply the imposition of a set of late twentieth-century preoccupations on the past. It is clear from the evidence that much of the print space in the nineteenth century thought to be neutral (i.e. securely male) is distinctly gendered in this way. Consequently, the discourses of two monthly magazines taken as 'neutral' are here identified as predominantly masculine and significantly gay respectively. A shift from the popular to the political is also detected in an early American woman's magazine, resulting in an identifiably feminist serial. All in all, media discourse in these studies appears as variegated granite, built up over centuries of cultural and linguistic pressure, shot through with gender patternings of different breadth, hue, and complexity.

The assumed distinction between the 'metropolitan' and the 'provincial', or in other terms 'centre' and 'periphery', is the dominant theme of the chapters in the final section. When it comes to defining nineteenth-century constructions of national and ethnic identities, surprisingly little attention has been paid to the manner in which the media set about formulating and shaping such constructs. Historians have often privileged the media as a primary source for recording events and contemporary opinions without regard for the geo-political centres in relation to which the mass media operate. In each instance, national or ethnic identities can be seen to some extent as responses to the political and cultural demands of the times. As these chapters make clear, Irish, Scottish, and Welsh cultural identities in the nineteenth century are traceable through a tension between their attempt to construct a separate status for themselves with regard to a larger British framework and, at the same time, their need to deal with the pressure to acquiesce. Attempts at the formation of Scottish national identity certainly found themselves engaging in just such a paradox throughout the nineteenth century, as they were called upon to celebrate symbols of past greatness within the context of a political union with a stronger, more economically powerful neighbour. How texts were made to participate in such cultural paradoxes is the subject of Toni Johnson-Woods's chapter on the adaptation and comparative marketing of cheap periodical literature in the different markets of London, New York, and Melbourne. The role played by the print media in the definition of a Welsh national identity is outlined by Aled Jones, who traces the development of an indigenous newspaper press in order to show how that same press helped throughout the late nineteenth and early twentieth centuries to formulate 'a new political vocabulary and a language of nationality'. Leslie Williams demonstrates in her essay on the British reportage of the Irish Famine how press reports and editorials written from the safe distance of London offered a representation for an increasingly peripheralized sector of the so-called United Kingdom that was, in the end, to have lethal implications.

Despite its breadth of concern, no volume such as this can offer a comprehensive treatment of media in the nineteenth century: the emphasis in the following pages is almost exclusively on the print media, and specifically the periodical press; others have explored, and continue to explore, aspects of the diverse spectacle of Victorian cultural life – architecture, music, the visual arts – of which media the printed text is one among many. Neither do we wish to suggest that the importance of media to the construction of identities begins in the nineteenth century. Work completed over the past few decades from the transformation of medieval reading practices, to the printing press as an agent of cultural change after Gutenburg, to the rise of the public sphere and the emergence of a new female reading public in the eighteenth century, demonstrate that many of the issues tackled in the following pages are not unique to the period but rather represent historical continuities traceable to other manifestations at other times. It might also be argued that as the first age of mass culture, many of the issues that confronted the producers and consumers of media in the ninteneenth century were not so very different from those that have come to be associated with more recent technological phenomena. The emphasis among today's practitioners of Media Studies on the electronic forms of radio, television, film, and the subsequent 'new technologies' of video and Internet, at the expense of these earlier modes of production, obscures the real continuities that exist in the History of Media.

LAUREL BRAKE
BILL BELL
DAVID FINKELSTEIN

Part I

Discourses of Journalism

1

George Newnes and the 'loyal Tit-Bitites': Editorial Identity and Textual Interaction in *Tit-Bits*

Kate Jackson

In 1890, nine years after establishing *Tit-Bits*, George Newnes assessed his contribution to periodical literature, when he separated from W. T. Stead over their joint venture, *The Review of Reviews*:

> There is one kind of journalism which directs the affairs of nations; it makes and unmakes cabinets; it upsets governments, builds up Navies and does many other great things. It is magnificent. This is your journalism. There is another kind of journalism which has no such great ambitions. It is content to plod on, year after year, giving wholesome and harmless entertainment to crowds of hardworking people, craving for a little fun and amusement. It is quite humble and unpretentious. This is my journalism.
>
> (Friederichs 1911, 116–17)

In this instance, drawing a distinction between the conventional nineteenth-century model of the press as an estate of the realm and his own seemingly less ambitious journalistic endeavours, Newnes conceptualized *Tit-Bits* as an extremely responsive medium, immersed in the rhythms of industrial life and realizing the humble needs of hardworking people. It was a social object, firmly rooted in contemporary culture: an extremely contextualized text. Moreover, Newnes identified closely with the paper's readers on a personal level. 'I *am* the average man', he always maintained, perhaps somewhat disingenuously, in explaining the success of *Tit-Bits*, 'I am not merely putting myself in his place. That is the real reason why I know what he wants' (Friederichs 1911, 188). Knowing what the reader wanted was, of course, a distinct commercial advantage, and *Tit-Bits* was one of the most lucrative publishing ventures of the 1880s.

Newnes was to become one of Britain's first media magnates, publishing a huge number and variety of publications including *The Strand Magazine, The Westminster Gazette, Wide World Magazine* and *The Captain*. But it was through his first publication, *Tit-Bits*, that he established his reputation as a periodical editor and publisher. Indeed, many historians and literary critics have commented that *Tit-Bits* revolutionized popular journalism, and have cited Newnes as the founder of modern journalistic practice. Such scholars have focused on Newnes's skill as a businessman and his use of promotional schemes and competitions to guarantee the commercial success of *Tit-Bits*, and a number have accused Newnes of undermining literary standards by subordinating journalistic quality to the transparent pursuit of profit.[1] Yet the success of *Tit-Bits* was as much the result of Newnes's inclusive, creative and dynamic personality as an editor – his editorial flair – as it was a product of his marketing genius. By 1891, when he established *The Strand* after editing *Tit-Bits* for ten years, Newnes ('Mr Editor') was, according to Reginald Pound, 'a vastly respected entity, infallibly wise and just, and always, in the people's fancy, benignly bearded' (Pound 1966, 7–8).

In the developing metacriticism of periodical research, significant emphasis has been placed on the uniquely interactive, open and self-referential form of the periodical text; the way in which it functions as social discourse rather than as direct social statement (see *VPR* 22. 3, 1989). This essay is an attempt to demonstrate the ways in which the weekly paper, *Tit-Bits*, functioned within this paradigm of social discourse, with special reference to its creator and editor, George Newnes. It examines the ways in which Newnes attracted a large circle of regular and loyal readers to this periodical by offering them entertainment, interaction, and creative participation and establishing a responsive editorial presence as the reader's friend and guardian, patron and pastor, adviser and representative. And it analyses the language and features of *Tit-Bits*, in relation to contemporary journalistic and cultural conditions as well as recent historiography, to clarify the ways in which the interactions occurring in the text represented a legal and moral contract between editor and readers; a bond of fellowship associated both with pre-industrial social models and new forms of collectivism; and an inclusive style of journalism which, whilst it was commercially motivated, addressed the needs and preoccupations of readers in a way which bound them closer to the authority of the editor and led them into identification with a discursive community of so-called 'Tit-Bitites'. Reader-response critics have created the term 'reading communities' to describe categories of readers linked together by a common experience or expectation of reading, and by common social, political, ideological or cultural objectives or bonds (rather than by physical proximity). Newnes's repeated characterization of the readership of *Tit-Bits* as 'loyal Tit-Bitites' is a transparent example of an attempt to exploit the concept of a community of readers.

Acknowledged by many historians and contemporaries to be the most popular penny paper of the late nineteenth century, *Tit-Bits* appeared on the publishing scene on 22 October 1881. It was essentially a paper of the miscellany variety that became popular in the late nineteenth century: a 16 page patchwork of advice, humorous anecdote, romantic fiction, statistical information, historical explanation, advertisement and reader correspondence. Regular columns and serials were interspersed with short jokes and sallies of the kind frequently furnished to newspapers and magazines by literary types. As the paper evolved it incorporated more original material. Serial fiction, a form which engaged reader response on a weekly basis in a way that allowed the audience to mediate in the production of the story, was introduced in 1889. Competitions and promotional schemes were a central feature, and notices relating to upcoming and current competitions appeared on the front page of each issue. Newnes also relied heavily on editorials and correspondence columns to create and maintain a bond of sympathetic intimacy with readers. A 'more personal tone' was, in fact, characteristic of contemporary journalism, as T. P. O'Connor observed (O'Connor 1889, 423).

The average weekly circulation of *Tit-Bits* during the years of Newnes's involvement (1881–1910) was 400,00 to 600,000 copies, but this figure fluctuated as circulation-boosting schemes came and went. By 1890, according to Geraldine Beare, its sales had exceeded half a million and by 1893 it was considered to be the world's most popular penny paper (Beare 1986, 20). *Tit-Bits* also maintained a substantial overseas circulation. This, the publisher stressed, was a *family* paper, catering to a variety of readers (male, female, and juvenile), and unerring in its support of traditional moral values. The popularity of the 'Tit-Bits Insurance Scheme' (a system of railway insurance which protected every commuter with a current copy of the magazine on his or her person whilst travelling on trains) was an indication that the paper's readership belonged to the expanding commuting public. Newnes's audience consisted of the lower-middle and upper-working classes. The lower-middle class, in particular, was rapidly expanding in the latter decades of the nineteenth century, and was one to which Newnes himself belonged (as a commercial traveller), before rising to the top of the professional middle class in becoming a self-made entrepreneur. This class was often the backbone of university extension courses and vocational evening classes, the taste for which was met by Newnes in the 'Tit-Bits Inquiry Column', and of new constituency organizations of both the major political parties, a fact which seems to have strongly influenced the language and editorial style of *Tit-Bits* (see Perkin 1969, 92–9).

Tit-Bits began as a collection of excerpts converted into 'text' purely by a process of creative editorial synthesis, and it was infused with a deep sense of editorial presence. Editorial interjection, often characterized by a familiar tone and containing references to readers as 'our friends', was the very essence of *Tit-Bits*: 'In consequence of the large numbers of queries we receive', claimed Newnes, referring to the 'Legal Tit-Bits' section, 'it is impossible to answer all we should like in this page'. Considerations of friendship, it was suggested, competed with the imperatives of economy:

> Those of our friends who find their questions omitted will please to understand that it is not from want of courtesy, but from want of space.
>
> (*Tit-Bits* 13. 313 (October 1887) 14)

A paper's 'personality', as Joel Wiener has pointed out, is partly a function of editorial persona (Wiener in Brake, Jones and Madden 1990, 155), and George Newnes, 'The Editor' was the very nucleus of *Tit-Bits*. Through *Tit-Bits*, Newnes evolved as an editor and a publisher, and the reader of his paper was closely involved in this metamorphosis by virtue of Newnes's characteristically interactive, self-conscious, self-referential, editorially transparent, personalized, dramatic, earnest, innovative approach.

The 'Answers to Correspondents' column, conveying an impression of editorial accessibility and reader involvement, was the linchpin of the interactive posture that Newnes adopted in *Tit-Bits*.[2] Known at the office as 'corres.', this column was one of the most popular amongst readers. 'For this page', according to Hulda Friederichs, Newnes's long-time friend, associate and biographer, 'the editor had a very special affection, together with very definite ideas as to the manner in which it should be conducted'.

> He held that, first and foremost, all answers should be given in a manner which would make each correspondent feel that he was treated with special consideration; that here behind this newspaper

> page, there was somebody to whom the inquirer's affairs were of real human interest; who sympathised, and tried to give his advice a practical turn. Secondly, Mr Newnes held that the answers should be couched in such terms, whenever this was possible, as to make them interesting to the general reader as well as to the individual correspondent.
>
> (Friederichs 1911, 106–7)

Thus, the text of *Tit-Bits* represented both a medium for personalized editor–reader interaction and a commercial product with broad narrative and journalistic appeal to a diverse audience. These were characteristics which it shared with other products of the New Journalism.

This column was one which Newnes, with his 'vivid imagination, his innate good sense, his ready wit, and above all his unfailing social tact' was ideally suited to managing, according to Friederichs:

> For years, he took these letters, in bundles twelve inches high, and higher, went carefully through each, and answered them so fully, so wisely and so well, that in course of time people belonging to every social class sought help and advice through 'corres'.
>
> (Friederichs 1911, 107)

Newnes's editorial persona was multiform. He was 'innovator and preacher' (Joel Wiener 1985), 'patriarch and pioneer' (Ogden Mahin 1924, xi), democratic representative, business partner, adviser and friend; sometimes upbraiding, sometimes cajoling, sometimes jesting, often avuncular. In a series of textual episodes, the editor was characterized variously as an 'august personage' of the press and public life, a slightly vague intellectual type, and a rather argumentative, difficult character whose volatility was only barely curbed by his diplomatic staff (*Tit-Bits* 15. 379 (January 1889) 237; 15. 381 (February 1888) 269; 5. 117 (January 1884) 203). The print culture of books, pamphlets, magazines and newspapers that developed during the nineteenth century under the new urban conditions (and spawned New Journalism), as Raymond Williams has pointed out, was very much interactive with a predominantly oral culture which encompassed such institutions as the theatre, the political meeting and the lecture, and such melodramatic forms as crime, scandal and romance (Williams 1961, 43–6). It could be argued, however, that *Tit-Bits* had closer connections with music hall entertainment than with the more structured environment of the nineteenth-century theatre. As Newnes fine-tuned his individual act, he resembled one of the great variety entertainers popular amongst the working class in the later nineteenth century. The audience applauded by buying the paper, week after week, and participating in its many competitions.

Yet Newnes took the relationship between himself and his readers very seriously, conceiving of it as a contract in which both parties were bound by legal and moral obligations. In the 'Tit-Bits Villa' Competition, for example, he went into endless detail regarding the conditions of the competition, answered the questions of prospective competitors in the weeks leading up to the judging, and reported on the entire process of awarding and bestowing the prize (a freehold house) as 'an absolute guarantee to the public that it has been given without favour'. 'We shall take pride in seeing that the house is one which all will admit is a *fair and reasonable fulfilment of our contract*', he announced.

> Let everyone remember that we have pledged our reputation in this matter... and all may be sure that we shall make this prize one which shall redound to the credit of *Tit-Bits*, and not one of which we shall be ashamed.
>
> (*Tit-Bits* 5. 112 (December 1883) 120)

The problem with journalistic competitions was that not only did they invite allegations that the publisher was undermining literary standards by employing cheap commercial ploys, but they left him susceptible to charges concerning the promotion of illegal gambling. 'We do not intend to have any more chance competitions', wrote Newnes in 1889, tackling the issue head-on, as was his wont. 'We never cared for them, and only instituted them because, unlike literary contests, they seemed to give an equal chance.' He continued in a pseudo-reformist vein:

> But it has been pointed out... that there is a danger of a demoralising effect from them, inasmuch as they encourage a tendency to gambling... they may be injurious to some, and in any case they are unworthy of *Tit-Bits* and of its subscribers.
>
> (*Tit-Bits* 16. 397 (May 1889) 97)

Aware of the decision that a general offer in an advertisement could be interpreted as an enforceable contract, Newnes made repeated reference to his legal obligations as publisher-promoter within the text of *Tit-Bits*.[3] He elaborated upon the terms of the contract as time went on. 'The greatest care is taken to give correct answers to inquiries', he explained in one issue, 'but at the same time our correspondents must understand that we disclaim any responsibility in giving our opinions'. In the same issue it was stated that correspondents were required to 'give their names and addresses, not necessarily for publication, but as a guarantee of good faith' (*Tit-Bits* 13. 313 (October 1887) 14). Every change in the format or content of *Tit-Bits*, including the introduction of advertising (*Tit-Bits* 17. 418 (October 1889)) was discussed at length. Thus Newnes developed an image as a collaborative editor, establishing a relationship of trust and loyalty – a kind of moral bond as well as a legal one – with his readers. Manchester readers took this bond very seriously. When Harmsworth introduced *Answers*, an imitation of *Tit-Bits*, they displayed their loyalty to Newnes by boycotting the newcomer (Friederichs 1911, 80–1).

Newnes's investment in the moral bond between editor and readers, in many ways a relic of the old journalism and an extension of his own strong sense of public responsibility, led him to insist repeatedly on the 'clean' quality of *Tit-Bits*. With an excessive degree of moral earnestness, he scrutinized every line of the paper right down to the joke 'fillers' at the foot of the columns. In one instance, he blue-pencilled a query about kissing in public, writing against it on the page proof: 'We should avoid any subject that may have an injurious effect on our readers' (Newnes as cited in Pound 1966, 25). His campaign to maintain his paper's impeccable reputation was obviously successful. The manager of W. H. Smith's bookstall at Victoria Station, Manchester, said that he decided to 'push' *Tit-Bits* because he thought that it might, as Newnes had intended, provide much-needed opposition to the 'blood-and-thunders' which had led Smith to consider compiling his subsequently effective blacklist (Pound 1966, 21).

The appeal of *Tit-Bits*, it could be argued, lay largely in its recreation of a pre-industrial kind of community of mutual responsibility, offering its readers personal involvement in its evolution as a text. In nineteenth-century Britain, industrialization and urbanization caused suburban spread, urban segregation and social fragmentation. The wealthy employing class tended to move outside the boundaries of the larger towns and cities, and thus removed themselves from the responsibilities and networks of the communities they left behind. Only some, like Newnes, continued to engage in what Harold Perkin has called 'competitive philanthropy' (Perkin 1971, 262), giving hospitals, libraries, public parks, town halls, and churches to the towns. In fact, Newnes's editorial image as a collaborative, benevolent and responsive paternalist paralleled his local reputation in Lynton, North Devon, where he built a country residence, financed the cliff railway, donated many buildings and facilities including the town hall, Congregational church, bowling green, golf course and cricket field, and hosted many public events. One West Country paper nominated Newnes 'Sergeorge the Giver', drawing upon a medieval social model (Friederichs 1911, 195–7). T. H. S. Escott observed that the donation of Lynton Town Hall and Putney Library (opened in 1899) were 'two specimens of conduct which made George Newnes the most widely popular as well as prosperous newspaper runner of the new era' (Escott 1911, 258). (They were, of course, good publicity as well as good deeds.) Newnes's father had belonged to a generation of Free Church pastors who brought to their work a strong sense of personal responsibility for and intimacy with the individual members of their congregations. Through the secular medium of the popular periodical, Newnes replicated the bond of human connection which the Church had previously provided.

In the 1880s and 1890s, many educated men and women acquired a practical personal commitment to the ideals of human fellowship and social unity. In pursuit of a common bond of citizenship, a number of middle-class students took up residence in newly-formed 'settlement houses' such as Toynbee Hall, which opened in 1885 and was followed by many more.[4] Such establishments provided a range of social services: health care, kindergartens, child-care classes, legal aid, lectures, libraries, concerts, and recreational facilities. J. A. Spender, who was to edit Newnes's liberal political weekly, *The Westminster Gazette* and become a great friend of its proprietor, was a 'citizen householder' at Toynbee Hall. The gospel of fellowship even permeated public policy, and the move towards collectivism in public opinion and the law was a manifestation of the same impulse towards the creation of a socially unified society based on mutual responsibility. A. V. Dicey argued, in fact, that these years saw a transition from individualism to collectivism (see A. V. Dicey 1905a, 281–300). Yet organic customary relations had been integral to earlier social models, revolving around a notion of paternalism that was as much based on mutual responsibilities as on hereditary rights. *Tit-Bits* was thus a manifestation of both George Newnes's attachment to old local and parish community traditions, and of his sense of new forms of collectivism. His paper, it is contended, was itself a kind of journalistic, discursive equivalent of a settlement house, offering mutual support, a wide range of social services, and a sense of citizenship and community to readers.

Newnes employed a variety of professionals to supply advice to and exert an educational influence upon his readers. A lawyer answered legal queries, a doctor answered medical questions, and the editor himself supplied answers to readers' general enquiries in correspondence columns and editorials. *Tit-Bits* also offered citizen-readers advice on emigration in a series entitled 'Intelligent Emigration', as well as an educational series of 'Literary Excerpts', and a plethora of informative 'Tit-Bits of General Information'. The editor-publisher introduced a column entitled 'Tit-Bits of Legal Information' in 1881, in response to requests to which he saw it as his 'duty' to respond:

> We have received so many applications from our subscribers for tit-bits of legal information on matters important to our correspondents, that we have secured the services of a legal gentleman to give the required answers.
>
> (*Tit-Bits* 1. 4 (November 1881) 14)

The column was a legal, textual version of the talkback radio programme. Over the early years of the paper's publication, a range of questions relating to urban, industrial life and to the legal preoccupations of late nineteenth-century society emerged within it. The intricacies of income tax and financial liability, the conditions attached to life insurance policies, the laws regulating apprenticeships and employment, and the obligations of patent-holders were all issues that concerned readers. Questions concerning the ability of a married woman to make a will were common. The Married Women's Property Act of 1883 had only just granted her that right, and had also enabled women to initiate civil law suits, hence the significant contingent of female correspondents to this column. The Legal Editor's advice on the laws governing marriage was supplemented by a series on 'How to Marry', appearing in 1883, which included a clear description (if one that was rather biased towards touting the advantages of recent Liberal reforms) of the 'Legal Effect of Marriage'.

A two-part series on 'Medical Questions' supplied 'an explanation of the causes and cures of the various ailments of which our querists have complained' as well as 'knowledge on the general principles of health'. Once again, Newnes combined personal interaction with the notion of general appeal. The medical conditions discussed included indigestion, headaches, 'noises in the ear', bunions and varicose veins, and the doctor's advice tended to take the form of a discourse on urban degeneration, social responsibility, and temperance, with phrases such as 'irregular living', 'deficient care' and 'bad cooking' figuring heavily. Indigestion, it was suggested, was a result of urban conditions:

> The hurried meals partaken of by a large number of clerks and warehousemen cannot but produce indigestion – all the laws of healthy digestion are violated. A run to a close restaurant, or a scramble to catch the bus or train, the food bolted, and then a smoke, and another scramble to get back to the office, is the daily routine of a large number of our townspeople.
>
> (*Tit-Bits* 5. 125 (March 1884) 333)

The column was almost reformist in tone. A later feature – a competition to produce the best list of 'Ten Long-Felt Wants' – similarly precipitated a disquisition on social reform and social responsibility (*Tit-Bits* 6. 130 (April 1884) 221, 250). The discourse of these columns was not partisan or directly political. But it created an atmosphere conducive to the voicing of critical views: to participation and representation.

The magazines established in the period of New Journalism were largely guided by the desire to take advantage of a lucrative commercial opportunity. *Tit-Bits* was no exception. 'Entrepreneurs such as Newnes and Lord Northcliffe stand out now', Stephen Elwell has thus commented, 'because they discovered in the 1880s and 1890s how to define and exploit the common interest of the middle class in *inclusive* rather than *exclusive* terms' (Elwell in Wiener 1985, 40).[5] After a period of class-based journalism, these years saw the emergence of a distinctly new editorial strategy. Magazines like *Tit-Bits* were able to speak to a broadly-defined reading community rather than a discrete class of readers. Newnes was an astute businessman, and he employed various competitions and promotional schemes to increase the circulation of *Tit-Bits* and thus attract advertising revenue. Yet such schemes were a mixture of business and benevolence, addressing contemporary cultural preoccupations, enhancing the editor's reputation for collaboration and beneficence by involving readers in an interactive textual relationship, and interweaving the language of commercialism with that of social conscience.

In May 1885, Newnes announced an innovative 'NEW SYSTEM OF LIFE ASSURANCE' in bold capitals:

> ONE HUNDRED POUNDS WILL BE PAID BY THE PROPRIETOR OF 'TIT-BITS' TO THE NEXT-OF-KIN OF ANY PERSON WHO IS KILLED IN A RAILWAY ACCIDENT, PROVIDED A COPY OF THE CURRENT ISSUE OF 'TIT-BITS' IS FOUND UPON THE DECEASED AT THE TIME OF THE CATASTROPHE.
>
> (*Tit-Bits* 8. 189 (May 1885) 97)

The idea for this scheme had actually been suggested to Newnes by the enquiry of a reader. The reader's husband, a devoted reader of *Tit-Bits* who was almost due to be rewarded for his constancy with a sum of money (in accordance with an incentive scheme for subscribers introduced by Newnes to raise circulation), had been killed in a railway accident. She applied to Newnes for some compensation, and was sent £100.

Insurance claims were always billed on the front cover of the paper, their visibility guaranteeing their impact. The first insurance claim was paid in August 1885. A 40-year-old coachbuilder with four children was killed when he was run over by a train after falling between the train and the platform at Hatfield Station. The Coroner's verdict was 'Accidental Death', and four witnesses testified to the fact that 'a copy of *Tit-Bits* had been found upon the unfortunate man at the time of the accident'. Thus the proprietor of *Tit-Bits* paid £100 insurance money to the victim's widow, and the recipient's receipt of payment was reproduced in *Tit-Bits* as proof of the transaction (*Tit-Bits* 8. 198 (August 1885) 241). A series of claims followed at the rate of about one every two months, and by September 1891 a total of 36 claims had been paid. An advertisement for *Tit-Bits* in *Under the Union Jack*, another Newnes publication, in 1900 stated that £12,500 had been paid in insurance claims (*Under the Union Jack* 1. 1, 1900). This was a vastly successful scheme that was soon copied by many other publishers.

All claims generally followed the same format, and were decided by 'the proprietor of *Tit-Bits*'. Appended to the eighteenth insurance claim was a moral:

> We trust that the publicity which the *Tit-Bits* Insurance Scheme has been the means of giving to this lamentable occurrence may to some extent put a stop to the habit of leaping out of a train while in motion, which is so prevalent in young men. To jeopardise life in this reckless fashion is an act of which no sensible person should be capable.
>
> (*Tit-Bits* 15. 387 (March 1889) 353)

In this manner, the editor-proprietor of *Tit-Bits* established himself as investigator and jury in cases of accidents occurring on the railways, taking evidence, hearing the testimonies of witnesses, and handing down a judgment. He set himself up as a guardian of social conscience and as a figure of paternalistic benevolence, fulfilling his responsibilities to his community of readers by dispensing insurance claims and offering sound advice. This scheme appealed to a commuting public preoccupied with the notion of life insurance and concerned with the frequency of railway accidents. But apart from anything else, the stories that emerged were gripping stories. Even agony columns, as Anne Humpherys has shown, depended upon the narrative gaps being filled by an audience with certain expectations about narrative form and convention. The power of such columns lay in the potential they represented for narratizing (Humpherys in Brake, Jones and Madden. 1990, 37). Thus the *Tit-Bits* Insurance System was structured in such a way as to produce a series of parables in which the grieving relatives of the victim were invariably saved (in a financial sense) by the money provided by the editor of *Tit-Bits*. There is something symbolic (if a little morbid) in the fact that all of these victims were found with a copy of *Tit-Bits* on their person. They were, it was amply demonstrated, 'loyal Tit-Bitites' to the end.

In November 1883, a revolutionary competition was announced. A seven-roomed freehold dwelling-house was offered as a prize to the person sending in the story judged by the *Tit-Bits* adjudicators to be the best. The main stipulation made by Newnes was that the winner should call the house 'Tit-Bits Villa'. The editor stressed that the competition was open to 'every member of every family in the Kingdom, from the highest to the lowest'. (The possibility of such inclusiveness was guaranteed by the extraordinary condition that entries could be selected from a published work rather than being original. Thus the competition required no great literary skill.) The purpose of the competition, of course, was to raise circulation, but those who joined the 'followers of *Tit-Bits*' could expect an intimate and responsive relationship with their leader, the editor, who throughout the competition answered the queries of competitors, explained, berated, commented, appealed to his readers' better natures, and publicized editorial dilemmas. At the conclusion of the competition, Newnes played police, prosecution and judiciary in passing sentence on the 'unprincipled miscreant' who had falsely informed a competitor that he had won the prize:

> we are of the opinion that an unnatural ruffian who would be guilty of such a dastardly act as this deserves punishment of the severest kind, and we offer
> £5REWARD
> to anyone who will disclose to us the name of this viper.
> If we unearth him we shall spare no pains or expense to prove that this act was forgery, and comes under penal laws.
> (*Tit-Bits* 5. 117 (January 1884) 205)

The role of the editor-leader of this discursive community extended, it seems, even to the policing of community interests. He addressed his audience in the language of popular melodrama ('unnatural ruffian', 'dastardly act', 'this viper').

Ultimately, Newnes published a 'REPORT UPON THE COMPETITION'. Over 22,000 letters were received at the *Tit-Bits* offices, some containing up to twenty entries. 'Tit-Bits Villa' became a place of pilgrimage for day-trippers and tourists; a kind of icon of the *Tit-Bits* community. It was open for inspection, and many hundreds took the trip by rail to visit. The villa was awarded in a public ceremony that must have resembled a political rally, and 100,000 photos of it were developed and sold as souvenirs of the competition. Even then, supply did not meet demand. Newnes had obviously entered successfully into his readers' shared dreams of domestic bliss.

The competitions, 'Inquiry Column', 'Legal Tit-Bits' and 'Answers to Correspondence' in *Tit-Bits* represented the possibility of reader participation, and in the terms of textual criticism, the creative freedom of the open text as opposed to the prescriptive reality of the closed text. Editor, sub-editors and readers mingled on the printed page. Thus, when Newnes erected a *Tit-Bits* Pavilion and Inquiry Office at the Paris Exhibition of 1889, describing it as 'a rendezvous for all readers of *Tit-Bits*, and a place for obtaining full information' (*Tit-Bits* 16. 394 (May 1889) 49), he was merely extending *Tit-Bits*'s textual function as the site at which editor and reading public interacted on the printed page to encompass the physical and tangible act of 'rendezvous'. Newnes himself appeared in *Tit-Bits* as a benignant administrator, conducting the affairs of his publishing house in an interactive fashion as if he were administering a democratic state. In the context of explaining the conditions of the Tit-Bits Villa Competition, for instance, Newnes articulated the democratic ideal by making the competition 'open to every one of our readers, irrespective of age, sex, nationality or colour'. Thus *Tit-Bits* was represented as a medium of participation for its readers, a pluralistic discursive sphere.

Newnes rarely spoke in parliament and became disillusioned after entering it with all the social and political idealism of an energetic young man. His journals *were* his public voice. Whilst *Tit-Bits* featured little that was explicitly political or partisan, its editor forged such strong links with his readers – 'the noble band of constant subscribers' – that they came to be something akin to constituents. As a result, perhaps, of the relationship that he developed with the reading public through *Tit-Bits*, Newnes's received not less than thirty-three requests from constituencies wishing him to represent them during the years of his retirement from politics (1895–1900). He had, in effect, already done so.

It is a curious fact that whilst the capitalistic development of the late nineteenth-century press entailed an increase in the distance between the organizational hierarchy of the publishing company and the mass reading public that fed its progress, the New Journalism was characterized by its personalized tone and its dependence upon the individual identities of editors, authors and illustrators. Vast increases in circulation meant that it was actually impossible for a publisher or editor to 'know' his audience in any real sense. But successful publishers like Newnes attempted to recreate the old communal relations of eighteenth- and early nineteenth-century Britain (Raymond Williams's 'knowable community') within various 'reading communities' with shared values and experiences, maintaining an interactive relationship with readers and manufacturing a community of interest, through editorials, correspondence columns, competitions and other features, in place of the organic neighbourhood in which rural people had enjoyed close personal involvement. In the pages of newspapers, as Piers Brendon has suggested, 'local gossip was writ large and the anonymous citizen became a living, breathing participant' (Brendon 1982, 17–18). *Tit-Bits* offered the security of a close community of loyal lower-middle-class readers, and the protection of a paternalistic and responsive editor-proprietor. It reflected and exploited the developments in the national community brought about by the extension of literacy and the popular press, the democratic extension of political suffrage, urban conditions and new notions of fellowship, cooperation and collectivism in politics and law.

The commercial success of *Tit-Bits* was underpinned by Newnes's editorial skill and by his deep sense of social responsibility and cultural awareness. Newnes, as Reginald Pound has acknowledged, 'might fairly have claimed that he was imbued by a desire for the betterment of his people' (Pound 1966, 23). As a publisher, by his own account, he had to possess 'the skill, the precision, the vigilance, the strategy, the boldness of a commander-in-chief'. As an editor, he was required to be 'a statesman, an essayist, a geographer, a statistician, and, so far as all acquisition is concerned, encyclopaedic' (*Tit-Bits* 1. 24 (March 1882) 5). As the editor-publisher of *Tit-Bits*, he was all things to all people. Above all, *Tit-Bits* was *inclusive*. The editor solicited the participation of readers in the creation of the text, and readers responded with enthusiasm. This publication represented a dynamic mechanism of communication; a carefully-balanced act of negotiation between editor-publisher and audience that was more than a timely but transient commercial triumph.

Through it, Newnes secured a loyal readership, and pioneered habits of journalistic discourse that foreshadowed many of his own subsequent publications and many later developments in New Journalism.

Notes

1. Historiographically, *Tit-Bits* has been enmeshed in the historical myth of the so-called 'Northcliffe Revolution'. Some contemporary social critics believed that journalism of the *Tit-Bits* mould perverted the aims of the 1870 Education Act and supplementary legislation that followed it. They argued that New Journalism offered information at the expense of knowledge and encouraged the kind of cultural Philistinism feared by Matthew Arnold. It was 'cheap', 'low', and 'trivial', they complained, and it undermined literary standards by playing to the lowest common denominator. See, for instance, Dicey 1905b, 917; Escott 1917, 368; Ensor as cited in Williams 1961, 196.
2. The 'Notices to Correspondents' Column had been a very popular department in family periodicals of the 1840s and 1850s such as *Lloyd's, London Journal*, and *Reynolds's Miscellany*. According to Anne Humpherys, *Reynolds's* acted as 'a stern adviser, a knowledgeable informant, an avuncular domestic manager' (Anne Humpherys, 'G. W. M. Reynolds: Popular Literature and Popular Politics' in Wiener 1985, 18).
3. As the use of competitions increased towards the end of the century, a number of cases involving competitions aimed at boosting newspaper circulations were brought before the courts as constituting unlawful lotteries. The key question, for judges, was whether the distribution of prizes in various promotional schemes was dependent upon chance, and thus constituted a lottery. Trade Marks Acts of 1875, 1883, 1888 and 1905 restricted the kinds of claims that could be made, particularly regarding rival products, and enforced protection of the trader's name. See T. R. Nevett, 133–7.
4. By 1913, twenty-seven settlement houses had been planted in the poorer areas of London (Kern 1983, 56). Polytechnics, launched at about the same time, also reflected a broad middle-class movement to decrease social distance and replace the paternalism of Victorian charity with a notion of people's participation in a common relationship of 'social citizenship' and entitlement to common social services.
5. James Curran has also pointed to the integrational themes stressed in the industrialized press (James Curran, 'The Press as an Agency of Social Control' in Boyce, Curran and Wingate 1978, 70).

Works cited

Beare, Geraldine, 'Indexing the Strand Magazine', *Journal of Newspaper and Periodical History* 11.2 (Spring 1986), 20.

Boyce, George, James Curran and Pauline Wingate (eds). *Newspaper History from the Seventeenth Century to the Present Day*. London, 1978.

Brake, Laurel, Aled Jones and Lionel Madden (eds). *Investigating Victorian Journalism*. New York, 1990.

Brendon, Piers. *The Life and Death of the Press Barons*. London, 1982.
Dicey, A. V., *Lectures on the Relationship Between Law and Public Opinion During the Nineteenth Century in England*. London, 1905a.
Dicey, Edward. 'Journalism, New and Old', *Fortnightly Review* 83 (May 1905b), 904–18.
Escott, T. H. S. *Masters of English Journalism: A Study of Personal Forces*. London, 1911.
—— 'Old and New in the Daily Press', *Quarterly Review* 227 (April 1917), 353–68.
Friederichs, Hulda. *The Life of George Newnes, Bart*. London, 1911.
Kern, Stephen. *The Culture of Time and Space, 1880–1918*. Cambridge, MA, 1983.
Nevett, T. R. *Advertising in Britain*. London, 1982.
O'Connor, T. P. 'The New Journalism', *The New Review* 1 (1889), 423–4.
Ogden Mahin, Helen. *The Editor and His People; Editorials by William Allen White*. New York, 1924.
Perkin, Harold. *The Origins of Modern English Society, 1780–1880*. London, 1969.
—— *The Age of the Railway*. Newton Abbott, 1971.
Pound, Reginald. *Mirror of the Century: The Strand Magazine, 1891–1950*. London, 1966.
Tit-Bits. vols 1–17, 1881–90.
Under the Union Jack. 1. 1 (1900).
Victorian Periodicals Review (Special Critical Theory Issue). 22.3 (Fall 1989).
Wiener, Joel H. (ed.). *Innovators and Preachers: The Role of the Editor in Victorian England*. New York, 1985.
Williams, R. *The Long Revolution*. London, 1961.

2
'A Simulacrum of Power': Intimacy and Abstraction in the Rhetoric of the New Journalism

Richard Salmon

In his autobiography, published in 1884, Edmund Yates, then editor of *The World*, claimed to have invented 'that style of "personal" journalism which is so very much to be deprecated and so enormously popular' (Yates 1884, I 278). The date given for this event is precise – 30 June 1855 – as is its location, in an article for Henry Vizetelly's *Illustrated Times*, entitled 'The Lounger at The Clubs'. However one might judge the accuracy of Yates's (characteristically self-promotional) claim, it serves to introduce both a significant shift in the history of journalistic practices and the terms in which this shift was commonly understood. By his use of the phrase '"personal" journalism', Yates offers a means of characterizing a form of popular journalistic discourse which extended, with a certain continuity, from the mid-nineteenth century to the 'New Journalism' of the 1880s and beyond.[1] Conversely, when recalling a later moment in his career, he accounts for the refusal of J. T. Delane, the editor of *The Times*, to be interviewed for his series 'Celebrities At Home' by invoking 'the old-fashioned theory that the editor of a newspaper should be an impersonal myth' (Yates 1884, II 168). While such oppositions between 'personal' and 'impersonal' modes of journalism are no doubt familiar to most historians of the nineteenth-century press, however, the normative (or ideological) function served by these terms has received little attention. In this chapter, I attempt to reassess the project of the New Journalism by focusing upon a number of contemporary debates in which the conceptual basis of journalistic rhetoric was directly addressed.

That the New Journalism was indeed conceived as a form of 'personal journalism' is evident from the comments of its most prominent advocates. In 'The Future of Journalism' (the second of two articles by W. T. Stead which appeared in the *Contemporary Review* in 1886), Stead, for example, commenced his famous editorial manifesto by insisting that, in journalism, 'everything depends upon the individual – the person. Impersonal journalism is effete. To influence men you must be a man, not a mock-uttering oracle' (Stead 1886b 663) Equally categorical was T. P. O'Connor's assertion (in his article 'The New Journalism', published in *The New Review* in 1889) that the 'main point of difference' between the New Journalism and its predecessor was 'the more personal tone of the more modern methods' (O'Connor 1889, 423). The consistency of such statements clearly suggests that a notion of 'personality' was central to the attempt to define the project of a 'new' journalism. What precisely was meant by this term, however, was a subject of varied and even contradictory accounts. Behind their apparent solidarity, this conflicting interpretation of the meaning of 'personal journalism' may be witnessed in the case of Stead and O'Connor. Whereas the former is concerned primarily with what we might term a personalization of the *subject* of journalistic discourse, the latter is occupied, for the most part, with personalizing its *object(s)*. Stead's polemic against 'impersonal journalism' targets the imposing rhetoric of a certain kind of 'high-Victorian' newspaper (*The Times* being the obvious example), and thus joins a longstanding debate about the prevailing use of a plural, anonymous subject – what Stead derides as the 'mystic "We"' – rather than a singular 'I', validated by the signature of the individual journalist. His advocacy of personal journalism is grounded in the belief that, since 'all power should be associated with responsibility', the identity of the journalist should not be effaced by the oracular voice of collective opinion (Stead 1886b, 663). While O'Connor also argued in favour of signed articles, as Tighe Hopkins reports in an 1889 article on the subject of journalistic anonymity (Hopkins 1889, 529), in 'The New Journalism', by contrast, O'Connor attempts to validate the interest of 'personalities' as such. The authenticity of the individual is to be located not in the charismatic figure of the editor, as it often seems to be for Stead, but rather in the 'great men' – Parnell and Disraeli being the examples given – whose 'strong personality' it is the task of the journalist to record (O'Connor 1889, 429).

These two distinct forms of 'personal journalism' are, of course, not entirely separable. Taken together, they may be seen to represent opposite poles of a shared desire to transform journalistic discourse into an intimate mode of communication. Tighe Hopkins revealed this aspiration in his insistence that the function of the newspaper was 'faithfully to reflect, the individual, personal life of its day': 'the Press', he claimed, 'has been more and more taking the tone of "a man speaking to a man"' (Hopkins 1889, 514). Nevertheless, it is important to recognize that this overarching project of the New Journalism was composed of differing lines of argument. As early as 1867, for example, John Morley – Stead's predecessor as editor of the *Pall Mall Gazette* – had suggested that the signing of articles might actually tend to preclude the introduction of 'irrelevant personalities in discussion' (Editor 1867, 290). In an editorial in the *Fortnightly Review*, Morley, like Stead, argued in favour of signed articles on the grounds that it would encourage 'personal responsibility' on the part of the individual journalist. As a consequence, the journalist would be properly exposed to the possibility of reciprocal communicative exchange, in which 'personalities may be in this case retorted, or their unbecomingness detected, whereas retort is impossible or futile against a journal' (Editor 1867, 290). The practice of signature, in other words, was not necessarily a corollary to the wholesale personalization of journalistic discourse, since, in theory at least, it could offer protection against the very phenomenon of which it appeared to partake: the fetishistic or intrusive veneration of 'personalities' of which later critics of the New Journalism were to complain.[2]

Despite these intentions, however, there was undoubtedly a sense in which this criticism became increasingly true. Personal journalism assumed an asymmetrical character in as much as its rhetorical form flagrantly contradicted contemporary developments in the organizational structure and technological resources of the newspaper and periodical press. In particular, Stead's emphasis upon the heroic figure of the individual editor in his *Contemporary Review* essays of 1886 can be seen as an oddly (or defiantly) anachronistic gesture in the context of the increasingly depersonalized system of corporate ownership which, as Raymond Williams and other, more recent historians have shown, began to emerge during this same period (Williams 1992, 199–206; Lee 1980, 79–93). It is surely not coincidental that at the very moment when the material basis of the press made it harder to locate an individuated source of authorial value, the discourse of journalism should so insistently declare its personalized character. This immanent contradiction between the form and the content of (so-called) 'personal journalism' was taken to extremes in monthly magazines of the 1890s such as *The Idler* and *The Strand*. At first sight, these magazines are entirely saturated with a rhetoric of authenticity: the value attached to the signature is literalized through facsimile reproductions and photographs in numerous interviews and related features on celebrities.[3] What this rhetorical strategy attempts to efface, however, is both the serialized form of its insistence upon the authenticity of the signature – literally, that is, such articles were published in lengthy serials – and its dependence upon the very forms of mechanical reproduction that it tacitly claims to disavow. In this context, Theodor Adorno and Max Horkheimer's critique of mass culture as a form of pseudo-intimacy seems entirely apposite, not because intimacy is to be devalued as a mode of public discourse, as conservative cultural critics have traditionally suggested, but, rather, because 'intimacy' in these journalistic texts is itself a form of abstraction (Adorno and Horkheimer 1986, 154–5).

While this is simply to offer a brief outline of a later and more familiar aspect of the New Journalism, for the purposes of this chapter it is perhaps more important to note that a similar contradiction may be observed within the theoretical projects of both Stead and O'Connor. What is characteristic of both editors is their simultaneous desire to utilize and to transcend the medium of print technology by making it into a vehicle of (apparently) unmediated speech. In 'The New Journalism', for example, O'Connor voiced his dissatisfaction with existing forms of parliamentary reporting. By reproducing parliamentary speeches through full and direct quotation, the 'old' journalism, he suggests, reveals all the more clearly the limits of its own mediation: the result being 'long, lifeless even columns' of print, which fail to reanimate the concrete event of speech (O'Connor 1889, 423). As a solution to this problem, O'Connor proposes to convey the context of the speech event and thus to rejoin the link between speech and speaker through the visualized representation of 'personal details with regard to public men'. 'Statesmen', he declares, 'are not ciphers without form or blood or passion. Their utterances and acts are not pure intellectual secretions' (O'Connor 1889, 428). Thus, paradoxically, it is only by devoting less space to the speeches themselves that the newspaper becomes able to render the truth of speech as such.

A similar attempt to transform journalism into a medium of oral communication may be discerned within the more ambitious schemes of Stead. Whereas O'Connor wishes to represent the individual 'personality' in all its concrete embodiment, Stead envisages the ideal newspaper as a medium capable, almost literally, of assembling the collective body of its readers. In 'Government by Journalism' (1886), for instance, he argues that technological advances have created possibilities of direct participatory democracy which undermine the representative function of parliament. Representative democracy is predicated upon the assumption that the sovereign body of 'the people' is no longer capable of being physically assembled, or made immediately present to itself. For Stead, however, this historical process of communal dispersion is reversible: 'The telegraph and the printing press', he insists, 'have converted Great

Britain into a vast agora, or assembly of the whole community, in which the discussion of the affairs of State is carried on from day to day in the hearing of the whole people' (Stead 1886a, 654). It is precisely on account of its utilization of modern communications technology that the newspaper gains the capacity to overcome distance and return the modern public sphere to its supposedly originary state: that of the 'agora' or 'Witanagemote', Greek and Germanic terms for the public space in which the 'whole community' assembles within a single oral and auditory horizon (Stead 1886a, 653). Stead, of course, is not so naive as to assume that this new technologically-mediated agora is, in truth, the product of a purely transparent mode of representation, one in which there is no longer any distance between the legitimate source and the effective instrument of political authority. Nevertheless, he does believe that the press, together with the extra-parliamentary political 'platform', constitute 'the most immediate and most unmistakeable exponents of the national mind' (Stead 1886b, 654). Unlike parliament, the representative function of the newspaper is ratified on a daily basis via the medium of its exchange as a commodity. Moreover, if the expression of public opinion mediated by the press is less spontaneous than that of the public which gathers together for the purpose of political debate, as Stead acknowledges, he is also able to observe that 'The growth of the power of the Platform', exemplified by the career of Gladstone, is itself 'largely the creation of the Press' (Stead 1886a, 656).

Like O'Connor, then, Stead seems to privilege speech over print as a source of political and moral authority, even though he recognizes that speech can only be rendered effectual through the mediation of print. This indicates an ambivalence towards the capacity of his chosen medium which might be seen as more generally characteristic of the New Journalism. The widespread adoption of the American technique of the interview in British magazines and newspapers during the 1880s and 1890s offers a concrete example of this aspiration towards the reproduction of oral forms, which Alan J. Lee has also observed (Lee 1980, 130). As a specific form of journalistic discourse, the interview, indeed, harbours a similar internal contradiction, since, on the one hand, it is concerned with 're-capturing' the authentic event of speech, often represented through the direct quotation of dialogue and the use of photographic illustrations, and yet, at the same time, this event is itself staged solely for the benefit of the text. Thus, in an attempt to efface their textual construction (which was rarely successful), interviews with celebrities were at first often disguised as mere 'dialogue' or informal 'chat'.[4] The form of the interview might well be described in terms which Stead applied to the newspaper as a whole. 'The press', he wrote in 'Government By Journalism', 'is at once the eye and the ear and the tongue of the people. It is the visible speech if not the voice of the democracy. It is the phonograph of the world' (Stead 1886a, 656). This oxymoronic collocation of sensory metaphors is indicative of Stead's populist desire to render even collective identity in a concrete, bodily form. His use of the term 'visible speech' offers an apt expression of the peculiar textual simulation of orality which is achieved both within the technique of the interview and the practices of parliamentary reporting advocated by O'Connor.

The significance of this form of 'personal journalism', however, can only be fully understood when it is set against the normative function previously ascribed to journalistic discourse. In a recent essay on the rhetorical construction of publicity, Michael Warner has argued that, in its classic bourgeois and civic republican forms, what is specific to public discourse is a principle of 'self-abstraction' or 'negativity' by which the subject is assumed to adopt a disinterested stance *vis-à-vis* his or her status as a private individual. In this view, the subject of public discourse is not simply a projection or amplification of the private self as the two spheres are thought to be discontinuous: 'What you say', according to Warner, 'will carry force not because of who you are but despite who you are. Implicit in this principle is a utopian universality that would allow people to transcend the given realities of their bodies and their status' (Warner 1993, 239). The adoption of a rhetorical posture of self-abstraction is thus central to the normative ideal of the public sphere since it preserves the formation of the 'common interest' against the infiltration of private or particular interests. For a number of critics, including Jürgen Habermas whose concept of the 'bourgeois public sphere' Warner is drawing upon, this model of publicity may be located paradigmatically in eighteenth-century periodicals such as *The Spectator*, where anonymous or pseudononymous authorship was explicitly rationalized in these terms (Warner 1993, 236–8; Habermas 1992, 42–3). In this context, anonymity was not perceived simply as an expedient way of concealing or suppressing the journalist's personal identity, although such reasons no doubt also played their part; rather, it belonged to a highly functional mode of rhetorical address in which the journalist wrote both within and to 'the public'.

From around the 1860s, however, this rhetorical convention of anonymity, and the principle of self-abstraction to which it gave support, came under sustained and systematic attack. Writing in *Macmillan's Magazine* in 1861, for example, Thomas Hughes anticipated a number of later commentators in the terms of his argument against anonymous journalism and, in particular, against the editorial use of an abstract collective subject, or 'mysterious "we"'. For Hughes, this mode of journalistic address practises a form of fetishism since it ascribes the products of human labour to what is merely an object, the journal itself, whereas, in fact, 'A thing can have no opinions' (Hughes 1861, 160). By subsuming the particular under the general, such rhetoric constructs a false totality that in turn generates a 'surplus weight and power' over and above any 'intrinsic worth' which a given article might possess (Hughes 1861, 158). Far from representing the legitimate authority, or 'voice', of public opinion as the convention of the journalistic 'we' implies, its function is dismissed as a strategy of domination over the public. Indeed, Hughes goes so far as to suggest that the very notions of '"The public", and "public opinion", are mere abstractions' (Hughes 1861, 162). The attempt to forge a homogenous collective identity, both for the journal and for its readers, can only be seen as an attempt to conceal the particular private interests which all journals inevitably represent. Whilst he rejects the prevailing practice of anonymity, however, it is worth noting that Hughes himself offers an interesting example of the rhetoric of self-abstraction. The extent to which this normative code of public discourse continued to shape the cultural horizon of mid nineteenth-century journalism is evident from the fact that Hughes challenges the system of anonymity on the very grounds of an appeal to the 'disinterested' sphere of public debate which underpinned that system:

> And so every Englishman, who values the freedom which we have, and is anxious that it should take no taint in our generation, ought to give as much spare thought as he can to the consideration of what a free press ought to be like in a free country... I do not ask this as an outsider, but as a member of the fourth estate myself, and one who has had much experience of public writing both in his own name, and anonymously. At any rate, I have this claim on their attention, that I am writing against my own interest; for, selfishly speaking, anonymous writing is to my taste by far the pleasantest, and, if I didn't believe that there are serious objections to it on public grounds, I most assuredly should never say a word against it.
>
> (Hughes 1861, 158)

What is striking about this passage is the way in which Hughes himself deploys all of the rhetorical signs and tokens of the discursive mode of bourgeois publicity, from his appeal to a common collective 'we' to his conspicuous disavowal of personal 'interest'. Such rhetorical devices function as a means of signalling the authenticity and probity of his own contribution to the debate on anonymity, even though the purpose of this contribution is to question the validity of the discourse which he invokes. Hughes's inability to recognize this contradiction is perhaps symptomatic of the disjunction between the normative function of public discourse and what appeared to be its merely rhetorical character, which became increasingly evident from the mid-nineteenth century onwards. It was during this period, in other words, that the rhetorical conventions of journalistic discourse no longer seemed to correspond transparently to any shared principle upon which their cultural legitimacy might rest (which is not, of course, to suggest that such norms had ever commanded universal assent). A similar scepticism towards the rhetoric of collective opinion, for example, was shown by John Morley in his 1867 editorial for the *Fortnightly Review.* Morley answered the charge that journalistic production was in its nature associative by arguing that the collective identity of the journal tended to enforce cohesion, rather than expressing cooperation: the 'Tone and Spirit of a journal, originally the creation of the contributors to it . . . assume[s] the mien and size and power of something outside of and superior to themselves, something to which they feel constrained to bend'. In this manner, he adds, the collective character of the journal – its use of an undifferentiated 'we' – is worshipped as the 'demi-god of an abstraction' (Editor 1867, 292).

During the 1880s, this line of argument was incorporated into the wider debate surrounding the New Journalism. A useful synopsis of the antagonistic positions adopted within this later discussion of the function of journalistic rhetoric can be found in Tighe Hopkins's article 'Anonymity?', which was published in two issues of *The New Review* in 1889 and 1890. While Hopkins declares himself to be strongly in favour of the introduction of signed articles, he also quotes the views of some thirty-five correspondents, who include such leading journalists and critics as Andrew Lang, R. H. Hutton, L. F. Austin, H. D. Traill, William Archer, as well as O'Connor. This collection of differing opinions shows, of course, that the argument against anonymity was by no means won, although by the 1880s, as Laurel Brake has observed (Brake 1994, 89–91), the debate was more narrowly focused on the newspaper press, signed articles having already become an accepted feature of many of the leading periodical reviews. What is interesting about Hopkins's position in the article, however, is the extent to which he seeks to discredit the very notion that the newspaper should function as a vehicle of collective opinion. Although he accepts the fact that the newspaper is comprised of many different kinds of texts, not all of which could or should bear the mark of an individual signature, he is nevertheless keen to extend this privilege to the remaining bastion of collective opinion, the leading article. Whereas Thomas Hughes had still, to some extent, recognized the power and indeed the pleasure of anonymous writing, even while attempting to resist it, Hopkins flatly denies the view that a collective utterance carries any more authority than that of an individual. The rhetorical force of the journalistic 'We' is stripped of any remaining connection to an actual or imagined community of writers or readers:

> The 'We' is not really 'we'; it is 'I'. If this be so, what or whose interest is served by the maintenance of the incognito? Not the interest of the writer, assuredly; and not, I think, the interest of the public. The writer suffers in his proper dignity by being compelled to withhold his name from his work in obedience to a custom which has its chief sanction in the credulity of *le gros public*. . . . As for the public, it ought to be a sufficient objection to the anonymous system that, whilst this system does not and cannot give any *real* force to the work of an honest writer, the dishonest may borrow from it a *simulacrum* of power which, if his mask were stripped from him, would vanish at once.
>
> (Hopkins 1889, 527)

For Hopkins, then, the aura of 'power' that is attached to the use of the collective subject is merely an illusion. Extending earlier perceptions of its ideological character, he further reduces the status of the 'we' to that of a fetish, the effect of which is 'derived' only 'from the willingness of the most credulous and ill-informed portion of society to accept and bow down to a huge and imposing symbol' (Hopkins 1889, 524). Once again, what characterizes this 'symbol' is its reified abstraction: the anonymous subject is, according to Hopkins, 'a bodiless thing, a poor negation' (Hopkins 1889, 526).

The striking recurrence of such bodily metaphors in the texts which I have been examining suggests, I think, that what is really at stake in the debates which surrounded journalistic anonymity from the 1860s to the 1880s is the legitimacy of a particular style of public discourse. If the bourgeois model of the public sphere, as it has been defined by Warner and, to some extent, by Habermas, may be said to have enjoined an essentially negative or disinterested relationship between the public subject and his or her private self, then these texts offer a series of more or less direct challenges to this normative separation between the spheres of personal and impersonal identity. Rather than stigmatizing the 'personal' as the merely particular, and thus barring it from the realm of public debate, the opponents of anonymity strived, in an almost literal fashion, to embody the language of journalism. In this respect, the project of the New Journalism, if not some of its practical manifestations, may be understood as a more radical cultural phenomenon than historians such as Lee have recognized (Lee 1980, 189–90). For as Warner points out, the value of abstraction is not in itself neutral. In historical terms, the virtue ascribed to the disembodied public subject marks the hegemony of a specifically bourgeois and masculine form of critical discourse (Warner 1993, 239–40). Hence, it is not surprising that the notion of 'disinterestedness' should have played such a central role in the critical thought of Matthew Arnold, the earliest and one of the most vehement opponents of the New Journalism. In 'The Function of Criticism at the Present Time' (1865), Arnold famously attacked the factional interests of the mid-Victorian press, and called instead for a means of 'combining all fractions in the common pleasure of a free disinterested play of mind' (Arnold 1907, 20). Yet by exalting his own critical position above the particularity of all others, he effectively reduced the status of 'disinterest' to that of another particular ideology. In the radicalism of Stead's 'personal journalism', Arnold was no doubt quick to sense an attack upon his own version of the theory of journalistic impersonality.[5]

This is not to suggest, however, that the terms of Arnold's evaluation of the New Journalism should be simply reversed. Rather, it is the very opposition between 'personal' and 'impersonal' journalism which must finally be questioned. As an expression of the character of journalistic discourse in the late nineteenth century, the notion of 'personality' is, as I suggested earlier, inherently problematic. By its very nature, the attempt to reproduce the immediacy of speech and the authenticity of the signature within the antithetical form of an increasingly commercialized mass-circulation press could only result in new ideological contradictions. Thus, in order to grasp these contradictions, it remains important to preserve those accounts of the conditions of journalistic labour which run directly counter to the rhetoric of personal journalism. In a guidebook to the profession entitled *Journals and Journalism* (1880), John Oldcastle (ironically a pseudonym for Wilfred Meynell), for instance, presents anonymous journalism in a very different light from Hopkins or Stead. The journalist, he writes:

> breaks up his individuality into a thousand separate fragments, not one of which bears the stamp of his name. The man is scattered and lost, the character of his work is dissipated by dissemination, and nothing remains but the influence of that work, falling as it may when sown broadcast.
>
> (Oldcastle 1880, 72)

Accounts such as this bear witness to the variety of cultural anxieties and pressures which surrounded the debate on anonymity. Rather than being figured as a merely rhetorical expression of collective identity, anonymity is here exposed as the real condition of journalistic authorship. As a heterogeneous textual form, what is peculiar to the newspaper is its natural resistance to the demand for individual authorial property: a demand which cannot be separated from the campaign to introduce signed articles. Yet whilst this dissolution of authorial identity is evidently disturbing, Oldcastle also acknowledges its necessity within an increasingly collective mode of production. This necessity continues to be felt in the case of the New Journalism. As a prototypical form of modern mass culture, the New Journalism extols the value of 'personality' whilst simultaneously extending the division of labour against which it protests. For this reason, the antagonism between intimacy and abstraction remains unreconciled.

Notes

1. The assumption that the New Journalism represented a sudden and dramatic shift from all preceding journalistic forms is one that has recently been challenged by Brake (83–103) and Wiener (47–71).
2. For an example of this common criticism, see 'Celebrities At Play' (145–9), an article satirizing the cult of 'personalities', which was published – ironically – in *The Strand Magazine*.
3. From 1891 to 1895, *The Strand Magazine*, edited by George Newnes, ran serials on 'Portraits of Celebrities at Different Times of Their Lives' (commencing vol. 1) and 'Illustrated Interviews' (vol. 2), as well as a number of articles on the signatures of famous authors written by J. Holt Schooling (vol. 7). Beginning in 1892, *The Idler Magazine*, edited by Jerome K. Jerome and Robert Barr, published many similar texts, and made greater use of modern techniques of photo-mechanical reproduction.
4. For examples of this strategy, see 'A Chat with Conan Doyle' (An 'Idler' Interviewer 340–9) and 'A Dialogue Between Conan Doyle and Robert Barr' (Barr 1894, 503–13), the American version of the same text. I discuss the form of the interview, and the problems which many journalists had in defining its proper function and status, in Salmon, 1997. For an historical account of its use as a journalistic technique, see Nilsson.

5. Arnold's disparaging remarks on the 'new journalism' are contained in his 1887 article 'Up To Easter', and were aimed specifically at Stead's *Pall Mall Gazette* (see Arnold 1977). For recent discussions of Arnold's contribution to the debate on the New Journalism, see Brake (1994, 83–103) and Wiener (1988, 47–8).

Works cited

Anon. 'Celebrities at Play', *The Strand Magazine* (1891), 145–9.
Adorno, Theodor W. and Max Horkheimer. *Dialectic of Enlightenment*, trans. John Cumming. 2nd ed. London and New York, 1986.
Arnold, Matthew. 'The Function of Criticism At The Present Time' in *Essays in Criticism*: First Series. London, repr. 1907, 1–41.
——'Up To Easter' in *The Last Words: The Complete Prose Works of Matthew Arnold*, vol. XI, ed. R. H. Super. Ann Arbor, MI, 1977, 190–209.
Barr, Robert. 'A Dialogue Between Conan Doyle and Robert Barr', *McClure's Magazine* 3.6 (1894), 503–13.
Brake, Laurel. *Subjugated Knowledges: Journalism, Gender and Literature in the Nineteenth Century*. Basingstoke, 1994.
Editor [John Morley], 'Anonymous Journalism', *The Fortnighty Review* 2.9 (1 Sept. 1867), 287–92.
Habermas, Jürgen. *The Structural Transformation of The Public Sphere: An Inquiry into a Category of Bourgeois Society*, trans. Thomas Burger with the assistance of Frederick Lawrence. Cambridge, 1992.
Hopkins, Tighe. 'Anonymity?', *The New Review*, 1.6, (Nov. 1889), 513–31.
——'Anonymity?', *The New Review* 2.10 (March 1890), 265–76.
Hughes, Thomas. 'Anonymous Journalism', *Macmillan's Magazine* (5 Dec. 1861), 157–68.
The Idler Magazine, 1–6 (1892–5).
An 'Idler' Interviewer [Robert Barr]. 'A Chat with Conan Doyle', *The Idler Magazine*, 6 (1894), 340–9.
Lee, Alan J. *The Origins of the Popular Press in England 1855–1914*. London, repr. 1980.
Nilsson, Nils Gunnar. 'The Origin of The Interview', *Journalism Quarterly* (Minneapolis) (Winter 1971), 707–13.
O'Connor, T. P. 'The New Journalism', *The New Review* 1.5 (Oct. 1889), 423–34.
Oldcastle, John [Wilfred Meynell]. *Journals and Journalism: With a Guide for Literary Beginners*. London, 1880.
Salmon, R. 'Signs of Intimacy: The Literary Celebrity in the Age of Interviewing', *Victorian Literature and Culture* 25 (1997) 159–77.
Stead, W. T. 'Government By Journalism', *The Contemporary Review*, 49 (May 1886a), 653–74.
——'The Future of Journalism', *The Contemporary Review*, 50 (Nov. 1886b), 663–79.
The Strand Magazine, 1–10 (1891–5).
Warner, Michael. 'The Mass Public and the Mass Subject' in Bruce Robbins, ed., *The Phantom Public Sphere*. Minneapolis, MN, 1993, 234–56.

Wiener, Joel H. 'How New Was The New Journalism?' in Joel H. Wiener, ed., *Papers For The Millions: The New Journalism in Britain, 1850s to 1914*. Westport, CT, 1988, 47–71.

Williams, Raymond. *The Long Revolution*. London, 1992.

Yates, Edmund. *Edmund Yates: His Recollections and Experience*, 2 vols. London, 1884.

3
Journalistic Discourses and Constructions of Modern Knowledge

Kate Campbell

Journalism's distinction as a writing form has turned on its frequency, regularity and reflexivity. Admitting the moment and ephemera, and related contingent phenomena – the partial and emotional, desire, daily matters, an immediate audience – it is a writing form obviously at odds with the traditional metaphysical strains and higher flights of Western culture.[1]

By the same tokens, however, it seems central to modernity – the era in which the claims of the surface and material world, the individual and contigency have been asserted as never before; and so the resistance its advance met with in the nineteenth century is hardly surprising. Too many substantial social, intellectual and political issues were at stake for its ready legitimation: much as it took time for persons to accept the Copernican sun, the journalistic sun necessarily took a long time to rise.

The contestatory import of journalism was recurrently registered and dismissed in terms of its superficiality, vulgarity and general dealing in 'trash'. The very introduction of the word from France in the early 1830s had a contestatory force, since 'journalism' yoked together higher and lower cultural forms, from the prestigious quarterly reviews to the daily paper: just to acknowledge commonality in such a range of writing challenged customary cultural distinctions.[2] Not surprisingly, then, its higher cultural practioners were apt to disavow their own involvement in the form.[3] Twentieth-century literary criticism has in the main reinforced this distance – most obviously in construction of modernism as a reactive formation to the development of the modern mass media (Williams 1989, 33; Huyssen 1986, 44–62).

This chapter argues that higher and lower cultural practices were, however, in important respects related. Focusing on the construction of a distinctly modern field of public knowledge, it explores how the relative claims of surface and depth knowledge were canvassed and exercised in nineteenth-century journalism. Following an exploration of the discursive construction of distinctly modern knowledge in higher cultural writers and journals, centering on the work of Walter Bagehot, it briefly considers particular forms of low knowledge in late-century news discourse and glances towards modernism. It argues that relatively high cultural writers were imbricated in changes associated with mass culture and modern journalism, and suggests that the similarities between high and low have been mostly overlooked in the negative coding of late-century journalistic developments in terms of 'trivia' and 'trash' in the press. Holbrook Jackson's awareness of the correspondence between the yellow press and yellow book literature in the 1890s has remained exceptional (Jackson 1927, 23).

The 'scraps' and 'trash' of late-century newspaper journalism in fact seem to comprise a neglected aspect of modernity's 'affirmation of ordinary life' (Taylor 1989, 13ff.). Both words refer to deprivation and negation – 'scraps' to remains, 'fragments (of food); broken meat', refuse; 'trash' to 'That which is broken, snapped, or lopped off anything . . . e.g. twigs, splinters, "cuttings from a hedge" . . . straw, rags; [again] refuse 1555', and in particular 'Worthless notions, talk, or writing' (*OED*). These very words of disposal and negation in journalistic contexts tell of inversion and refusal, a world where fragmentation is becoming ontological and waste is not entirely worthless. The ironies are accordingly rich when conferences now address 'The Nineteenth-century Media and the Construction of Identities', in that journalism was widely viewed in this negative light – as a destructive form ending depth, assaulting reason and integrity, affronting coherence and connection.

Accordingly, in Bagehot's essay discussed shortly, in newspapers – which are like London – 'everything is there, and everything is disconnected' (Bagehot 1879, 194). By the end of the century, with their failure to live up to an enlightenment role particularly lamented after the 1870 Education Act, their deficiency in offering 'reasoning understanding' and proper knowledge had become a common theme, with journalism widely denigrated as 'an agency for collecting, condensing and assimilating the entire trivialities of the entire human existence' (Stillman 1891, 689).[4]

The impression was based on the increased range of affairs covered in late nineteenth-century journals and their apparently reprehensible new practices generally, not least the growth of investigative and descriptive writing. Innovatory series such as 'How the Poor Live' and 'Celebrities at Home' illustrate the general shift alongside more spectacular instances headed by W. T. Stead's investigation into child prostitution in 1885, 'The Maiden Tribute of Modern Babylon'.[5] In such journalism a broad sea-change was discernible, whereby 'the journalistic eye is now of far greater importance than the journalistic pen' (Watson 1906, 84) – not so much in respect of illustrations as such as the way in which the visual and other senses were permeating daily journalistic prose.

Most ominously, newspapers' key liberal office, their serious political opinion-making function, appeared in abeyance beside entertainment as reasoned political analysis and verbatim reports of politicians' speeches gave way to interviews, interpretative summaries and sketches. Locally, features such as Henry Lucy's 'Peeps at Parliament' offered 'graphic scraps' and sentiment; on a broader cultural front such representations contributed to 'The Fall of Public Man' (Sennett 1986, 150–218, 259–68), the ascendancy of the image, the growth of symptomatic readings, the aestheticization and putative deterioriation of political and daily life through the daily paper. Rather more was at stake than dismissive responses to the journalistic changes concede.

Coming to terms with trivia: constructions of modern knowledge in higher journalism

The phrase 'graphic scraps' appears in Bagehot's essay on Charles Dickens which first appeared in October 1858 (Bagehot 1879, 192). A prolific producer of higher journalism, like other higher journalists Bagehot recurrently reflected on the form. In as much as to take seriously literature's educative and political obligations for most of the century involved promoting right opinion among 'the many', Bulwer Lytton, both Mills, Bagehot, Matthew Arnold and others suggested its pre-eminence as a modern writing form. The reduction and abolition of the stamp duty were obvious occasions for such reflections.[6]

More than for most writers, however, Bagehot's perception of journalism as a key modern form exceeds instrumental considerations and recognizes its consonance with complex and unresolved realities. The utilitarian aspect is famously voiced in his early essay, 'The First Edinburgh Reviewers': this notes how 'Some things, a few, are written for eternity; some, and a good many, are for time', and urges 'if you wish a person to read, you must give him short views, and clear sentences' (Bagehot 1879, 26, 4). But in it Bagehot soon adumbrates a style veering to intricacy and indeterminacy, in the manner of

> the manifold talker, glancing lightly from topic to topic, suggesting deep things in a jest... expounding nothing, completing nothing, exhausting nothing... fragmentary yet imparting what he says.
>
> (Bagehot 1879, 4)

Whereas the discursive construction of journalistic writing as writing 'for time' customarily signalled a teleological undertaking – putting the present times on the right tracks – here this sense is complemented by writing 'for time' as writing in time: in the thick of things, that is, where all life counts towards knowledge and deserves not to be run dry.

Three years later Bagehot like others sees Dickens in a journalistic light encouraged by his early mode of publication – a sharp-sighted, street-walking reporter, 'a special correspondent for posterity' (Bagehot 1879, 197). The much-quoted-from essay reflects on understanding and elaborates an intellectual opposition between literature proper and journalism that is recurrent in nineteenth and twentieth-century writing. More specifically it seems to take up the *Saturday Review*'s attack on the novelist earlier in the year for his deficiency in philosophy and 'pert, flippant frame of mind' (8 May 1858: quoted Bevington 1941, 162). So, presenting 'graphic scraps' lacking 'the crystalline finish characteristic of the clear and cultured understanding', Bagehot's Dickens shows a

> deficiency in those masculine faculties.... the reasoning understanding and firm, far-seeing sagacity... component elements which stiffen the mind and enable it to protect itself from the rush of circumstances.
>
> (Bagehot 1879, 192–3; 220)

System, coherence, wholeness, balance, perspective, distance: bristling with 'perceptive sharpness', Dickens lacks these indices of 'reasoning understanding' and Bagehot displays as much propriety as the elitist *Saturday* towards him.

Matthew Arnold's famous criticism of 'a new journalism' thirty years later belongs in this critical line, charging cognitive inadequacy: the Home Rulers and the 'new journalism' of W. T. Stead who supports them in the *Pall Mall Gazette* do not look to the end and forecast consequences. Echoing Bagehot's and others' gendering – centring on 'deficiency in those masculine faculties' – Arnold's 'new journalism' is 'feather-brained', somewhat feminine, born along by the wind and the senses, lacking mental rigging, those 'elements which stiffen the mind'; consequently it fails to 'get at the state of things as they truly are' (Arnold

1887, 638–9). The indictment is intensified by the ostensible compliment of its being 'full of ability, novelty, variety, sensation, sympathy, generous instincts': appertaining to surface and sensation, these are somewhat two-edged.

Bagehot seems more profoundly divided, however: if he is clear what Dickens lacks, what he in the end makes of him is not. For punctuating his high line is a defence of Dickens's difference – 'the bizarreness of his genius' is strikingly commensurate with city life, journalism and modern life itself, in which 'everything *is* there, and everything *is* dis-connected' (Bagehot 1879, 194, 197). Bagehot in fact concedes the claims of this writer's 'seeing observation': the street-walking writer (or flaneur) who possesses it, 'the meditative idler amid the hum of existence is much more likely to know its sound and to take in and comprehend its depths and meanings than the scholastic student intent on books' (Bagehot 1879, 218). The 'seeing observation' has a synaesthetic freight, encompassing sensory cognition in general as against 'reasoning understanding'; and the 'depths' of this modern environment tend to the material, suggesting cellars and underground life rather more than sophisticated insights.

Literature seems incompetent faced with such modern realities and 'depths', the knowledge its 'scholastic student' can offer distinctly limited beside that of the street writer. Its poverty relates not least to an 'intent' mode deaf to the hum of things: only a casual, receptive approach can 'take in' the way the world is. Not just what is known but ways of knowing are at issue here – a downgrading of intellectual cognition, oriented towards intensive clarity and summary, system and mastery, and elevation of aesthetic cognition, in which clarity has to do with extension and fidelity to the disorderly way things actually are (see Gasche 1987, 142–50). An 'intent' mode in modern circumstances may be obtuse.

The ontological and epistemological claims of an uncrystalline and formless actual world – a dense material world that tends to defeat the 'reasoning understanding' – appear even less diffidently in Bagehot's ironically entitled *Spectator* essay 'The Trash of the Day' a few years later. Against those who disparage such journals besides book literature, this argues that *The Times* is necessary reading as 'a vast map of embodied events and living detail' (Bagehot 1861, 976).

The journalistic idiom here, with 'embodied events' and 'living detail', is most striking. It impresses what is at stake: a contingent unstable material order, whose sensory data are essential for the distillation of less inchoate and proper sorts of knowledge – knowledge 'reduced into human shape', in Bagehot's question-begging formulation (Bagehot 1861, 977). The newspaper's epistemological and political claims in fairly indiscriminately presenting low fragments are reiterated in this short piece. So it appears both a necessary warehouse for the modern empiricist, and the writing form of modern democracies: 'much of the most valuable part of human knowledge lies there half known, half realized, half said'; 'It is as the many-sided reflex of the many-sided aggregate that what seems valueless trash is practically invaluable' (Bagehot 1861, 977). The newspaper's admission of plurality makes reading the vulgar form oddly circumspect.

In summary, Bagehot is of special interest here because of his foregrounding of low knowledge and the low form of journalism and their conjunction – for instance rehearsing a standard intellectual opposition between journalism and literature over the corpus of the 'special correspondent' Dickens, where the 'seeing observation' and 'graphic scraps' of street life contrast with 'the crystalline finish... of the clear and cultured understanding'. A similar drama of surface and depth is enacted in his and others' writing on the journalist-historian Macaulay, and on Gladstone – a notable political analogue of journalistic superficiality and reflexivity.[7] John Davidson provides one of many late-century variants of the theme in his *Fleet Street Eclogues*, where journalism is imaged in gaudy birds – presumably jays and magpies that seize what the moment brings, gaudy surface knowledge-dealers given to picking up bright things.[8]

However, beyond offering fairly standard analysis Bagehot is significant as a respectable apologist of the sort of low knowledge newspapers increasingly dealt in: acknowledgement of the low extends to its validation. Rather than merely acknowledging the paper's trivia, that is, he concedes its knowledge claims – the cognitive and political importance of journalistic 'trash' and superficiality in complex modern societies whose democratizing norms jar with selective procedures. For if systematic, 'reasoning understanding' involving abstraction and distance has been the traditional high road of knowledge in organic societies, we may now need to abandon such hierarchical order and square with fragmentation, disconnection, surface – proper knowledge now being predicated on an epistemological plunge, suspending 'reasoning understanding' for the 'seeing observation' capable of knowing cellars.

Thus were the vagaries of an aesthetic disposition and trivial newspaper items validated for public knowledge. Similar claims appeared among champions of the relatively new movements of realism in fiction and of empiricism in history, and of course, as indicated already, daily newspaper journalism itself increasingly embraced a common phenomenal world. The new historical mood can aptly be glimpsed in a survey of 'The Modern Newspaper' in 1872: 'The details of history, in truth, can only be gathered from the immense and varied surface' presented by newspapers (Anon. 1872, 348). The underlying concept of 'mirroring' the world and curtailment of selective procedures informed realism and modern journalism as well as newer, more 'informal' history. Henry James casually commends these, seconding Balzac's complaint that

> the official historians have given us no information about manners that is worth speaking of... future ages will care much more for the testimony of the novel, properly executed, than for the writers who 'set in order facts which are about the same in all nations'.
>
> (James 1878, 99)

James's *Century* article on Alphonse Daudet is particularly illuminating concerning the 'new' school of fiction and its overlap with journalistic discourse. So it makes the broad point, that a writer's work may be richer precisely through knowing things lightly, in their surface aspects; and it insists on the modernity of such an approach. This same essay, in a striking proto-modernist move that also harks back to Bagehot's 'seeing observation', deprecates understanding and rational explanation for exposition and the 'pictorial form'. Respecting this, untypically for James but appositely here, with graphic scraps-turned-snaps it nonchalantly proposes the supplementarity of the modern novel and newspaper: Daudet

> held up the mirror to contemporary history, and attempted to complete for us, by supplementary revelations, those images which are projected by the modern newspaper and the photographic album.
>
> (James 1883, 507)

Bagehot's friend and long-serving editor of the weekly *Spectator*, R. H. Hutton, is lastly notable in this context as a persistent exponent of 'seeing observation' and mundane low knowledge, opposing Arnold on precisely this ground. So he repudiates Arnold's notion that

> the thoroughly clear apprehension of a moderately rich experience contributes much more to what is properly the 'modern' element in

literature, than a half-clear apprehension of a very much richer experience.

(Hutton 1869b, 222)

In effect deprecating the intensive clarity of Arnold's intellect beside the extensive clarity due a 'rich and complex' world, Hutton's criticism and own journalistic practice constitute an epistemologically-motivated defence of journalism in terms of the complexity peculiar to modern times (see Woodfield 1986).

Where Arnold approaches journalism on standard intellectual grounds as a vehicle for the best thinking, and so falls in with conventional measures whereby it is mostly found wanting, Hutton perceives it more sensually and sympathetically as a form where such intellectuality may be earthed and less shallow knowledge arise through attention to surface matter. Going against the grain of much 'cultured understanding' he thus elaborates the importance of personal details in gauging the meaning of words: 'Is not the deed or sign, the action or the smile or frown an essential element in the truth itself?' (Hutton 1877, xi).

The question conjures a proto-Bakhtinian dialogizer moving beyond the objective referential axis of words in insisting on the semantic value of the 'living mouth'; and Hutton's animosity towards Arnold as the 'cut-and-dried' man of culture given to dogmatic utterance further suggests an opponent of detached knowledge and monologism.[9] Populist variants of such hostility appeared in many 'new' journalists. Most quoted is T. P. O'Connor's manifesto in the evening *Star*, promising public figures would be presented 'living, breathing, in blushes or in tears – and not merely by the dead words that they utter' (*Star* 17 Jan. 1888, 1).

The notion of some sorts of words and accounts being 'dead' haunts the writers considered so far, the corollary of claims for 'seeing observation' and 'graphic scraps' (these were indeed imaged as ending greyness and bringing colour). Such terms, like Bagehot's 'embodied events' and 'living details', and their high cultural defence, press the newspaper provenance of such knowledge and the possibility of positive constructions of new newspaper practices involving them.

Shifts in public knowledge in late Victorian newspapers

The positive terms offered by higher journalistic discourses for reading changes in late Victorian newspapers – say their 'embodied events' – cannot be systematically followed through here. It is only possible to touch on the epistemological significance of some of the new low traits including the 'graphic scraps' and 'seeing observation' that particularly obviously align such journalism with modernism. Here the point already made about 'seeing observation' needs reiteration: in the increasingly synaesthetic environment of late-century newspapers, visual references mostly appertained to sensory cognition generally.

More broadly, we can note how 'new journalism' extended beyond its most flagrant locations, in the *Pall Mall Gazette* under W. T. Stead in the 1880s and O'Connor's *Star*. A much contested term, it seems best apprehended loosely, with reference to a growing emphasis within the main London daily papers, traceable to the mid-century, on how things present and life's whole surface are evident in the growth of features such as headlines, written sketches, interviews, illustrations, investigations and the 'personal' and 'human interest' note that still inform newspapers. For convenience Stead's work will be most mentioned but it is important not to overlook this broad identity leading to a modern package of 'the news' before the century's end.[10]

Among the most widespread of journalistic revisions of public knowledge, operative for instance in the institution of lobby correspondents from the 1880s, was the already-mentioned shift from official reporting of political utterance to contextual details – from the what of public articulation to the how, from transcription to description and images. This shift from content to manner, signified to signifier, extended beyond political affairs in all matters of 'human interest' and lay behind Hutton's dictum that 'the modern newspaper and the novel are beginning to assimilate, and are becoming very much alike' (Hutton 1869a). The interview form, promoted by Stead and enjoining and parading a being *en rapport* above the simple report, was in particular testimony to a shift from the final word to the 'living mouth': growing impatience with the 'cut-and-dried' words of public figures and infusion of 'personal' tones (see Stead 1884).

In this environment, Stead's description of the newspaper press as the 'great inspector with a myriad eyes, who never sleeps' was increasingly apt – as with an early 1880s meeting at the Reform Club to consider 'paragraphs' in the *Echo*, *Standard* and *Scotsman* that had 'made public' certain private proceedings, making it seem that 'Everywhere, newspapermen were on the prowl' (Stead 1886, 673; Koss 1981, I. 239). 'The Truth about...' as title, as in the *PMG*'s 'The Truth about the Navy Campaign' (September 1884 – June 1885), and the frequent signature 'By One Who Knows the Facts', were the up-front and underhand faces of this revelatory office gravitating to 'graphic scraps', the nefarious and hidden. Allon White's analysis of early modernist obscurity inadvertently points to the consonance of high art and this lower journalistic environment, remarking of George Meredith, Joseph Conrad and Henry James, 'There is in their work the feeling that inner privacies are being cruelly probed and exposed in a way that was quite new' (White 1981, 53). It seems to me this 'feeling' and modernism's 'age of suspicion' had some journalistic basis, not least in 'graphic scraps' telling unofficial stories.

The shift away from the routine 'cut-and-dried' occurred not just in the increasing use of new journalistic practices in themselves. For these rebounded on readers, inviting them to question a fairly 'cut-and-dried' hierarchical order in itself – specifically, to evaluate the authority and utterance of public figures. Relating their public words to their positions and personalities – the 'living mouths' – could make the distinction between signifier and signified and the what and the how just invoked problematical, for the what might well be revised in the knowledge of how and why, and then constructed quite differently, with resultant changes in meaning – and possibly in the credibility of authority too.

An appropriate example is the *Pall Mall Gazette*'s frequent undermining of Arnold under Stead's editorship. The elitism and remoteness of his pronouncements were stressed in occasional notes on his less-than-enthusiastic reception in America, attention generally to disaffection in his audiences and the dotting of i's that Arnold himself avoided (so on 1 Nov. 1883 the paper declared the gist of one of his speeches: the minority are right, the majority wrong). Given the sometime animosity and undermining of his 'meaning' it seems not fortuitous that Arnold somewhat carelessly charged the 'clever and energetic' Stead with 'invention' of 'a new journalism' (Arnold 1887, 638).

Broadly speaking, the increased reliance on 'graphic scraps' and 'seeing observation' in even the respectable papers seems to have fostered a sign-reading, semiotic approach *avant la lettre* and in tension with long-standing public knowledge. In support of this we may note Dorothy Richardson remarking the knowingness of late-century papers in *Pilgrimage*; and within the period, widespread awareness of Gladstone's theatricality alongside simple-minded veneration of the Grand-Old-Wood-Cutter (see Hamer 1978). Here liberal conceptualization of politics as an arena of debate was liable to erosion by intimations of it as a performing art to see through.

Particularly among journalists themselves the knowingness appears in awareness of the politics of style and presentation. So, for instance, Stead repudiated the 'social conventionality' accompanying 'well-balanced sentences' (Stead 1886, 667); and the ex-*PMG* and *Cornhill* journalist,

Frederick Greenwood, was not alone in apprehending the performative aspect of language which so concerned modernist writers, suggesting how unintentionally the semblance of truth entailed writing having the force of it: because speculative articles from abroad

> are telegraphed, and because they are printed with news as news, the writer's remarks are invested by most minds with the importance due to a statement of facts... the reader often takes that for veiled information which is merely speculation.
>
> (Greenwood 1897, 714)

Lastly in this brief run-through, the late-century papers' epistemological significance includes their disaffection with a single order of knowledge altogether. Thus in the *PMG*, alongside headlines suggesting the opposite – proclaiming 'The Truth' – ran features subtitled 'the point of view of...' and the admission of hitherto unheard voices – women, anarchists, socialists, match-girls (for instance, 'Streetsellers', 3 Jan. 1887, 4–5). Though the admission of new subjectivities was extremely limited and superficial it seems that, however nominal, the incorporation of difference, as in a 'many-sided aggregate', went some way to unsettling a singular world view and contributed to the currency of a 'principle of radical doubt' – a principle seen as a hallmark of modernity (see, for instance, Giddens 1991, 3).

In conclusion, in this and other respects the features I have noted are related to the mid-century construction of 'modern knowledge' and roughly correspond to epistemological changes associated with modernist writing: deprecation of straightforward referentiality, 'rational understanding', abstraction and authority; admission of plurality and fragments; prioritization of the signifier, point of view and image, with attendant growth of symptomatic readings. The homologies could be further elaborated, undermining the common, fairly strict polarization of high art and low mass culture at the *fin-de-siècle*. The consonance between the knowledge offered at different levels is crudely highlighted in H. G. Wells's indictment of Henry James for foregrounding 'an egg-shell, a bit of string' – erstwhile trivia, that is; and in Georg Lukács's indictment of modernism, for failing to distinguish the incidental and fundamental in life – a charge of trivialization routinely levelled at modern newspapers (Wells 1915, 100; Lukács 1963, 17–39).

There is growing appreciation of different 'modernisms' now, as against 'Modernism', and my account of overlap respecting knowledge obviously fits some versions better than others (more James's and Woolf's than Imagism, say). The fact that 'seeing observation' and pictorial accounts were upheld in terms of distinctly modern knowledge against 'rational understanding' by some Victorian high cultural writers suggests 'modernities' may be an equally necessary revision. Their particular constructions of 'modern knowledge' argue the continuities as well as rupture in modernisms' emergence.

While much of the new 'human interest' and low, often 'graphic' knowledge found in late-century English newspapers certainly was reductive, facetious and sensational, and many modernists most assiduously distanced themselves from the mass media and journalism, I have suggested how such accounts should not be seen as the whole significance of changes in newspaper journalism. Rather there seems to have been no absolute line between superficial and profound knowledge, and similar modes of cognition advanced at quite different cultural levels. What for long has passed as trivial stuff in newspapers may have been conducive to and even incubated modernist sensibilities.[11]

Notes

1. See for instance how 'Men are divided broadly into journalists and eternalists, ephemera and immortals' (Stillman 1891, 688).
2. The *OED*'s first citation for the word is in a *Westminster Review*, Jan. 1833; though note also 'Journalism, by a Newspaper Editor', May 1831, a privately printed pamphlet anticipating this by two years.
3. The phenomenon of disavowal and Walter Pater's in particular are addressed in Brake 1992, 59–86.
4. On the liberal vision and subsequent disillusion see respectively Lee 1976, 21–6, 213ff.
5. Before book compilation Sims's forrays featured in the *Daily News*. It was the late 1880s's weekly the *World* that featured 'Celebrities at Home' but on many fronts daily papers were disrupting privacy too – not least in divorce court reporting.
6. For instance, Bulwer Lytton's *England and the English* (1833), with its exposition of the importance of the press in liberal thinking, was in part an argument for the reduction of the stamp duty (passed in 1836).
7. See for instance Bagehot 1879. Respecting Gladstone, Arnold was not exceptional in identifying a certain speciousness that broadly accords with the analysis of Hamer 1978.
8. Thus a carrier of literature's grey distaff at one point remonstrates to his journalistic shepherd companion: 'gaudy birds/Gay flowers and jewels like me not' (Davidson 1896, 32).
9. Mikhail Bakhtin, apparently unfamiliar with 'new journalism', sees journalistic writing as creating 'the least favourable conditions for overcoming monologism', in contrast to Dostoevsky's – reluctant 'to separate the thought from

the person, from a living mouth, in order to bind it to another thought on a purely referential and impersonal plane' (Bakhtin 1989, 95).
10. On the broad changes in metropolitan newspaper journalism see Brown 1985. Descriptions of 'new journalism' are too numerous for citation here; Brown offers some, Lee also.
11. Consider for instance the low knowledge which Henry James gravitated towards in reading the papers, evident in his *Notebooks*.

Works cited

Anon. 'The Modern Newspaper', *British Quarterly Review* 55 (April 1872), 348–80.
Arnold, Matthew. 'Up to Easter', *Nineteenth Century* 21 (May 1887), 629–43.
'Charles Dickens', [1858], 'Thomas Babington Macaulay', *Literary Studies*, II. 184–220; 221–60.
'The Trash of the Day', *Spectator* 34 (7 September 1861), 976–7.
Bagehot, Walter. 'The First Edinburgh Reviewers', *Literary Studies*, ed. R. H. Hutton, 2 vols. London, 1879. I. 1–40.
Bakhtin, Mikhail. *Problems of Dostoevsky's Poetics and Prose*. Manchester, 1989.
Bevington, M. M. *The Saturday Review: 1855–1868*. New York, 1941.
Brake, Laurel, 'Aesthetics in the Affray', *The Politics of Pleasure*, ed. S. Regan. Buckingham, 1992. 59–86.
Brown, Lucy, *Victorian News and Newspapers*. Oxford, 1985.
Davidson, John. *Fleet Street Eclogues*. London, 1896.
Gasche, Rudolf. 'Of aesthetic and intellectual determination', *Poststructuralism and the Question of History*, ed. Attridge, D., Bennington, F., Young, R. Cambridge, 1987.
Giddens, Anthony. *Modernity and Self Identity*. Cambridge, 1991.
Greenwood, Frederick. 'The Newspaper Press', *Blackwood's Edinburgh Magazine* 161 (May 1897), 704–20.
Hamer, D. A. 'Gladstone: The Making of a Political Myth', *Victorian Studies* (Autumn 1978), 29–50.
Hutton, R. H. 'The Empire of Novels', *Spectator* 22 (9 January 1869a), 43–4.
——'Mr Matthew Arnold on the Modern Element in Literature', *Spectator* (20 February 1869b), 222–3.
——*Theological Essays*. London, 1877.
Huyssen, Andreas. *After the Great Divide: Modernism, Mass Culture, Postmodernism*. Bloomington, CA, 1986.
Jackson, Holbrook. *The Eighteen Nineties*. London, 1927.
James, Henry. *French Poets and Novelists*. London, 1878.
——*The Complete Notebooks of Henry James*, ed. Leon Edel and Lyall H. Powers. New York and Oxford, 1987.
——'Alphonse Daudet', *Century* 26 (August 1883), 498–509.
Koss, Stephen. *The Rise and Fall of the Political Press in England*. vol. 1. London, 1981.
Lee, A. J. *Origins of the Popular Press in England, 1855–1914*. London, 1976.
Lytton, Bulwer. *England and the English*. [1833] London, 1835.
Lukács, Georg. *The Meaning of Contemporary Realism*. London, 1963.
O'Connor, T. P. *Star* 1 (17 January 1888), 1.

Sennett, Richard. *The Fall of Public Man*. London, 1986.
Stead, W. T. 'Interviewing versus Bookmaking', *Pall Mall Gazette* (16 August 1884), 1.
——'The Maiden Tribute to Modern Babylon', *Pall Mall Gazette* (May 1885).
——'Government by Journalism', *Contemporary Review* 49 (May 1886), 653–74.
Stillman, W. J. 'Journalism and Literature', *Atlantic Monthly*, 68 (November 1891), 687–95.
Taylor, Charles. *Sources of the Self*. Cambridge, 1989.
Watson, E. H. L. *Hints to Young Authors*. London, 1906.
Wells, H. G. *Boon*. London, 1915.
White, Allon. *The Uses of Obscurity: The Fiction of Early Modernism*. London, 1981.
Williams, Raymond. *The Politics of Modernism*. ed. Tony Pinkney. London, 1989.
Woodfield, Malcolm. 'Victorian Weekly Reviews and Reviewing after 1860: R. H. Hutton and the *Spectator*', *Yearbook of English Studies* 16 (1986), 74–91.

4
A Centre that Would not Hold: Annuals and Cultural Democracy

Margaret Linley

Writers for journals and reviews in the first quarter of the nineteenth century imparted to middle-class readers an unparalleled power of reading, as Jon Klancher demonstrates. By ceaselessly mapping interpretive strategies, journals forged a distinctive identity against those above and below them (Klancher 1987, 50–2). This print boom in the early nineteenth century had a tremendous impact on 'the way knowledge was conceptualized' (Butler 1993, 122). Increasing fragmentation, compartmentalization and specialization of information involved a negotiation of gender as well as class relationships in terms of nationality. While the journals were working to establish certain forms of theoretical and political knowledge as masculine and middle-class, publishers began defining the women's market in the 1820s and the juvenile market at the end of the decade through gift books and annuals. From the appearance of Rudolph Ackermann's *Forget-Me-Not, a Christmas and New Year's Present* in 1823 to the thirtieth and final issue of the *Keepsake* in 1857, the annual market was fiercely competitive, with sixty-three gift books making an appearance in 1832 and more than two hundred by the end of the decade.

The separate spheres ideology implicit in the fragmentation of the print market should not prevent us from observing the overlap between readers and writers of annuals and journals. Not only did women, such as Maria Jane Jewsbury, Felicia Hemans, Caroline Norton and Letitia Landon, contribute articles and poetry to journals, but men regularly wrote for and edited annuals. The interdependence of the two markets is obvious in the regular appearance of advertisements and reviews of annuals and gift books in journals, indicating that despite the implicit gendering of audiences, readership was to some extent shared. Annuals, moreover, often contain prose and poetry that refer to political issues analysed in journals.

Though annuals could run as high as a guinea in England and $5 in America, they were priced generally in the ten to twelve shilling range, less expensive than a year's subscription to a quarterly, which Klancher says cost four to six shillings each (Klancher 1987, 50–1). Consequently, while annuals, like journals, determined the outer limits of their audience on an economic basis, they also shared with journals a rather imprecise delineation of readership, blurring social differences among readers while uniting them within a framework characterized by style, visuality, and, above all, a desire for cultural self-improvement. Designed to take their place in the drawing room, annuals, often gilt-edged, came in ornate paper or cloth covers, and silk or embossed leather bindings; they usually offered between ten and thirty plates (mostly steel engravings of either original drawings or famous paintings) and up to four hundred pages of poems, sketches, tales and travel essays by the best-known writers of the day. If the journals attempted to define how a middle-class person should think, annuals laboured to establish the *look* of middle-class leisure and that look was especially centred around a self-conscious display of a taste for beautiful things.

Given their ostentatious association of an appreciation of beauty with aristocratic sensibility, the annuals would appear to illustrate Pierre Bourdieu's point that aesthetic disposition functions as a marker of class distinction and is therefore inseparable from the history of social conflict, not least because cultural competence is available only to those who have access to the codes into which a work of art is encoded (Bourdieu 1984, 2). Yet while the annuals evince an especially volatile moment in the struggles for the definition of what Bourdieu calls 'cultural nobility,' registering differing ideas within the middle and upper ranks about socially recognized hierarchies of art and the corresponding status of consumers, they also call in question his opposition between popular and traditional aesthetic proclivities. Although the aristocratic trappings of these fancy gift books encourage us to perceive them as 'modelled on aristocratic assumptions and priorities' (Hofkosh 1993, 205) or as selling 'social snobbery' (Reynolds 1995, xxvi), their massive, if not precisely mass, audience suggests another possibility: not that the annuals are organized in opposition to all that is popular, vulgar and ignoble, but that they are constituted, rather, at the paradoxical juncture of popular and traditional, high and low taste.[1]

Just as the proliferation of journals foregrounds the incoherence of the communal identity they attempted to compose, annuals necessarily announce, both materially and textually, the ambivalent effects of the beautiful forms they celebrate. While at first glance the annuals may appear simply to mirror aristocratic values, a closer look reveals a fluctuating cultural market whereby social positions, and the political outlooks they imply, are in the process of being formed and differentiated through continual negotiation. Organized around an economy of gift-giving, the 'annual inundation,' as one reviewer phrased the yearly increase in supply (Anon. 1831a, 704), significantly expanded the merchandising and consumption of cultural artifacts produced in part by and aimed especially at the growing middle ranks of Victorian England.[2]

Early each fall, annuals aggressively competed to capitalize on the spirit of Christmas giving by representing value as more than monetary worth, a surplus ambivalently concentrated in the notion of the gift. When the engraver Charles Heath launched *The Keepsake* for the 1828 Christmas season, the annuals had been flourishing in the Christmas market for five years. Leigh Hunt's well-known essay, 'Pocket-Books and Keepsakes,' which functions as an extended advertisement for the volume, provides an insightful analysis of the value of keepsakes generally and of the book as keepsake specifically. Outlining economic conditions governing the bestowal of a keepsake in the broadest terms possible, as being 'neither so priceless as to be worth nothing in itself, nor yet so costly as to bring an obligation on the receiver' ([Hunt] 1828, 14), Hunt draws attention to the coexistence of monetary and symbolic significances in the economy of cultural production, whereby one system of value does not necessarily cancel another. Economic value of a keepsake is negotiable precisely because surplus or symbolic value translates the object into a gift through a process not of metaphor, or exchange, but of metonymy: 'something that has been about a friend's person', especially a lock of hair, stands in for the whole person because it is a 'part of the individual's self' ([Hunt] 1828, 15).[3]

Establishing the special status of the book as keepsake becomes for Hunt a question of redefining the exchangeable commodity as 'part of the individual's self'. A book is important in this regard, Hunt argues, because 'like a friend, it can talk with and entertain us' ([Hunt] 1828, 16). A book given as a keepsake can represent the absent giver, appearing to close the temporal gap that separates giving, and hence the giver, from the receiver's pleasure in using the gift; by this process, the purchase of the beautiful commodity is translated socially and symbolically from a commercial transaction into a communal act, while the receiver's solitary act of reading and viewing becomes a form of communication. Moreover, the receiver's consent to a shared aesthetic sensibility, demonstrated by accepting the book, converts spectatorial desire for the beautiful object, which might inspire a sense of dispossession and alienation, into sympathetic identification between individuals.[4]

A book specifically designed as keepsake allows for more, however, than the communication of sympathy and the display of taste. Bestowing a present of a gift book allows for the self-conscious staging of sympathy and affection as a semiotic dynamic at play in the performance and intervention in the fantastic world of beautiful things. Hunt suggests that the giver can manipulate the visual effects of the gift by marking 'his or her favourite passages throughout (as delicately as need be), and so present, as it were, the author's and giver's minds at once' ([Hunt] 1828, 16). The giver can make a further impression on the impact of the gift by adding a personal signature or note: 'one precious name, or little inscription at the beginning of the volume, where the hand that wrote it is known to be generous in its wishes, if not in its means, is worth all the binding in St. James' ([Hunt] 1828, 17). Finally, the gift book can be combined with non-monetary keepsakes, especially the 'most precious of all the keepsakes – hair' ([Hunt] 1828, 18); hair not only supplements the symbolic value of the book, but is itself, when braided, recast through combination with the book into a useful object, a bookmark.

In contrast to the identified paintings and engravings in *The Keepsake*, all the textual contributions are published anonymously. Hunt inscribes this exigency of publication in the irony that those very strategies of intimacy and individuality which he recommends to the giver of the gift book simultaneously eviscerate personalization. It is equally ironic that although individualizing techniques enable the giver to become a writer of sorts, the self-authorization and self-possession authorship implies are incomplete without the coincidence between 'author's and giver's minds' or between giver's and receiver's sympathies. This decentring of the supremacy of the self in the act of expressing affection and generosity is repeated in the material composition of the multi-authored mixed-media gift book as well as in the performative dimension of its social utility. Moreover, by encouraging the reader to assume the creative position of the author (and actor), indeed of various authors of the annual, the ritual of gift-giving becomes a dialogic process whereby new aesthetic and political possibilities for representation can find articulation (and action) precisely through the meeting and mingling of different perspectives.

In terms of the politics of cultural production, the annual phenomenon thus represents a link between eighteenth-century notions of the idea of the liberal state ordered and bound together by affection instead of by force and the later nineteenth-century view of culture as an aristocracy within, a hidden source of value that can redeem the masses from themselves. Arising alongside the reform movement of the second quarter of the century, the annual phenomenon shares, in fact commodifies, that socially oriented revisionist and unifying aim, cultivating and legitimizing a taste for the beautiful, the civil and the refined which marks off members of a class, and affirms and naturalizes their differences from those with alternative aesthetic dispositions, not only vertically, but also horizontally across other social categories such as gender, religion and race. At the same time, however, they simultaneously imply the production of a desire for self-representation that might go beyond the civic rules they are trying to establish, a desire which potentially upsets hierarchies of authority, generates radically autonomous subjectivities and encourages independent cultural identities.

Although the amorphous public this aesthetic democracy produced was primarily composed of consumers, not voters, an edge in the annual market nevertheless required constant redefinition, increasingly through the merchandising of famous names, aristocratic titles and celebrated authors, artists and engravers and always through material innovations, the most important being the commercially viable steel engraving.[5] Not surprisingly, the competition that drove the production process left its mark on the text. On introducing embossed leather bindings into the marketplace, for example, the 1830 *Friendship's Offering* facetiously proclaimed its superiority with the following prologue, which announced, among other things, the primary importance of gender, and of femininity in particular, in the shaping of aesthetic sensibility:

> That the meek prudish AMULET
> With bitter jealousy will fret;
> That KEEPSAKE, GEM, FORGET-ME-NOT,
> And some whose names I have forgot,
> Who dress themselves in silk attire,
> For very envy will expire.
>
> (F. O. 1830, vi)[6]

Here we appear to encounter an example of what Anne Mellor identifies as a 'systematically constructed... hegemonic ideal of feminine beauty,' whereby the annuals promote 'an image of the ideal woman as specular, as the object rather than the owner of the gaze' (Mellor 1993, 111). However, there is a highly competitive, indeed downright desirous, gaze attributed to these feminized and anthropomorphized gift books (*Amulet, Keepsake, Gem, Forget-Me-Not*), suggestive, moreover, of the possibility that free competition unsettles any easy assumption of pristine virtue and purifying disinterest supposedly inhering in either the beautiful object or its perception. Transformed into female jealousy, competition, like femininity, appears to be as trivial as it is beautiful, while the ideal of beauty becomes a mere sign of fashion. Not only are feminine beauty and the serious masculine business of competition simultaneously levelled by the parodic gesture of the gendered commodity text, but the very notion of the separation of the spheres is itself the stuff of comedy.[7]

We see the transformative and appropriative figuration of social relations at work throughout this self-advertising prologue to the *Friendship's Offering*. 'Competition' between the 'pretty *Bijou*,' the 'coquettish *Keepsake*,' and the 'sweet *Souvenir*,' the poem goes on to suggest, 'brings much good' to the nation as a whole, for ''Tis owned that our proud Native Land / Alone can boast so fair a band' (F.O. 1830, vii). Dressed up in this extended feminine metaphor, competition and beauty masquerade in different directions across gender and class boundaries. The feminine guise enables a naturalization of competition in the feminine sphere, as female biological reproductivity becomes the hopeful analogue for the annual's wish for prosperity and posterity: this 'band of bright compeers' will teach their 'offspring to inherit / A generous RIVALSHIP IN MERIT' (F.O. 1830, vii). Ideal beauty and patrilineal inheritance are transformed into the children or products of feminine competition. Coded in the polite tones of chivalry, middle-class competition gains a semblance of aristocratic disinterest while aristocratic privilege is recast, renewed and subsumed within middle-class notions of merit and virtue.

Distinction of taste, and the superior knowledge it implies, are played out within masculine terms as well and, in so doing, further mark a struggle for the meaning of cultural competence, but in noticeably different ways. The introduction by Alaric Watts to the 1831 *Literary Souvenir* directly challenges the power of the critic as arbiter of aesthetic judgement. Responding to a litany of complaints published in journal reviews, he returns a volley of his own charges against 'a certain class' of critics who condemn entire volumes of annuals on the faults of a few pieces, who debase popular writers wholesale when one of their contributions is deemed faulty, who then turn around and criticize annuals for using less popular writers, who harp, moreover, on the triviality of annuals but who also publish pieces they have rejected and, finally, who have been known to praise precisely those works when published in a single-authored book (and under another name) that they have condemned earlier in annuals. This attack on the capriciousness of reviewers is not simply an attempt to establish the superior judgement and careful selectivity of the annuals. It also points to a fundamental irony in traditional critical practices: critics favour single-authored texts, yet they themselves not only publish for the most part in widely circulating multi-authored media, but in so doing maintain their stronghold on the politics of taste. By implication, then, Watts suggests that critics, knowing all the while that their publishing medium commands a wider audience than that of the authors they review, invest in the supremacy of the individual author and the propriety of the single-authored book out of a self-interested desire for cultural power.[8]

For Watts the battle line is drawn across gender. The masculine preserve of critical objectivity is, in his view, an irrational pretence, arbitrary and despotic toward the more refined sensibility of a feminized taste for beauty. Not only did the 1831 *Athenaeum* free supplement on the annuals confirm Watts's position by claiming that one effect of the gift books is 'to smooth and boarding-school down all manliness and vigour' (Anon. 1831a, 704), but it also mobilized gendered discourse to answer his charges directly in a review of the *Literary Souvenir*: 'we think it extremely bad taste in Mr. Watts to make this display of temper before the ladies, and in our drawing-rooms' (Anon. 1831a, 705). Whereas the paradoxical and reversible nature of femininity played upon in the self-advertisement to the *Friendship's Offering* exemplifies the conventions of material production and reception of the annuals themselves (uniting socially diverse readers around a counterfeit aristocratic beauty which it simultaneously mocks), the *Literary Souvenir*, in contrast, situates itself as a protector and guardian of that same feminine realm. Constructing interference from a 'certain class' of male critics as tyrannical, misguided and self-aggrandizing, the annual editor takes up the position of the defender of civility and refinement, thereby contending for an alternative cultural authority based on the feminization of taste.[9]

Watts's confidence in the possibility that a male might construct an alternative cultural authority in the feminized domain of the annuals was by no means always shared by other male professional writers. The conflict and contradiction recognized by Watts in the attitudes of some male critics toward the feminized commodity gift book is announced in boldface, as it were, in the difference between the actual practice and the extreme opinions of regular contributors to the annuals, such as Robert Southey and Charles Lamb. Southey, who as poet laureate could command fifty guineas for a contribution to the 1828 *Keepsake* (which, he recorded, sold 15,000 copies), called the annuals 'picture-books for grown children.'[10] Similarly, Lamb, after agreeing to contribute to *The Bijou* in 1828, wrote to Bernard Barton complaining how 'I shall hate myself in frippery, strutting along, and vying finery with Beaux and Belles, with "future Lord Byrons and sweet L.E.L.'s."'[11] Southey was deeply resentful of competition from well-paid aristocratic amateurs, and Lamb, who could not win the high sums pocketed by Southey, seemed contemptuous of anyone who earned more.[12] Vociferously denigrating a rapidly growing market in which no one could afford not to participate, these authors both express and simultaneously attempt to resolve the displaced and increasingly marginal status of the male poet in a new world (dis)order of feminized and commercialized literature.

Hunt's *Keepsake* essay suggests that the gift book giver could virtually usurp the position of author by leaving personal marks on the text, and Watts's introduction to the *Literary Souvenir* registers a general perception that the collective nature of annuals might threaten (what we now call after Foucault) the 'author-function,' as property, sovereign name and authentic individual. Aesthetic democracy further threatened the supremacy of creative genius by eroding the distinction between talent and title: inside the covers of the annual, the gifted author competed for recognition alongside those who were traditionally his patrons. Despite attempts to stave off criticism, such as the *Friendship's Offering*'s typographical shout proclaiming that annuals promote a 'RIVALSHIP IN MERIT,' some of the strongest disapproval was directed toward the levelling effect of the annual market on the status of creative genius. Thus an 1831 *Athenaeum* review dismissively remarks that the *Keepsake* for 1832 is 'to all intents and purposes, a book belonging to rank, claiming the same exclusive homage to its literature, which the titled contributors claim for their station', while also hinting that this very circumstance is calculated precisely to ensure 'the "Keepsake" will sell' (Anon. 1831b, 736).[13]

Five years later, Thackeray's review of annuals for *Fraser's* unequivocally links the commercialism of the gift book industry to 'a kind of prostitution', with the conclusion that 'a little sham sentiment is employed to illustrate a little sham art' ([Thackeray] 1837, 758). While Thackeray displays the false pretenses of 'exclusive homage' all the more vividly by mapping it in moral and sexual coordinates shared by *Fraser's* readers, he does so striking an indignant pose undoubtedly calculated to offend anyone ever having anything to do with annuals – staging, in other words, a performance more inimical to bourgeois decorum than the prostituted 'little sham art' he mocks. Likewise, by the end of the decade, Carlyle situates himself outside the middle-class world of simulacra and similitude in order to 'take the *spectacles* off his eyes and honestly look, to know' (Carlyle 1966, 175); in the process, he associates the modern catastrophe of vision with a failure of authority: 'Heroes have gone-out; Quacks have come-in' (1966, 174).

Connecting prostitution and performance to a crisis in autonomy, authority and authenticity, these authors understand the problem of the male professional writer in the same terms as the 1831 *Athenaeum* quoted earlier, as a threat to 'manliness and vigour'. The feminine domain of the annuals is, for the male professional writer, simultaneously an effeminate and effeminizing one, where, in Lamb's words, titled males stage a 'barefaced sort of emulation' (Lamb 1891, 2. 212) of both masculine genius and feminine beauty itself. Assertions of 'heroic masculinity' that bring the 'hero as spectacle' into prominence confirm, as James Eli Adams argues, a persistent affinity between dandy and poet-prophet, the one making a spectacle of himself in the world, the other remaking himself as a spectacle of estrangement from the world (Adams 1995, 215). Aesthetic democracy thus potentially dethrones the supremacy of poetic and critical vision by exposing the performative basis of its gender and genius; aesthetic democracy allows, moreover, for the unruly significance of 'true no-meaning' in linguistic transvestism by cross-dressing gender-bending males, who perpetually destabilize any normative definition of masculinity to which notions of merit are subject.

While the 'dandy' visibility of the annual world provided a forum for struggles for cultural power between males,[14] the very attempt to define this world as masculine, however effeminate, signals a simultaneous conflict between genders, as some males strenuously objected to women's increasing participation in the literary marketplace. Thus Southey's infamous comment to Charlotte Brontë that 'literature cannot be the business of a woman's life, and it ought not to be!'[15] and Lamb's that 'if [L.E.L.] belonged to me, I would lock her up and feed her on bread and water till she left off writing poetry. A female poet, or female author of any kind, ranks below an actress.'[16] Such histrionics aside, annuals, far more so than journals, made possible a steady income for professional women writers from the upper and middle ranks of society and provided a forum for their self-representation on an unprecedented scale. The spirit of reform animating the aesthetic democracy of the annual fervour no doubt contributed momentum to women's increasing demands for broader economic opportunities as well as for legal and political representation in the second half of the century. Policed more by popularity than by paternalistic decree, women as both editors and writers of annuals earned financial independence and a degree of cultural authority through a highly developed and often critically acclaimed femininely coded poetics.

Letitia Landon, who published under the signature L.E.L., was one annual editor who specifically called attention to and manipulated the feminine coding of the annuals. Under Landon's editorship from 1832 until her death in 1838, *Fisher's Drawing Room Scrap-Book* made a distinguished entry into the annual fray by introducing the quarto format, a size larger even than the *Keepsake*'s octavo (established five years before), and by presenting a substantial number of plates occupying nearly half of every volume.[17] A typical volume of roughly sixty pages could thus count as many as thirty or more engravings. In addition to this generous proportioning of visuals and text, the *Scrap-Book* further differentiated itself by specializing almost exclusively in three thematic areas: portraits of aristocrats and famous people, picturesque scenes of English landscapes and monuments, and oriental exotica. Consequently, a representative index (for 1833) lists the following juxtapositions: 'The Queen of Portugal,' 'The Collegiate Church, Manchester,' 'Mohamed Shah's Tomb, Bejapore,' 'Sir Thomas Lawrence,' 'Liverpool,' 'Sarnat, a Boodh Monument,' and so on.

Although the style of the text, its twenty-one shilling price tag, aristocratic themes, and target audience of a 'younger gentler class of readers' (Landon 1832, 3) suggest a concerted effort both to reflect and compose a certain nobility of taste, those efforts are nonetheless disrupted by the very discontinuity and arbitrariness intrinsic to the scrap-book form itself. At the material level, the *Scrap-Book* is a collection of remnant plates reprinted from Fisher's other illustrated publications, which Landon recontextualizes and embellishes with fragments of history (also often culled from Fisher's library) as well as with her own trademark spontaneous-styled verse. The *Scrap-Book*'s frequently pronounced disjunction between word and image thus has a source in material constraints determined by Fisher's publishing format, a source which Landon acknowledges in the preface to the opening volume: 'It is not an easy thing to write illustrations to prints, selected rather for their pictorial excellence than their popular species of composition' (Landon 1832, 3). Sounding somewhat desperate, Landon admits to adopting 'any legend, train of reflection, &c. which the subject could possibly suggest' and to accepting 'gratefully' two poems from a friend (Landon 1832, 3).

Through such a confession of limitation Landon strategically demonstrates the requisite conformity to conventional standards of feminine modesty necessary for a woman's legitimate access to cultural power. At the same time, however, she registers the irony that readers' 'literary luxury' (Landon 1832, 3) is generated out of the professional woman writer's economic necessity, a circumstance which caused Landon to complain in private that 'a horse on a mill has a better life than an author' and to long for 'oblivion and five hundred a year!' (Blanchard 1841, 1.225 and 54). By thus inscribing the *Scrap-book*'s material contradictions, prompted by its exploitation both of the author-editor's labour and of Fisher's own previously published archives, into the formal features of the book, Landon destabilizes and displaces the publisher's own broadly defined unifying subjects through competing gestures toward the discontinuities and wastes that 'scrap' also implies. Landon's prefatory remarks thus function to expand the connotative resonance of the book's very title by including inflections of the tension and conflict marking the writer's struggles to fill the blank page beside the picture as well as of the contest between word and image that transforms the *Scrap-Book* into a multivalent struggle over aesthetic values signifying power and knowledge.

Although Landon often appears to have little affinity with engravings Fisher supplied her of picturesque scenes of English rural life, her textual responses demonstrate an astonishing ability to turn the adversities of the book trade into creative material. In the 15-line lyric, 'The Vale of Lonsdale, Lancashire,' Landon asserts a lack of sympathy with the picturesque aesthetic itself, first through the opposition expressed in the twice repeated statement, 'I could not dwell here it is all too fair,' and secondly through the ambivalence of the final lines, 'For the heart feels the contrast all too much,/Between the placid scene, and its unrest' (Landon 1832, 30). By thus constructing a split within the viewing subject, Landon assigns to the picturesque a fundamental inability to bridge the gap between seeing and feeling. In thus presenting her verse as an exemplary response to the image, Landon attempts to mediate the viewing experience as constituted by alienation and contradiction: an identification on the one hand with the static nature of the 'placid scene' and a desire on the other for independence and distance. While generalizing the confrontation between word and image as a conflict within the visualizing self, Landon ultimately dismisses the picturesque aesthetic as both incomplete and uninspiring: its 'too fair' portrayal of nature cannot accommodate the female writer's and reader's emotional and creative 'unrest' within its all too captivating and constraining frame. Landon's verse interpretation thus complicates the picture, rewriting the scene of domesticated nature as one of confinement, isolation and discontent, a potential parallel, in other words, of domestic femininity.

Figure 1 'Waterfall and Stone Quarry, near Boscastle', engraving. After painting by Thomas Allom, engraved by Wm. Le Petit (*c.*1832), 1833.

In the 1833 'Boscastle Waterfall and Quarry,' Landon rewrites the image by guiding the visual perspective away from the desolate beauty of nature depicted in the painting (see Figure 1) to a celebration of art and culture. Devoting more than two-thirds of this 36-line ode to the glories of the city, Landon accepts neither waterfall, nor quarry, nor even the people in the image as her focal point; rather, the poem imagines, with almost parodic irreverence, Westminster Abbey as its centrepiece. Refusing to recognize any value in the quarry as aesthetic object in and of itself, Landon contemplates the utility of the quarry's stones in the construction of monuments of historical and cultural significance located in London, 'the city vast and grand and wonderful' (Landon 1833, 22). While the poem apostrophizes the 'gloomy quarry' only to hurry the reader's 'eye' quickly away to the cosmopolitan spectacle well beyond the frame of the rural setting, the prose note at the bottom of the page confides directly to the reader: 'we talk of the beauties of nature, I must own I am more pleased with those of art. I know no spectacle more impressive than a great street in a great city' (Landon 1833, 22). Taking the reader on an imaginative tour through Piccadilly, Green Park, Westminster Abbey and Hyde Park, Landon not only attempts to rewrite the story of human destruction that the picturesque scene of the quarry reveals, but she also celebrates human creativity in a rejection of the exploitation of nature that makes up picturesque pleasure.[18]

Landon's dispute with the empire of nature in Romantic aesthetics answers those males who perceived the reign of literary commercialism as a threat to masculine authority. In seizing the opportunity to exercise her own access to cultural power, Landon counterposes the aesthetic appeal of the country scene with that of the city. For the romantic attribute of solitude, she suggests 'you may walk for days [in London], and not meet a creature you know'; for tranquillity, 'our fine old Abbey'; for variety, 'the immense variety of faces that hurry past'; for sublimity, 'Westminster Abbey rising in dim dusky grandeur'; for space, 'the open space behind' Hyde Park 'shrouded in unbroken darkness' (Landon 1833, 22). Landon maintains, furthermore, that the city surpasses the country as the prime inspiration of the art of nature by virtue of being the very embodiment of the nature of art, encompassing past, present and future forms of culture. Thus she merges the creativity of humanity with commercial abundance itself, celebrating 'the shops, where every article is a triumph of ingenuity – some curious, some beautiful' (Landon 1833, 22). Most importantly, a cosmopolitan aesthetic is able, she implies, to recognize the social origins of the poverty that also composes its spectacular pleasures. Whereas the rural scene appears to naturalize destitution and deprivation, overwhelming and diminishing human relations by its very vastness, Landon's cosmopolitan aesthetic takes account of 'the grades of society; the wealth few pause to envy, the poverty fewer pause to pity' (Landon 1833, 22). By thus drawing the reader's attention to the social contingency of poverty, Landon emphasizes the potential for communal sympathy in her revisionist aesthetic and thereby locates her work within the growing liberal impulse for social reform.

In addition to disrupting the rural picturesque with a cosmopolitan perspective, Landon often rewrites English scenes so as to resituate them within the British empire. The 1833 poem 'Liverpool,' for example, accompanies an illustration of the harbour from on board a ship called 'The Mercer'; the view of the shore is intersected by small boats gathered for a royal parade celebrating the launch of an 1832 mercantile expedition to central Africa (see Figure 2). Rather than describe the scene, the poem embarks on an expedition of its own. Thus the dark streams of smoke trailing out of the factory stacks in the painting are displaced in the poem by the 'pure intent' of imperial commerce, science and the pursuit of knowledge, which in turn have superseded the 'former days' of 'wretched' slavery and the 'false light' of military conquest. In a prose note, Landon connects the commercial success of the expedition with

Figure 2 'Liverpool, from the Mersey Commencing at the Prince's Parade'. Engraving by Samuel Austin engraved by Robert Brandard (*c*.1832), 1833.

the advance of 'geographical knowledge,' while in the poem itself she claims a far 'noble[r] aim': 'to civilize' and 'by traffic to reclaim'; 'to draw the savage and unknown / Within the social pale' (Landon 1833, 13). In constructing 'a deep and ardent sympathy' with 'the bold' adventurers who aspire to 'traffic' with, rather than in, African peoples, Landon transforms a capitalist enterprise into a heroic mission of empire. This rhetoric of imperial benevolence functions to deflect class-based fears that the visual images of industrialism may provoke, while the language of racial superiority works to ameliorate anxieties of empire that the poetic images of encounters with the unknown might simultaneously raise. Landon thus simply reinscribes these contradictions at the structural level in the relationship between word and image when she establishes the possibility of identification with the event portrayed in the picture by ironically turning the gaze imaginatively outward.

Landon's revisions to the Liverpool picture underline the fundamental necessity of discourses on empire in the constitution of national identity, a necessity which supplies a linking thread between the *Scrap-Book*'s many portraits of celebrated heroes and aristocrats. Perhaps the most exemplary of these is the 1837 reproduction of George Hayter's 'The Princess Victoria,' a portrait which went into mass circulation with royal assent, epitomizing the commodification and domestication of

Figure 3 'Victoria, 10 August 1835' engraving. After George Hayter, Esq., 1835 portrait 'Princess Victoria', engraved J. Cochray 1836, 1836 for 1837.

aristocracy for middle-class consumers (see Figure 3). The economic, cultural and social significance of this traffic in gifts was not lost on one of Landon's most supportive critics, Christian Isobel Johnstone, literary editor of *Tait's Edinburgh Review*. According to Johnstone, commercialization of the Christmas spirit is of valuable national utility:

> We know that thousands of Annuals now travel regularly down into the provinces, either as gifts sent *home*, or as town presents given in interchange for the turkeys, goose-pies, and hams which London levies every winter from those kindly rural neighbourhoods where a new book or a picture hardly ever went before.... There are few rings or broaches, given as *souvenirs*, that we would compare in value with 'Finden's Tableaux'; or London caps and turbans with 'Fisher's Drawing Scrap-Book.' Yet it is precisely this kind of trinketry, and millinery, and dainty cakes and bon-bons, of which these modern elegancies in art and literature are taking place as friendly gifts and tokens. In the natural march of refinement, the change was inevitable.
>
> (Johnstone 1837, 678)

Economic progress is, Johnstone suggests, most effectively achieved through the cultural commodity of the book as gift circulated through a consumerism bound by feminine sympathy. Such a marriage between production and consumption seems to promise the incorporation of geographically disparate locations and diverse social identities into a centralized yet boundless cosmopolitan democracy, forming a consensual middle way.

While luxurious books, beautiful bodies and exotic geographies all stand as metonyms for an endlessly expansive commodity aesthetic, Landon is not unaware of the paradoxes unsettling her collaboration with the project of empire. 'The Princess Victoria,' which warns that 'a nation's hopes and fears blend with thy destiny' (Landon 1837, 22), concludes with an ambiguous acknowledgement of that same constriction Landon challenges in the English picturesque. Implying a highly tentative 'future' that '*may*' be 'blessed in thee' as well as a qualified 'peace' that '*may*' 'smile on all [who] be round thy throne', Landon renders the window framing the Queen's innocently open face both a hopeful sign of empire's breadth and a more sinister intimation of the young monarch's potential vulnerability, exposure and entrapment under the pressure of the public gaze.

Landon's commentary on the constraints of romantic and domestic convention noted in her poems on the English picturesque, receives its broadest scope in Orientalizing contexts. In the 1836 'Immolation of a Hindoo Widow,' composed for the plate, 'A Suttee' (see Figure 4), Landon expresses an affiliation with the colonized woman that sensationalizes and universalizes the condition of female oppression.[19] By emphasizing gift-giving in the ritual of sati, criminalized in Anglo-India in 1829, Landon suggests a potential lawlessness at the very heart

Figure 4 'A Suttee': Preparing for the Immolation of a Hindoo widow, engraving. Reprint of 'Immolation of a Hindoo Widow at Baroda, in Guzerdt' by Capt Robert M. Grindlay from his *Scenery, Costumes and Architecture Chiefly on the Western side of India*. London: R. Adelman, 1826. Engraver J. Redway, 1835 for 1836.

of that English national identity which centres its legitimacy precariously upon a paradoxical outside within – whether that be the excluded female, marginalized poet, remote provincial region or colonized subject.

Emphasizing the heroism of the Indian woman's self-sacrifice, Landon points to the analogical connection between the colonial heroine and the self-abasing domestic English woman, and suggests, in so doing, that representation itself is a highly political theatre expressing both oppression and resistance. Willingly dispersing her 'garland' and 'gems,' indeed all 'love-signs,' the exotic heroine's self-fragmentation incorporates 'every gazer' in the desire for ritualized destruction of the scene's very centre, for picture and poem alike hinge on an anticipated consummation which entails the self-consumption of the woman holding the gaze. Attempting to revise the opposition between subject and object, Landon tries to propel the static figure into motion by making the sati's gift-giving her final assertion of unassimilable identity. Torn between erotic identification with the heroine's saintly purity and fascination with her pure rejection of signification itself, the narrator interprets sympathetic gift-giving as a proclamation of independence from the world's 'pleasant vanities'. If death signifies the end for *the* representative feminine paradox of autonomous nobility and dependent sympathy, then Landon indicates how representation of that death might also mark its political beginning somewhere between the static poles of domination and subjection.

Although Landon initially presented the *Scrap-Book* as 'addressed chiefly to a younger gentler class of readers' (Landon 1833, 3), her final preface transfers the emphasis from readers to the poetics itself, describing her writing as a process of 'giving words to those fancies and feelings which constitute, especially, a woman's poetry' (Landon 1838, 3). That 'women's poetry' can be described as a national literature claiming legitimacy through commercially viable 'fancies' of its nobly self-sacrificing wastes; the aesthetic excesses of such a visual feminine sympathy torn between heroically eating its heart out and giving that heart away may well have produced boundless identities, a democracy of identity, out of the necessary illusion of cultural distinction and political coherence. The very popularity of the annuals and the expansive empire of culture they staked out simultaneously ensured finally that their centrality, indeed the idea of one centre, could not hold.

Notes

I would like to acknowledge the support of Simon Fraser University's President's Research Grant and the Social Sciences and Humanities Research Council of Canada during the research and writing of this chapter.

1. According to Altick (1957, 362), 'in one season, 1828, it was estimated that 100,000 copies were produced, at a retail value of over £70,000. Smith, Elder's *Friendship's Offering*, priced at 12s, alone sold between 8000 and 10,000 a year. These annuals... were priced too high to affect the mass audience'. However, A. A. Watts, son of an annual editor, claimed in 1884 that 'the influence of these annuals was eminently healthful and refining, highly conducive, as they were, to the awakening and stimulating among the *great mass* of the middle classes, especially in the country, of the taste for landscape art in England' (my emphasis; cited in Siemans 1978, 143).
2. See Richards (1990, 40–53), and Stephenson (1995, 126–55).
3. Note that the economic value of Heath's *Keepsake* was *not* negotiable; priced at a guinea, it was furthermore the most expensive annual on the market in 1828.
4. See Jaffe (1994, 254–65) for an analysis of the way in which spectacular forms of cultural representation evoke desire and compel sympathetic identification while precluding participation.
5. For a brief history of the introduction of the steel engraving into general use, see, for example, Anthony Dyson, *Pictures to Print* (Dyson 1984, 113–44), and Percy Muir, *Victorian Illustrated Books* (Muir 1989, 59–62). On other crucial

innovations concerning covers, embossing and printing processes, colour and size, see especially Renier (1964), Faxon (1973), and Jamieson (1973).

6. F.O. stands for *Friendship's Offering* of course, but the poem was likely written by the then editor, Thomas Pringle.
7. On the history of the ideology of separate spheres and its relation to middle-class institutions and culture, see Davidoff and Hall (1987).
8. Lee Erickson argues that the annuals significantly cut into the market for verse by individual authors, competing directly for most of the poetry readership (Erickson 1996, 26–39).
9. Watts's position may have been tempered by the fact that his wife was also an annual editor, whom the *Athenaeum* critic accused in the next column for lifting 'substantially word for word' from a French guidebook a story for her *New Year's Gift* (Anon. 1831a, 705).
10. Letter to Caroline Bowles, February 1828 (Southey 1965, 2.324). Many of Southey's complaints are instigated by the 'sad diminution' in sales of his books, which he claims averaged at one time around £200 but by the late 1820s 'scarcely produce anything' (Southey 1976, 228 and 1965, 2.335). See also letter to Allan Cunningham (Renier 1964, 13).
11. Letter to poet and annual editor Bernard Barton, 28 August 1827 (Lamb 1891, 2.217). A footnote indicates that the quotation is from verses by Lamb himself, written to answer the question, 'What is an Album?'
12. Complaining about late payment, Southey remarked that Fredric Mansel Reynolds, editor of the *Keepsake*, likely discovered that 'he has paid young men of rank and fashion somewhat dearly for the use of their names' (letter to Allan Cunningham; in Renier 1964, 14). Lamb, likewise concerned with payment, wrote: 'For God's sake do not let me be pester'd with Annuals.... I get nothing out of any of 'em, not even a Copy' (Lamb 1888, 2.362). Arthur Hallam makes similar remarks on Tennyson's unprofitable foray into the annuals (Hallam 1981, 460), and although Wordsworth was also a regular contributor he blamed 'the ornamented annuals' as the cause of poor sales of his books in the 1820s (cited in Jack 1963, 175).
13. This, despite Fredric Mansel Reynolds's announcement in the previous year's *Keepsake* that he, as editor, has introduced 'a few *anonymous* articles, for the satisfaction of those who may desire to judge of the merit of a work, undazzled by the prestige attached to an illustrious name' ([Reynolds] 1831, iv). Of course, the reviewer was correct; as noted above, this annual sold 15,000 copies in its first year (Southey 1965, 2.324).
14. The word 'dandy' is used by Lamb: 'I hate the paper, the type, the gloss, the dandy plates.... In short, I detest to appear in an Annual' (Lamb 1888, 2.204).
15. Southey made this comment when Brontë presented him with her poetry. Letter to Charlotte Brontë, March 1847 (Gaskell 1987, 127).
16. P. G. Patmore, *My Friends and Acquaintance* (Patmore 1854, 1.84, cited in Leighton 1992, 46).
17. Not only did Landon edit and write most of the contributions for *Fisher's Drawing Room Scrap-Book* from 1832 to 1839, but she also edited *The Easter Gift* (1832) as well as the first number of *Heath's Book of Beauty* (1833), *A Birthday Tribute* (1837), and *Flowers of Loveliness* (1838). Landon's first biographer claims 'she derived [from the annuals] sums considerable enough to show

that it was no immutable decree of fate by which poetry and poverty had been made inseparable companions' (Blanchard 1841, 1.65); see also Stephenson (1995).

18. See Michasiw's (1992, 76–100) careful distinction between the gaze of the tourist and that of the proprietor. Landon's representation of the gaze as that of the dispossessed yet mobile tourist displaces habitual patterns of viewing and asserts the superiority of migrancy and movement over fixity of vision. Her disruptive aesthetic marks a significant intervention in the struggle between classes and genders for the production of cultural power.
19. Landon's most complex exploration of the relationship between femininity and oriental exoticism occurs in 'The Zenana, An Eastern Tale', comprising most of the 1834 *Scrap-Book*.

Works cited

Adams, James Eli. 'The Hero as Spectacle: Carlyle and the Persistence of Dandyism,' *Victorian Literature & the Victorian Visual Imagination*, eds Carol T. Christ and John O. Jordan. Berkeley, Los Angeles and London, 1995, 213–32.

Altick, Richard D. *The English Common Reader: A Social History of the Mass Reading Public 1800–1900*. Chicago and London, 1957.

[Anon.]. 'The Literary Souvenir, for 1832,' *Athenaeum* 209 (29 Oct. 1831a), 704–05.

[Anon.] 'The Keepsake, for 1832,' *Athenaeum* 211 (12 Nov 1831b), 736–7.

Blanchard, Laman. *Life and Literary Remains of L.E.L.* 2 vols Philadelphia, PA, 1841.

Bourdieu, Pierre. *Distinction: A Social Critique of the Judgement of Taste*. Trans. Richard Nice. Cambridge, 1984.

Butler, Marilyn. 'Culture's Medium: The Role of the Review,' *The Cambridge Companion to British Romanticism*, ed. Stuart Curran. Cambridge, 1993. 120–47.

Carlyle, Thomas. 'The Hero as Man of Letters: Johnson, Rousseau, Burns,' *On Heroes, Hero-Worship and the Heroic in History*, ed. Carl Niemery. Lincoln, 1966, 154–95.

Davidoff, Leonore and Catherine Hall. *Family Fortunes: Men and Women of the English Middle Class, 1780–1850*. Chicago, IL, 1987.

Dyson, Anthony. *Pictures to Print: The Nineteenth-Century Engraving Trade*. London, 1984.

Erickson, Lee. *The Economy of Literary Form: English Literature and the Industrialization of Publishing, 1800–1850*. Baltimore and London, 1996.

F.O. [Pringle, Thomas]. 'Prologue,' *The Friendship's Offering*. London, 1830, v–ii.

Faxon, Fredrick W. *Literary Annuals and Gift Books: A Bibliography, 1823–1903*. Middlesex, 1973.

Foucault, Michel. 1977. 'What Is an Author?' *Language, Counter-Memory, Practice: Selected Essays and Interviews*. Ithaca, NY 1977, 113–38.

Gaskell, Elizabeth. *The Life of Charlotte Brontë*. London, 1987.

Hallam, Arthur Henry. *The Letters of Arthur Henry Hallam*, ed. Jack Kolb. Columbus, OH, 1981.

Hofkosh, Sonia. 'Disfiguring Economies: Mary Shelley's Short Stories,' *The Other Mary Shelley: Beyond Frankenstein*, ed. Audrey A. Fisch, Anne K. Mellor, Esther H. Schor. New York, 1993, 204–19.

[Hunt, Leigh]. 'Pocket-Books and Keepsakes,' *The Keepsake*. London, 1828, 1–18.

Jack, Ian. *English Literature, 1815–1832*. Oxford, 1963.

Jaffe, Audrey. 'Spectacular Sympathy: Visuality and Ideology in Dickens's *A Christmas Carol*,' *PMLA* 109.2 (1994), 254–65.

Jamieson, Eleanore. 'The Binding Styles of the Gift Books and Annuals,' *Literary Annuals and Gift Books*, ed. Fredrick W. Faxon. Middlesex, 1973, 7–17.

[Johnstone, Christian Isobel]. 'The Books of the Season,' *Tait's Edinburgh Magazine*, 8.4 (1837), 678–88.

Klancher, Jon P. *The Making of English Reading Audiences, 1790–1832*. Madison, 1987.

Lamb, Charles. *The Letters of Charles Lamb, Newly Arranged, with Additions*, ed. Alfred Ainger. 2 vols. London, 1891.

Landon, Letitia Elizabeth. *Fisher's Drawing Room Scrap-Book*. 8 vols. London, 1832–1839.

Leighton, Angela. *Victorian Women Poets: Writing Against the Heart*. Charlottesville, VA, 1992.

Mellor, Anne. K. *Romanticism and Gender*. New York, 1993.

Michasiw, Kim Ian. 'Nine Revisionist Theses on the Picturesque,' *Representations* 38 (1992), 76–100.

Muir, Percy. *Victorian Illustrated Books*. London, 1989.

Renier, Anne. *Friendship's Offering: An Essay on the Annuals and Gift Books of the Nineteenth Century*. London, 1964.

[Reynolds, Fredric Mansel]. *The Keepsake*. London, 1831, iii–v.

Reynolds, Margaret. 'Introduction I,' *Victorian Women Poets: An Anthology*, ed. Angela Leighton and Margaret Reynolds. Oxford, 1995, xxv–xxxiv.

Richards, Thomas. *The Commodity Culture of Victorian England: Advertising and Spectacle, 1851–1914*. Stanford, 1990.

Siemans, Lloyd G. 'The Poet as Huckster: Some Victorians in the Toyshop of Literature,' *English Language Notes* 16 (1978), 129–44.

Southey, Robert. *Letters of Robert Southey to John May, 1797–1838*, ed. Charles Ramos. Austin, TX, 1976.

Southey, Robert. *New Letters of Robert Southey*, ed. Kenneth Curry. 2 vols. New York and London, 1965.

Stephenson, Glennis. *Letitia Landon: The Woman Behind L.E.L.*. Manchester and New York, 1995.

[Thackeray, William Makepeace.] 'A Word on the Annuals,' *Fraser's Magazine*, 16 (1837), 757–63.

Watts, Alaric A. 'Introduction,' *The Literary Souvenir*. London, 1831, vii–x.

Part II

The Reader in Text and Image

5

A Paradigm of Reading the Victorian Penny Weekly: Education of the Gaze and *The London Journal*

Andrew King

This chapter continues the debate about the study and definition of Victorian periodicals that was raised by the 1989 'Theory' issue of the *Victorian Periodicals Review*. It is a case study, experimentally applying a pragmatic concept of how meaning is produced to illustrations taken from one of the best-selling magazines of the mid-nineteenth century, the penny fiction weekly *London Journal* (1845–1906) (Anderson 1991, 3; Mitchell 1987, 1; Ellegård 1957, 37; Altick 1957, 358; de Clarigny 1857, 228).

There are several reasons for studying a popular Victorian magazine and its illustrations in this way. First, it can contribute to the growing concentration of press historians on areas that have hitherto received comparatively little attention, in this case the mass-market penny fiction weekly and more particularly the non-verbal material in it. Second, by employing a conceptual framework derived from the discourse of linguistic pragmatics, rarely associated with the study of Victorian periodicals, it can take up some of the challenges for periodical research raised by Lyn Pykett (in 1989), namely, the need for interdisciplinarity, close reading of texts, deconstruction of a text/context hierarchy. Finally, it allows for interrogation of an anti-theoretical press history that is produced from positions that claim, usually implicitly, to transcend their own historical specificity, refusing to problematize referentiality and absolute meaning. My use of pragmatic theory is intended to avoid the production of transcendent meanings by concentrating on the extremely local conditions under which meanings are produced.

What exactly 'pragmatic theory' comprises in linguistics is much debated (Levinson, 1983), but common to most definitions is a concern with utterance in a defined context, where the latter term is understood to cover the identities and relationships of the participants in the communicative event, together with its temporal and spatial parameters. What meanings, if any, will be produced from the utterance will depend upon these. It follows that the same utterance in a different context will produce different meanings: 'It's hot here' can mean, for example, 'I am so happy feeling the Mediterranean summer air after spending a long time in a cold climate', 'I am a person of a certain social group addressing someone I believe belongs to the same, and am enjoying this party', or even 'Please open the window'. In the importance it gives to context, pragmatics may seem at first to resemble speech act theory, which is rather more well-known in literary studies because of Derrida's engagement with it in 'Limited Inc.abc.' (1977). It differs from speech act theory, however, in that it does not depend upon what Derrida regarded as its feet of clay, the 'good faith' of the participants in the communicative event. Furthermore, while acknowledging, like speech act theory, the importance of conventions, lexicons and rules for the production of meaning, pragmatics by no means privileges them above the parameters of context. In these senses too it differs from the best-known application of linguistics outside its own field, Saussurean semiotics in structuralism, which seemed to strive for universality rather than specificity.

Real-life conversations have been the principle 'texts' that pragmatics has been applied to, and the contextual parameters of these are often known from sources external to the actual text analysed. The traces of contextual parameters are then sought in the text itself through an often very elaborate kind of close reading. In literary/historical studies it would be possible to apply an approach like this to culturally powerful material such as *Blackwood's* or novels by George Eliot without too much difficulty. An abundance of commentary written by the cultural élite survives, along with a great deal of information about them, which enables the various parameters to be set independently of the text. In the case of nineteenth-century popular periodicals such as *The London Journal* we do not have such a plethora of commentary, since both text and readers' interaction with it existed below the cultural threshold which marked what was considered worth preserving. A Critical Heritage volume on *The London Journal* would be very slim indeed (Patricia Anderson [1991; 1992] refers to most of the commentary available). While the *Journal* ran a 'Notices to Correspondents' page in every issue, it is interesting that these provide almost no direct information on how the rest of the magazine was read. A very few early readers offered comments on the serial novels *The London Journal* ran, but I have found nothing in 'Notices to Correspondents' that gives any indication of how the illustrations were understood. Again the position of the popular under the 'threshold of preservation' must be posited as the reason for this. How illustrations were understood was not considered culturally important enough to be printed even by the magazine itself (assuming in the first place that letters were written about them). Even their origins were considered unimportant: most were unsigned and their artists remain unknown (as in the case of those reproduced here).

One parameter that governs meaning which can be deduced from these pages, however, is a generalized idea of the social category of readers – 'Mill-girl, Apprentice and Clerk' as Anderson puts it (Anderson 1992) – although the magazine's readership actually covered a much wider social spectrum than that, from the £500 (and more) a year downwards. A more consistent parameter than income or any simple notion of social class based on occupation is readers' marginal position as regards cultural power. Their marginality is expressed in their questions about the law, etiquette, morality, history and correct linguistic usage. They are excluded from the centre yet acknowledge its power, wishing to model themselves upon it or at least know its rules. This is as much the case with 'Dan Florestan', a solicitor with £400 a year, as it is to 'Maria M. . . . in her twenty-third year but portionless' (*The London Journal* 14 (1852), 32). Buying *The London Journal* meant cultural marginality: it was not bought for discussions of Comte.

A culturally marginal position is confirmed by those few accounts of reading *The London Journal* that do survive. Those dealing with its early years come from well outside the readership it was intended for: their survival indeed depends largely upon their proximity to cultural power. There are two that I want to mention here, one by Margaret Oliphant and a slightly later and more famous one by Wilkie Collins. Although dating from over a decade after the illustrations I shall examine, appearing very close together in the late 1850s, they are nonetheless the earliest extended readings of the magazine. Oliphant starts by recounting how she read her copy of *The London Journal* (along with several other similar publications) to a bored little girl within the precincts of a cathedral on a hot summer's day. This can be regarded as part of a narrative strategy intended to attract and maintain the attention of *Blackwood's* readers through the pleasures of scandal: the paradox of placing a low-status object within a high-status location, with the attendant possibility of its corruption of innocence. Oliphant will later diffuse the scandal by revealing that the offending object is in reality a harmless toy of impeccable moral tone, but still dismisses *The London Journal* and its companion magazines as culturally invalid: they do not properly follow through the rules of high status literature or art, although she does remark how they gesture towards them. The article in sum justifies and confirms the assumed shared cultural preferences and practices of her readers. The parameters of place, time, identity and relationship of participants she assigns *The London Journal* are clearly subordinate to a narrative aimed at amusing and validating a high culture market. What is very important here is her realization that the two lexicons of high and low cultures are not hermetically sealed off from one another. She suggests *The London Journal* is offering a sort of failed copy of the high.

Wilkie Collins's witty exploration of mass-market reading in *Household Words* treats *The London Journal* as though it were an exotic artefact from a dark continent ready for cultural colonization and exploitation, 'The Unknown Public'. Like Oliphant, the parameters he assigns the *Journal* are conditioned by the location of his own communication, but his article offers parameters validated by other sources. He first encounters and then keeps encountering the magazine in shop windows ([Collins] 1858, 439; see also [Wright] 1867, 194; [Rands] 1873, 163; Stevenson 1922, 334–5). In so doing he locates the text not only spatially but also temporally (the moment just before the magazine is purchased) and furthermore indicates the relationship involved in the communicative event between viewer and text, that of appeal or advertisement. This is when the commodity status of the text is most to the fore, and when the text is trying to seduce and transform the potential purchaser-reader into an actual one. These three parameters, time, space and relationship of participants, shall form the core coordinates of my reading, and their traces in the text shall be examined. In choosing them I am not only being faithful to a contemporary source, but am also privileging the commodity status of periodicals in an attempt to consider some ways in which meanings are produced under capitalism. A fourth parameter considered will be the identities of the participants. While I shall regard the page as not varying significantly (itself an ideal situation), the identities of its viewers will: how they decode it will depend upon their access to cultural codes and therefore upon their social category.

The parameters I have set for the magazine have a further implication for the study of periodicals that again recalls the *VPR* 'Theory' Issue. Margaret Beetham raised the issue of the 'definition of the [periodical] text', asking us to consider whether 'the single number constitute[s] our text or the whole run of numbers' (Beetham 1989, 96). In restricting my parameters to those of the shop window I am taking an even more extreme line, fragmenting the periodical into 'texts' which comprise single pages. At the same time, however, the text will be shown by no means to be a hermetic unit but made constantly to refer to signs and

THE

LONDON JOURNAL;

And Weekly Record of Literature, Science, and Art.

No. 1.] FOR THE WEEK ENDING MARCH 1, 1845. [Price One Penny.

THE CASTLE OF ASPIROZ;
OR, THE SPECTRAL BANQUET.

Figure 5 *London Journal*, 1 March 1845.

discourses outside itself, in a deconstruction of the text/context dyad that is basic to pragmatics. My methodology revisits Beetham's article in another way in that by analysing non-successive pages which nonetheless share the same basic parameters, a masterfully single and linear argument can be avoided, returning the reading of periodicals at least partially to the dispersion, disruption and seriality which are characteristics too often elided in our retrospective panoptical surveys.

The first problem any commodity faces when it comes on the market is to initiate the communicative event it is economically dependent upon. It must create its addressee by identifying potential purchasers. It must make potential purchasers give it an identity and persuade them that it relates to themselves. In order to do this it has to get the public to differentiate it from competing products while at the same time it must give the appearance of being already well-known to them (Williamson 1978, 24ff). The commodity thus suffers the dilemma of having to be new-and-not-new.

When *The London Journal* first appeared in the shop window on its first Saturday of publication, it would have struck the passer-by most by its use of a print and type on the same page. This identifies it as 'modern' (since it was enabled by the comparatively new stereotype process) and designed for sale to the culturally marginal. This is in contrast to the old-fashioned and élitist look that, for example, Dickens and Browne, and later the elegant *Cornhill*, cultivated through the comparatively expensive steel etching on separate plates.

In the shop window, after the *gestalt* of the page layout, the picture would have been perceived rather than the words, if indeed any of the minutely printed verbal text of *The London Journal* would have been decipherable at all. This temporal location of print in the sequence of perception undermines the priority often given automatically to words when textual location is not concretely considered (e.g. 'the primary function of the illustration of literature is to realize significant aspects of the text', Hodnett 1982, 13). It reveals such privileging as the product of a focus on textual origins rather than on reading processes: the implied location of the text is a vague moment of production. Figure 5 is a picturesque vignette with strong narrative elements. Despite the modernity and popular layout of the page, it is in an old-fashioned style: the heyday of the picturesque vignette had taken place fifty years previously and by 1845 it had become a markedly conservative art-form (Bermingham 1988; Andrews 1989). In periodical illustration it had appeared frequently in the 1820s and 1830s in genteel miscellanies such as *The Portfolio* and the *Penny Novelist* (Maidment 1992), and *The London Journal*'s associations with such publications can also be seen in the typography of its headpiece, which is very similar to theirs. Connection with commodities no longer on the market relies on the good memory of the viewer: a more immediate comparison, which was published on a Saturday like *The London Journal*, was that eminently respectable publication, *The Penny Magazine* (1832–1845), which enjoyed a circulation of around 40,000 at the time (Bennett 1984, 236; Anderson 1991, 81). *The Penny Magazine* had previously printed numerous picturesque vignettes, though by 1845 it preferred narrative pictures of various types which illustrated series such as 'The Year of the Poets', and 'The Canterbury Tales'. By placing a print that was both narrative and picturesque and combining it with type on the front page, then *The London Journal* is able to begin solving the problem of market differentiation, projecting a recognizable image of itself.

So far the 'not-new' has been covered, but it is also necessary for the commodity to differentiate itself from competitor products. *The London Journal* differentiated itself from *The Penny Magazine* in two main ways. The first is through size: the 16cm × 13cm vignette in *The London Journal* on a 28cm × 21cm page was noticeably larger than most of the images on the 26cm × 17cm *Penny Magazine*. Of course, *The London Journal* was not competing with the luxurious folio *The Illustrated London News* which, at 6d, was in a totally different market, but in accordance with a sales ploy common in Victorian verbal advertising, its size was declaring that it offered quantitatively more for a penny. Second, although *The London Journal* will find, after experimenting with various kinds of image, that offering modes of illustration that had cultural capital to validate them brought the largest profits, these modes had to avoid threatening the public with too great a display of their hegemony. Unlike in *The Penny Magazine*, direct reproduction of gallery art is rare, as Anderson (1992) has rightly pointed out. When it is offered in *The London Journal* its exclusivity is played down either by claims that good art is democratic art by definition (see e.g. *The London Journal* I, 290; IV, 166) or by ripping it out of its auratic context altogether (e.g. a reproduction of Géricault's *The Raft of the Medusa* in volume VI, 113 is used to illustrate a sensational marine adventure with no mention of its origin). This can be contrasted with *The Penny Magazine's* front page that appeared on the same day as Figure 5, showing a medallion of Arcite and Palamon from 'The Knight's Tale' in classical garb with an elaborate symbolic frame, prompting a reading of the picture from within the exclusive discourse of History Painting, the highest form of High Art. Anderson is wrong, however, in failing to appreciate how extremely important gallery art is to *The London Journal*. For, as Oliphant realized in her own way, the magazine offers a *simulacrum* of gallery art, not only in Figure 5 but in almost all of its prints: that is, the rules and the lexicon of culturally powerful art are consistently employed as models (cf. Baudrillard 1988). While employing these models, however, *The London Journal* acknowledges only partially, if at all, their function as social excluders (cf. Bourdieu 1984).

What such simulation offers is a fantasy deconstruction of 'high' and 'low', an instantaneous and temporary social mobility without any of the threats that actually challenging or changing cultural hierarchies would involve. Through the relatively non-exclusive (but still genteel) codes of the magazine, comparatively culturally marginalized readers can gain imaginary access to a centre of the world splendid with wealth yet voided of the myriad symbolic mechanisms designed to keep them out. The same relationship to cultural power is at work in the enquiries and replies of 'Notices to Correspondents' that I mentioned above. The distance of the questions and replies from the *res ipsa* of élite culture may be obvious and amusing to Oliphant and Collins, but to the more marginal such simulations seem to have given at least the illusion that élite culture was accessible. Whereas *The Penny Magazine* tried to control by imposing cultural power through direct reproduction, *The London Journal* offers simulacra, which being tangential to cultural power, are less obviously violent. For the simulacra do not *raid* the culturally powerful in the resistive manner described by de Certeau (1984). It is important to stress the relative and not absolute distance of the magazine from the cultural centre. This can be seen in its negative relation to publications only a few years older than itself and which started publication more recently than *The Penny Magazine*, whose visual language directly opposed the hegemony of gallery art.

When *The London Journal* first came out on 1 March 1845, there were, as far as I have been able to discover, no penny illustrated fiction weeklies quite like it. Its closest rival would have been Edward Lloyd's *Penny Sunday Times and People's Police Gazette* (1840–49?) which comprised fiction and embroidered police reports, with a sensational woodcut on its front page. The *Penny Sunday Times* was very similar in tone and iconographic style to the penny dreadfuls that Lloyd also published. Illustrated on their front page, stereotyped, costing a penny a part, and issued on Saturday, they were in many ways rivals to *The London Journal*. But the illustrations to Lloyd's publications offered different kinds of pleasures from Figure 5. By markedly refusing culturally powerful styles available in magazines like *The Penny Magazine* in favour of non-naturalistic iconography linked to the older broadside, to the penny gaff and melodrama (James 1963, 163; James 1976a, 86), they offered resistance to hegemonic culture. Figure 5, by contrast, proffers (imaginary) integration for the same price as a penny dreadful rejects it, and operates *in opposition* to such cultural rebellion.

Confirming its rejection of the non-genteel viewer, Figure 5 offers an extra pleasure to those with the appropriate cultural lexicon to formulate it. By making use of a *repoussoir*, a figure often found in the picturesque who directs the viewer's gaze and with whom the viewer is supposed to identify, a visual allegory of the viewer's own relationship to the magazine can be understood. The *repoussoir* is resting before the open gate, just as the viewer has stopped to look at *The London Journal*. Like the viewer next to the shop door, he has arrived at a liminal point in his narrative: whether or not to enter. The lance of the rider to whom the

THE

LONDON JOURNAL;

And Weekly Record of Literature, Science, and Art.

No. 3. Vol. I. FOR THE WEEK ENDING MARCH 15, 1845. [Price One Penny.

THE SECRET CHAMBER IN THE GENERAL POST OFFICE.

SAINT MARTIN'S LE GRAND.

Figure 6 *London Journal*, 15 March 1845.

repoussoir directs the gaze acts as a secondary *repoussoir*, an arrow pointing into the journal itself, instructing the reader to open it. Meanwhile the castle gates open inwards, inviting the viewer inside, and of course into *The London Journal* itself. The background comprises a series of corridors that does not seem to end, and allows a choice of entry, just as the magazine comprised different sections that the reader could begin with at will. Such mirroring of the ideal consumer's relation to the commodity is a well-known characteristic of twentieth-century advertising (Williamson 1978), but here it is available in the 1840s, suggesting both the continuity of strategies in visual advertising in capitalism and the close implication into it of supposedly pure artistic elements (allegory, the picturesque) (cf. Berger 1972).

The choice of subject matter in the third issue is interesting for what it can tell us about *The London Journal*'s appeal to different kinds of reader in issues that closely succeed one another. At this stage in its run neither the identity of the magazine nor its readership has been fixed. If issue one had tried to catch the respectable reader, in its third issue *The London Journal* pursues the reader interested in political scandal. Unlike the picturesque, which was not closely identified with either sex, Figure 6 more specifically genders the reader as male, since female franchise was not seriously considered by anyone at this time: the 1848 Charter is notorious now for making no mention of it.

The caption tells us that the picture shows 'The Secret Chamber in the General Post Office' where letters of suspected persons were opened. News of its existence had broken in June the previous year and the matter having been reopened in Parliament in February, was still being discussed in March. The topic had already proved itself saleable to the public in a variety of forms: there were even Parian ware figurines being made of a central figure of the story, the Italian revolutionary politician Mazzini (Smith 1970). Figure 6 is therefore a commercially safe choice, as in a different way was Figure 5: *The London Journal* was, in both, capitalizing on what had already proved its worth in the market. But the publication of news or political comment was in 1845 still theoretically a legal (rather than a commercial) risk for a magazine that was neither a stamped nor a 'class publication'. *The London Journal* is certainly appealing to a 'radical' reader in Figure 6, but it dare not identify itself too definitely as a political magazine. If anyone was able to read the tiny print through the thick glass of a Victorian shop window they would find the disclaimer that 'the policy of any measure of government [is] not [a] subject for discussion in the columns of a journal devoted to literary and scientific purposes'. These words seem to renege on the promise of the picture, yet *The London Journal* is able to continue discussion of politics by quoting at length from the best-selling novel, *The Mysteries of London*, which was currently coming out in penny parts. By quoting a fictional text (even though this section obviously refers to real events) the magazine can claim not to be reporting news but to be reviewing a novel. The style of the image in fact is very similar to that of the early volumes of *The Mysteries of London*, suggesting the reader make a further link with it.

This picture also shows how varying the parameter of the reader's identity varies the decodings that are possible. Besides any public meaning, the story would have had a significance available only to readers who belonged to *The London Journal* coterie. What the public could not have known at this stage was that the editor of the magazine and the writer of *The Mysteries of London* were the same man, G. W. M. Reynolds, since at this stage the editor's name is nowhere printed. Whether the public noticed from the inconspicuous colophons or cared that the publisher of *The Mysteries of London* and *The London Journal* was the same is debatable, but the social group connected to the producers of the text would certainly have recognized the puffing here. They would also have known that the scandal of the Secret Room had been discovered by the engraver W. J. Linton (Smith, 1970) who worked for the *Illustrated London News*. He would thus have been known to the owner of *The London Journal*, George Stiff, himself recently connected to the *ILN*. Furthermore, Linton was a friend of Mazzini after whom G. W. M. Reynolds named one of his sons (James and Saville 1976, 150), and only Mazzini out of all those involved in the scandal is named in *The London Journal*. Such exclusive meanings, only part of which I have been able to recover here, often form part of the function of an advertisement. Given the marked failure of almost all advertising as sales pitch (Fiske 1989, 30–2), it cannot be assumed, as it usually is, that the direction of the advert is exclusively out towards the purchasing public. The advert exists also to create and maintain group solidarity for those who produce the commodity advertised. Figure 6 therefore offers different meanings to two distinct categories of readers and fulfils two kinds of social function while the parameters of time, place and relationship remain constant. As an advert to the public, it is intended to sell both *The London Journal* and *The Mysteries of London* for Vickers, and identifies the former as a publication with radical tendencies which is trying to avoid prosecution. As an advert to readers associated with the magazine's producers it performs the social function of consolidating a corporate identity through knowledge available only to that group.

Figure 7 offers a further refinement of advertising techniques. It is clearly a torture scene, but what exactly is happening cannot quite be determined from perusal of the front page alone. The magazine's written decoding of the print will not take place until the third page of this issue, which means that just looking at the front page in a shop window will only ever raise, not answer, questions. It is necessary to buy in order to find the words that will anchor the image. This is the practice that *The London Journal* front-page fiction illustration will now almost always follow, becoming a key part of the seduction of the potential purchaser.

Retention of elements in successive prints seems to me an indicator of communicative success, and if the above aspect of the print proved successful, a less subtle one, the sadistic subject, does not seem to have been so, for *The London Journal* never again repeats this kind of image on its front page. Indeed, even on inside pages there are very few prints that show suffering like this. Certainly there was a market in Britain for

THE

LONDON JOURNAL;

And Weekly Record of Literature, Science, and Art.

9 Vol. I. FOR THE WEEK ENDING APRIL 19, 1845. [Price One Penny.

THE MYSTERIES OF THE INQUISITION.

Figure 7 *London Journal*, 19 April 1845.

images of violence: Fox's *Book of Martyrs*, murder broadsides, penny dreadfuls and the sensational *Penny Sunday Times* all made use of them, but none used a detailed naturalistic style. Figure 7 does so, employing the lexicon of French romantic-terrific art such as can be found in Delacroix or Horace Vernet. It is in fact directly copied from (not only modelled on) the illustrations to the French novel that was being translated below it, *Les Mystères de l'Inquisition* by 'Victor de Féréal', the pseudonym of Madame Suberwick (see Vicaire 1896). That *The Mysteries of the Inquisition* was a much reworked translation of the Suberwick was unlikely to have been realized by the British public at this time (*The London Journal* (1. 256) only admits to it two months later in a reply to a correspondent). What is of more interest here than the attempt to pass the novel off as an original work is the assumption that an iconographic lexicon is universal whereas words have to be translated into the local language. Intriguing is how another translation of *Les Mystères*, published simultaneously by Geoffrey Pierce only a few doors from *The London Journal*'s printer, did not make this mistake, offering much tamer and home-grown images that are recognizable through a more respectable British iconographic lexicon.

We can recover several reasons for the universalist assumption behind Figure 7. First, art was often considered to transcend location, an idea disseminated for example in Mrs Jameson's discussions of the Italian renaissance in *The Penny Magazine* (1844–5). Secondly, the editor, G. W. M. Reynolds, straddled two cultures and had command of two cultural lexicons. He had lived in France for eight years, even becoming a French citizen and serving in the French army (Reynolds 1841, ii). He knew the French literary market well, as his *Modern Literature in France* shows. He would have had no difficulty in accessing his French art lexicon for Figure 7: current theorizations of the universality of images would have encouraged him to believe that others would automatically have the same capacity. The third reason is likewise concerned with the bridging of two cultural lexicons. *The London Journal* was at this time also being published in Paris by Galignani, as well as in London by Vickers, so the magazine seems to be considering two publics simultaneously, in Britain and Paris. As Reynolds would certainly have known, *Les Mystères* was a great success in Paris in late 1844: *La Presse* (29 January 1846, 4) notes how the whole of Paris knew it. A translation and copied images must have appeared as a way to address the two publics simultaneously. What it does not take into account is the local specificity of codes, for Figure 7's simulation of what for Parisians would have been recognizable as modelled on 'Art', would have registered for respectable British viewers as vulgar and disreputable.

Despite what we can assume to be the failure of its address to the respectable British public, the print nevertheless brings up another important point about how different meanings are formulable through different relationships to the magazine considered as a run of issues. While Figure 7 and the chapter it illustrates have been moved from the centre of *Les Mystères* to the beginning of *The Mysteries*, strongly implying a conviction that the cruel image would prove popular, I suggest that a *purely* sadistic reading was intended to be produced by a viewer only in a certain relationship to the magazine. If viewers have seen and read earlier numbers of *The London Journal* they will have been encouraged to formulate a different meaning. In the previous issue the English state had been called 'the Inquisition' in an article that accompanied the front-page illustration of a secret chamber concerned with tax collection. That print is very similar to Figure 6 above, to which the reader is in fact verbally referred there. In both, groups of men in authority are perpetrating outrages in enclosed spaces. If Figure 7 is read through these earlier illustrations it is possible to construct an allegorical protest against English state oppression *as well as* a sadistic spectacle. If political allegory is considered too sophisticated a mode of reading for the popular market, it is as well to note that it is documented by Mayhew: he records an occasion when an assembled crowd decoded in exactly this way a passage from Reynolds's *Mysteries of the Court of London* which we should understand today as primarily sexually sadistic (Mayhew 1985, 27).

This double reading of political allegory and sadistic spectacle requires two kinds of relationship to the magazine, which are not discreet but can exist simultaneously in the same reader: that of the 'First Time Viewer' and that of the 'Constant Subscriber' (a frequent pseudonym of readers in 'Notices of Correspondents'). The ideal First Time Viewer will relate the front page only to other products or cultural codes in general (as in my reading of the first page of the first issue in Figure 5), while the Constant Subscriber will relate it to previous issues of the same periodical. Thus the First Time Viewer will see the sadistic while the Constant Subscriber will be encouraged to read allegorically as well. Obviously, the magazine must cater for the First Time Viewer in all its numbers, since it is not sold purely on a subscription basis, especially so early in its lifetime, but the Constant Subscriber is much more important for a periodical's continued survival. *The London Journal* encourages Constant Subscription by offering the pleasures of exclusive decodings available only to those who know previous issues. The First Time Viewer, on the other hand, must not be put off by hermetic discourse: hence the importance of making available different decoding practices at the same time. In general terms, the Constant Subscriber will recognize the faces and names in the picture and be *au fait* with the requisite reading strategies. The illustration will offer some new development of the plot that will either confirm or challenge reader expectations (both have their pleasures). The First Time Viewer on the other hand will see an ambiguous or 'strong' situation which requires explanation and can partially be read through more generalized cultural codes.

Conclusions

This chapter has applied a pragmatic theory of meaning production to three nineteenth-century popular visual texts, using parameters of space, time, identity of participants in the communicative event (with their cultural lexicons), and their relationships. As has been shown, these parameters are by no means reducible to finite or single positions: each may and will usually be multiple, and even contradictory. Even within the relationship of 'advertisement' which I stated at the outset would be constant in this chapter, several distinct locations have been constructed. It will also be evident that parameters are not discrete: the borders between 'relationship' and 'identity' of participants blurs several times in the above analysis, for example. But this is less important than how, if only one parameter changes while all the others remain stable, the meanings produced will change. In this chapter, it has been assumed that one participant (the page), time and place (the glimpse in the shop window) have remained constant whereas only the relationships of the participants and the identity of one participant (the viewer) have varied.

The result is that the page comes across as semiotically malleable. This can be regarded as key to the communicative strategies of *The London Journal* (though it is not of course particular to it). To reader /viewers, *The London Journal* offers some interpretative freedom, teasing them into active participation with the text through the offer of implied meanings they are free to supply or not as they want. The interpretative freedom of the reader, encouraged by the use of generalized models or simulacra, has the corollary that it is decodable by a broad spectrum of social categories. *Respectable marginality* is what they have in common, a sense of non-rebellious inferiority which will part with its penny for a fantasy trip to a simulated centre. The offer of temporary and fantasmatic centrality to the marginal constitutes one of the secrets of *The London Journal*'s success: never quite getting where you want, you keep buying it in the hope that next time you will arrive, but also in the comforting knowledge that you won't.

Works cited

[Anon.] 'Cours et Tribunaux', *La Presse* (29 January 1846), 4.

Altick, Richard D. *The English Common Reader: A Social History of the Mass Reading Public*. Chicago, IL, 1957.

Anderson, Patricia. 'Mill-girl, Apprentice, and Clerk', *VPR* 25 (1992), 64–72

Anderson, Patricia. *The Printed Image and the Transformation of Popular Culture 1790–1860*. Oxford, 1991.

Andrews, Malcolm. *The Search for the Picturesque: Landscape, Aesthetics and Tourism in Britain 1760–1800*. London, 1989.

Baudrillard, Jean. 'The Masses: The Implosion of the Social in the Media', *Selected Writings*, ed. Mark Poster. Cambridge, 1988, 207–19.

Beetham, Margaret. 'Open and Closed: the Periodical as a Publishing Genre', *VPR* 'Special Issue: Theory', ed. Laurel Brake and Anne Humpherys, 22 (1989), 96–100.

Bennett, S. 'The Editorial Character and Readership of the Penny Magazine: An Analysis', *VPR* 17 (1984), 126–41.
Berger, John. *Ways of Seeing*. London, 1972.
Bermingham, Ann. *Landscape and Ideology: the English Rustic Tradition 1740–1860*. London, 1988.
Bourdieu, Pierre. *Distinction: A Social Critique of the Judgement of Taste*, trans. Richard Nice. Cambridge, MA, 1984.
[Collins, Wilkie]. 'The Unknown Public', *Household Words* (21 August 1858), 439–44.
De Certeau, Michel. *The Practice of Everyday Life*. Berkeley, CA, 1984.
De Clarigny, Cucheval. *Histoire de la presse en Angleterre et aux États Unis*. Paris, 1857.
Derrida, Jacques. 'Limited Inc. abc', *Glyph* 2 (1977), 162–254.
Ellegård, Alvar. *The Readership of the Periodical Press in Mid-Victorian Britain*. Göteborg, 1957.
Fiske, John. *Understanding Popular Culture*. London, 1989.
Hodnett, Edward. *Image and Text: Studies in the Illustration of English Literature*. London, 1982.
James, Louis. *Fiction for the Working Man 1830–1850*. London, 1963.
James, Louis. *Print and the People 1819–1851*. London, 1976a.
James, Louis and John Saville. 'Reynolds, George William MacArthur', *Dictionary of Labour Biography*, ed. Joyce M. Bellamy and John Saville, III, 146–51, London, 1976b.
Levinson, Stephen C. *Pragmatics*. Cambridge, 1983.
Maidment, Brian, *Into the 1830s: Some Origins of Victorian Illustrated Journalism. Cheap Octavo Magazines of the 1820s and their Influence*. Manchester, 1992.
Mayhew, Henry. *London Labour and London Poor*, ed. Victor Neuberg. Harmondsworth, 1985.
Mitchell, Sally. *The Fallen Angel*. Bowling Green, 1987.
[Oliphant, Margaret]. 'The Byways of Literature', *Blackwood's* (1858), 200–16.
Pykett, Lyn. 'Reading the Periodical Press: Text and Context', *VPR* 'Theory Issue' ed. Laurel Brake and Anne Humpherys, 22 (1989), 101–8.
[Rands W. B.] 'Penny Awfuls', *St Paul's Magazine* 12 (1873), 161–8.
Reynolds, G. W. M. *Modern Literature in France*. London, 1841.
Smith, J. B. 'British Post Office Espionage, 1844', *Historical Studies* 14 (1970), 189–203.
Stevenson, Robert Louis. *Random Memories and Other Essays*, *Works* XII. London, 1922.
Vicaire, George. *Manuel de l'amateur de livres de XIXe siècle 1801–1893*. Paris, 1896.
[Wright, Thomas]. *Some Habits and Customs of the Working Class by a Journeyman Engineer*. London, 1867.
Williamson, Judith. *Decoding Advertisements*. London and Boston, 1978.

6

From Street Ballad to Penny Magazine: 'March of Intellect in the Butchering Line'

Michael Hancher

Early in the nineteenth century much laughter was directed by the British middle and upper classes at the intellectual aspirations of the lower classes. A grandiloquent slogan, 'the march of intellect' – sometimes 'the march of mind' – was draped over those aspirations, usually to mock them. The first use of this phrase registered by the *OED* appeared in a satirical poem that *The Gentleman's Magazine* published near the end of 1827; the *OED* notes that the phrase was 'very common (esp. in ironical allusion) between 1827 (the date of the foundation of the Society for the Diffusion of Useful Knowledge) and 1850'.[1] In fact, however, the phrase 'march of intellect' had become a ridiculous byword at least five years previously. Near the end of a long, anonymous poem called 'The Press, or Literary Chit-Chat: A Satire' published in 1822, the phrase is used to connote a general prodigality of the press, and refers more specifically to the growing craze for travel literature, a genre that was identified as both product and cause of 'the travelling mania that pervades / Both wives and husbands, bachelors and 'maids'!'

> All glorious age! the march of intellect
> Produces curious wishes to inspect
> All that is foreign: Cunningham in vain
> Caution'd his countrymen, and may again.
> Hobhouse, and Leigh, and Matthews, Mrs. Grahame,
> The deaf, the dumb, the blind, and eke the lame
> Publish their travels – lo! the passion spreads,
> And each, delighted, foreign regions treads.[2]
>
> (Anon. 1822, 120)

This early march of intellect proceeds from a decidedly middle-class base; if any satire on class pretension is involved here, it has to do with middle-class apings of the aristocratic grand tour. But the march of intellect became most threatening and most ridiculous as it enlisted the lower classes. On 30 September 1826 *The Times* published three squibs under the already-ironic rubric, 'PROOFS OF THE MARCH OF INTELLECT': these made fun of the intellectual failings of 'a clownish scholar' catechized by an 'irritated schoolmaster in the west of England'; also, the epistolary illiteracy of a customer in arrears to 'a respectable tradesman'; and the solecisms of a posted advertisement offering the services of a day-school teacher(3). *The Times* continued to run paragraphs under a 'March of Intellect' headline at least until 1850; these gradually shift from ridiculing the intellectual limitations of the semi-literate, to satirizing their intellectual ambitions. In 1829 (21 March: 3) *The Times* reported under that general rubric the 'True Story' (in scare-quotes) of a fat, hospitalized butcher intent on willing his body to medical science for dissection: the tension between his gross corporeality and his desire to advance abstract knowledge produces, in the telling, a long series of puns. By 1846 (26 Jan: 4) readers of *The Times* could contemplate the 'March of Intellect in Africa': specifically, the pretentiousness of an Algerian potentate who, mimicking 'European usages', had 'ordered visiting cards . . . engraved . . . in letters of gold, "The Kalifa of Constantine Ali-Ben-Ba-Hamet," and under it also in gold, the Cross of Commander of the Legion of Honour'. Here the march of intellect is clearly a matter of social-climbing, and racial social-climbing at that. What we call social-climbing the Victorians called snobbery.

For nearly half a century the march of intellect provoked the comic muse; the most common cause for concern was the insubordination of the lower classes. In *The March of Intellect: A Comic Poem* (1830), the popular farceur, W. T. Moncrieff, devotes most of his wit to this theme:

> The scullion acts by mental rule,
>   Soars 'bove her situation, –
> Boasts, brought up at the parish school,
>   A *liberal* education.
> . . .
> Mechanics' Institutions
>   At each second step we meet;
> And Birkbeck's resolutions
>   Stare us in every street.
>
> The *barber* takes you by the nose,
> And talks about *nosology*;

And *Thames Street* warehousemen disclose,
  Their art in *crane*-iology.

Last-dying speeches beggars sell,
  And prate about *buy*-ography;
While *journeymen* take walks and well
  Improve them in *toe*-pography.

And so on, in bad puns worthy of Thomas Hood, whom Moncrieff invokes as a kind of muse at the outset. Though Moncrieff asserts that the ridiculous effects of the march of learning are universal – 'Nor stand nor stall does it neglect, / It every where is shown' – they are most conspicuous in those who aspire to rise above 'their station' (Moncrieff 1830, 20, 29–31).

The following year, in 1831, W. H. Harrison published a poem called 'The March of Intellect'; it elaborated on a Rowlandson drawing, posthumously published, in which a butcher boy and a clergyman are seen shopping at the same bookstall (Figure 8). In this poem the distinctions of body from mind, corporeality from spirituality, lower class from upper, are threatened by the social convergence that was being impelled by print consumerism.

As, t'other day, I bent my way
  Through Holborn, I espied
A butcher, at a book-stand, by
  A prebendary's side.

'*Pares cum paribus*,' cried I,
  'For each has got a stall;
This cuts up in the Quarterly,
  And that in Leadenhall.'

My cousin Ned, who heard me, said,
  'Now, Harry, only look!
How gravely yonder butcher cons
  His newly purchased book.

'Such from the march of intellect
  Results – 't is really dreadful,
To see one in his class of life
  With learning stuff his head full.

Figure 8 W. H. Harrison, 'The March of Intellect'. Engraving after Thomas Rowlandson. *The Humourist*, London, 1831.

'The vulgar set more learn'd will get
  Than many of their betters:'
'Dear Coz,' quoth I, 'they may do that,
  And yet scarce know their letters.'

But Ned the strain took up again:
  'You can't, 't is quite horrific,
Address your servant, but you get
  An answer scientific.'

(Harrison 1831, 24)

All sorts of labourers as well as servants were tempted by literacy to put on airs. Robert Seymour ridiculed such vanity in two lithographs published around 1834, which were reprinted as engravings in 1838. (Seymour is now best remembered for having initiated *Pickwick Papers* with Charles Dickens in 1836.) One of them, titled 'Coalheavers in the Byron Coffee House', shows two grimy navvies debating the interest of newspaper reports of parliamentary debates, as against 'the scientific notices' also published there; they ignore the fact that both the affairs of parliament and the affairs of science are above them. Another, titled 'Literary Coalheaver and Butcher in a Coffee House', shows a labourer and a butcher debating the usefulness of newspapers as against the sentimental uplift to be gained from books (presumably fiction or poetry); again, both domains are above their class.[3] An anonymous woodcut, 'The Literary Dustman', perhaps published around 1835, amplifies the theme (James, 1976, 149). It shows a middle-aged dustman lounging at his ease on a sofa in a parlour crowded with furniture and decorations, while his son plays and sings at a piano – the new piece of furniture most closely identified with domestic gentility. To round out his pleasure the father not only savours a pipe but grasps a copy of the *Penny Magazine*, which the Society for the Diffusion of Useful Knowledge had published weekly since 1832, with the express intention of improving the intellectual life of artisans.

The persistent moral to all these satires is that if workers know how to read they cannot also know their place. Mandeville's century-old warning against educating the lower classes, put forward in his *Fable of the Bees* (1714), was coming true:

> When Obsequiousness and mean Services are required, we shall always observe that they are never so chearfully nor so heartily perform'd as from Inferiors to Superiors; I mean Inferiors not only in

> Riches and Quality, but likewise in Knowledge and Understanding. A Servant can have no unfeign'd Respect for his Master, as soon as he has Sense enough to find out that he serves a Fool.
>
> (Mandeville 1924 1. 289)

An anonymous broadside ballad of the early nineteenth century, 'March of Intellect In the Butchering Line', offers a special perspective on the growing problem (Figure 9).[4] If the usual point of view in 'March-of-Intellect' satires is that of the literate bourgeois whose turf is being trespassed upon by the marchers, the point of view in this broadside ballad is that of someone whom the march leaves behind in the dust. The speaker is a butcher, who suffers from the march of intellect in his own family: his wife and his daughters are caught up in it, eager to progress through literacy to a superior social class. His coarse labours support their lives and genteel aspirations; their literary fantasies make him déclassé. He seems to identify with penny ballads (certainly he is constructed by one); but they identify with 'the penny publications'. These publications probably do not include the socially improving *Penny Magazine*, nominally addressed to artisans, which from the moment of its appearance in 1832 made all penny broadsides a doubtful bargain; there is not much evidence that the *Penny Magazine* appealed to female readers. Rather, the butcher's wife is likely to have 'taken in' what Patricia Anderson calls 'the second generation of illustrated penny miscellanies', that is, *The London Journal* (begun in 1845), *Reynolds's Magazine* or *Reynolds's Miscellany* (from 1846), or even *Cassell's Illustrated Family Paper* (from 1853) (Anderson 1991, 84). These did cater to, and construct, middle-class feminine taste.

Like most broadsides, 'The March of Intellect In the Butchering Line' is not dated. Mention of the fashionable purchase of 'a bran new PYE [ANE] R' suggests a time not much before mid-century (some 25,000 pianos were manufactured in Britain in 1851) (Drain 1988, 605). The ballad-broadside business, though it may have subsided after the Napoleonic wars, persisted on a substantial scale at least until the 1860s.[5] A plausible date for this text would be 1850.

The ballad is headed by a woodcut showing a gentleman holding a carriage door for a lady – both of them dressed in Regency-style clothing. Presumably the block itself survived from earlier in the century; prolonged reuse was common. By thematizing fashionable taste the cut is less irrelevant than most such ballad illustrations were. But of course the gentleman holding the door is no butcher. And it is the butcher who speaks:

I keep a snug shop, which once had a good stock in,
  But the life I now lead is indeed very shocking;
I contrive to get money by industry's plan,
  My family spend it as fast as they can,
My spouse, who once work'd hard as any wife going,
  By this 'march of intellect's' so genteel growing.
She dresses herself and her daughters up fine,
  Although I am but in the butchering line.

Spoken – She takes in all the penny publications though she can't read without spelling the hard words – makes poetry though she can't write, and as to blank verse, makes nothing of it – she has made herself a HALBUM out of an old day book, and my eldest daughter write down all the good things they can scrape together – if she goes into the shop to serve a quarter of a pound of suet, or a pennyworth of lights, she puts on a pair of white kid gloves, with the fingers cut off – and its all through the march of intellect.

As this ballad continues to shift back and forth between verse and prose the bill of complaint accumulates elaborate detail. The basic antinomy is the body versus the intellect: projected upon it are male versus female, lower class versus upper class, labour versus leisure, penny ballad versus penny magazine, the economics of shop-keeping ('day book') versus literary aesthetics ('HALBUM'), porter versus madeira, English versus French, and meat versus language ('the French teacher having got on the [account] books, "For sundry legs of mutten [*sic*] and beef," we takes it out in lessons'). The two daughters, Georgiana Matilda and Isabella Caroline, are pampered by their mother, dressed in 'frocks' that look 'like gowns' (the mother is dressed in gowns that look like frocks) – no doubt inspired by the magazines; 'the girls are all the mother's delight – while the poor little boy, Augustu[s] Henry William runs about in ragged breeches; and his mother don't like him at all, because he never wipes his nose – and its all th[r]ough the march of intellect'. The father suffers a similar rejection:

> She wants me to order a pair of false whiskers for Sundays, and 'cause I wont she never gives me a civil word – and what d'ye think? though we've been married 18 years, she says it's very vu[l]gar to sleep together, and so we have separate beds, and it's all through the march of intellect.

March of Intellect
In the Butchering line.

I keep a snug shop, which once had a good stock in,
But the life I now lead is indeed very shocking ;
I contrive to get money by industry's plan,
My family spend it as fast as they can,
My spouse, who once work'd hard as any wife going,
By this "march of intellect's" so genteel growing.
She dresses herself and her daughters up fine,
Although I am but in the butchering line.

Spoken—She takes in all the penny publications, though she can't read without spelling the hard words—makes poetry though she can't write, and as to blank verse, makes nothing of it—she has made herself a HALBUM out of an old day book, and my eldest daughter write down all the good things they can scrape together—if she goes into the shop to serve a quarter of a pound of suet, or a pennyworth of lights, she puts on a pair of white kid gloves, with the fingers cut off—and its all through the march of intellect.

She dresses herself and her daugnter so fine,
Although I am but in the butchering line,
I get back from market each morning at seven,
But wifey ne'er rises till after eleven,
She don't condescend to take breakfast with me,
For chocolates much more genteeler than tea,
She quarrels with what she calls my vulgar manner;
She's just order'd home a bran new PYE-ANNER;
Of course we must have a music-master so fine,
Although I am but in the butchering line.

Spoken—We've got two daughters and one son—Georgiana Matilda learns the pye-anner and singing, 'cause she's got a woice; and there she is a strumming and so-fa-ing from morning till night, enough to drive all the customers out of the shop—Isabella Caroline, she learns French, and parlez vous like a good un, only we don't understand her. The music-master has hard cash for his notes ; but the French teacher having got on the books, "For sundry legs of mutton and beef," we takes it out in lessons—the girls are all the mother's delight—while the poor little boy, Augustus Henry William runs about in ragged breeches ; and his mother don't like him at all, because he never wipes his nose—and its all through the march of intellect.

The mother and daughters together combine,
And cock up their noses at the butchering line,
In vain 'bout extravagant whims I do rate her.
'Tis useless, for if I go to the theathre,
In dress circle boxes her feather she nods,
While I has a sixp'n'worth along with the gods,
Though my daughters are young, they have each got a lover ;
They wear long frill'd trousers, their ankles to cover,
Their mother's determin'd to make them both shine,
Although I'm but in the butchering line,

Spoken—She scolds me for drinking porter, 'cause its so vulgar; drinks Cape Madeira at 18d the bottle—she puts all the washing out 'cause the steam's unwholesome—all her gowns are made like frocks, and all the girls frocks like gowns—milliner's bills come in by the dozen, she has a new front from the barber's ev'ry month, 'cause the fashion changes so, and she wants me to order a pair of false whiskers for Sundays, and 'cause I wont she never gives me a civil word—and what d'ye think ? though we've been married 18 years, she says it's very vulgar to sleep together, and so we have separate beds, and its all through the march of intellect.

These genteel ideas may be very fine,
But she'll soon make an end of the butchering line.

Figure 9 Anon. 'March of Intellect In the Butchering Line', broadside.

'These genteel ideas may be very fine', the butcher concludes, 'But she'll soon make an end of the butchering line'.

There is some pathos to this comic situation, the pathos of the one left behind, like the crippled child in Browning's 'Pied Piper of Hamlin'. The butcher himself is too robust to feel the pathos: outrage and exasperation are the emotions he expresses instead. However, all these emotions differ from the conventional response to the march of intellect, which was viewing-with-alarm from a superior position. It is tempting to think that this is a different kind of text from those already discussed, less condescending, more engaged.

Of course this ballad is not quite an 'inside view' of the literacy crisis. Its anonymous author was not likely to have been a butcher himself; his fondness for puns is conventionally literary, in Hood's and Moncrieff's style. The badly paid hacks who wrote street ballads were threatened by the march of intellect not as consumers but as producers: it was not their taste that was made déclassé, but their product, gradually edged out by the penny magazines – which promised more value for the penny. Their situation is not quite the same as the butcher's, but it is similar enough; and it is different from the usual view.

'March of Intellect In the Butchering Line' illustrates an important generalization about class and gender that Elizabeth Eisenstein ventures in *The Printing Press as an Agent of Change*. Summarizing' centuries of social adjustments that concluded in the nineteenth century, she notes that, thanks to the medium of print:

> the 'family' was not only endowed with new educational and religious functions, especially in regions where the laity was trusted with printed Bibles and catechisms; but the family circle also became the target of a complicated literary cross-fire. As book markets expanded and divisions of labor increased, feminine readers were increasingly differentiated from masculine ones, and children were supplied with reading matter different from their parents.
>
> (Eisenstein 1991, 133)

In 'March of Intellect In the Butchering Line' 'the family circle' has indeed become 'the target of a complicated literary cross-fire'. Some of these shots are older than any reading public constructed by and for print: complaints against extravagant wives form a genre as old as the fabliaux on which Chaucer drew. But the recurrent use in this ballad of the epithet 'genteel' marks a newer anxiety. The descent from the medieval word *gentil* (meaning 'noble') to the nineteenth-century word genteel (meaning 'pretentious'), a descent that matched the ascent of the social aspirations of the lower classes, was enabled by print, as print popularized those aspirations and fixed the new spelling;[6] and it was confirmed by the magazines. Patricia Anderson, in *The Printed Image and the Transformation of Popular Culture*, gives a good account of the process of cultural ascent as it worked in early nineteenth-century periodicals. 'March of Intellect In the Butchering Line' registers one contemporary objection to the process, plausibly on behalf of many, especially male, members of the working class, as well as the hack writers who catered to that class.

The penny street ballads had many faults: often they were crudely written, in halting or slap-dash meters; crudely printed on cheap paper, in battered type, carelessly set; and crudely read and performed, to indulge a popular taste for sex and violence. Instead of gentility they presented a kind of authenticity. Their fate was to succumb to the March of Intellect, driven as it was by the steam-powered printing press, and drawn by the allure of 'genteel ideas'.

Notes

1. The Society was actually founded in November 1826 (Stewart 1986, 188).
2. The word 'maids' is represented by a blank space. 'Cunningham' is glossed on p. 129: 'Mr. Cunningham of Harrow hath written an Address to Britons on the danger likely to arise from indulging the travelling mania'. Regarding 'Hobhouse, and Leigh, and Matthews, Mrs. Grahame': 'The names of different travellers whose works have lately been published. The list might be increased by a hundredfold'.

   'The Press' has been reprinted in photofacsimile as part of a volume collecting three works supposed to be by John Hamilton Reynolds: 'The Garden of Florence', 'The Press', and 'Odes and Addresses to Great People'. In his Introduction, Donald H. Reiman gives reasons for doubting the attribution of 'The Press' to Reynolds.
3. Seymour, from the series 'Every Day Scenes': scenes 4 and 18 respectively; the titles are taken from the list of illustrations in Bohn's edition of Seymour.
4. This broadside is uncommon in another sense. Although broadsides were published in very large numbers in the nineteenth century – hundreds of thousands if not millions of copies, in some cases – they quickly disappeared. Apparently no copy of 'March of Intellect In the Butchering Line' survives in the large collections at the British Museum or the St Bride Institute. The copy I have used (Quarto 820.1 C683) is pasted into an album owned by the University of Minnesota Library. Recently the Bodleian Library published another copy online in facsimile – apparently the same setting of type (<http://www.bodley.ox.ac.uk/ballads/>, Harding B17 [189a]). Neither copy carries a printer's imprint, though many others in the Minnesota album are from John Catnach's establishment. For accounts of Catnach see Hindley and Shepard.

5. Henderson cites Douglas Jerrold's comment on the apparent decline in 1840; for a later date see p. 10.
6. See, for example Jonathan Swift's satiric manual of *Genteel and Ingenious Conversation* (1738).

Works cited

Anderson, Patricia. *The Printed Image and the Transformation of Popular Culture, 1790–1860*, Oxford, 1991.

Anon. *The Press or Literary Chit-Chat: a Satire*, London, 1822.

Drain, Susan. 'Piano', *Victorian Britain: An Encyclopedia*, ed. Sally Mitchell. New York, 1988.

Eisenstein, Elizabeth. *The Printing Press as an Agent of Change*. 2 vols. Cambridge, 1979; rpr. 1991.

Harrison, W. H. *The Humorist: A Comparison for the Christmas Fireside*. London, 1831.

Henderson, W., ed. *Victorian Street Ballads: A Selection of Popular Ballads Sold in the Street in the Nineteenth Century*. London, 1937.

Hindley, Charles. *The History of the Catnach Press*, London, 1887; rpr. Detroit, 1969.

James, Louis. *Print and the People*, London, 1976.

Mandeville, B. 'An Essay on Charity and Charity Schools', *The Fable of the Bees*, ed. B. F. Kay. Vol. I. Oxford, 1924.

Moncrieff, W. T. *The March of Intellect: A Comic Poem*. London, 1830.

Reynolds, John H. *'The Garden of Florence', 'The Press', and 'Odes and Addresses to Great People'*, ed. Donald H. Reiman. New York, 1978.

Seymour, Robert. *Seymour's Humorous Sketches*, ed. Henry G. Bohn. London, 1868.

Shepard, Leslie. *John Pitts, Ballad Printer of Seven Dials, London. 1765–1844*. London, 1969.

Stewart, Robert. *Henry Brougham, 1788–1868*, London, 1986.

Swift, J. [pseud. 'Simon Wagstaff']. *A Complete Collection of Genteel and Ingenious Conversation*, London, 1738.

7

'Penny' Wise, 'Penny' Foolish?: Popular Periodicals and the 'March of Intellect' in the 1820s and 1830s

Brian E. Maidment

1

One of the aims of mass-circulation graphic satire – cartoons, caricatures, and the like – is that of representing social anxieties and complexities through simplified visual codes which draw on (and construct) stereotypes, emblems, and repeated images. Nowhere is this process more obvious than in the considerable number of caricatures and prints which delineate and describe the massive cultural and educational developments within the artisan and working classes in the 1820s, 30s, and 40s, developments which were given shorthand titles like 'the march of intellect', 'the march of mind', 'useful knowledge', or 'the pursuit of knowledge', sometimes 'under difficulties'.[1] The social and ideological complexities of these developments are obvious enough. Did thinking, literate, and increasingly articulate working men and women challenge the political status quo in presumptuous or even dangerous ways, or was cultural advance a key method through which political challenge could be dispersed or assimilated into the dominant culture? Whatever the dangers, the advantages of a literate, reflective, and socially ambitious artisan class for the development of an industrial economy were immediately apparent to employers, investors, and social theorists. For artisans themselves self-education might provide, in addition to literacy, articulacy, and education 'for their own sakes', improved career prospects and better access to technical, instructional, and recreational literature.

Issues concerning the cultural advance of the working classes ran so deep within the culture that a wide range of the visual resources developed and deployed by eighteenth-century caricature – emblem, stereotype, cultural allusion, captions and other verbal amplifications and explanations – was deployed to describe them in the 1820s and 1830s.[2] At this time, many of the structures of eighteenth-century caricature remained in place. Caricaturists in the 1830s still maintained the characteristically unaffiliated stance of previous generations. They were by and large opportunistic in their choice of social and political targets and felt no pressing need to represent consistent ideological or political views in their work. But, largely as an outcome of the development of a mass market for printed texts, changes were beginning to take place in the forms, codes, and discourses of visual satire. Many of the complex images and representational codes developed by genteel caricature (including physiognomy, posture, and gesture)[3] were becoming available to a wider audience in the first three decades of the nineteenth century. Cheap, mass-produced woodcuts or wood engravings, for long associated primarily with vernacular texts (especially romance and gothic narratives and play texts), began to share many of the mannerisms, forms, and stereotypes of more sophisticated single plate caricature.[4]

But, even with these emergent developments, up to the late 1830s at least (one emblematic way to date the change would be the launch of *Punch* in 1841) single plate caricatures with their traditional satiric codes remained one important site for the depiction of complex social issues.[5] By the end of the decade, quite apart from the emergence of comic journals like *Figaro in London* and *Punch*, comic or satirical graphic images had begun to be assimilated into what were essentially verbal texts. These texts themselves represented new literary forms. One important genre was the non-picaresque narrative with illustrations (the realist novel), which Dickens, among others, was formulating in the late 1830s. These narratives were being taken up by monthly journals like *Bentley's Miscellany, Ainsworth's Magazine*, and *The New Monthly Magazine* in the early 1840s. Another fast-developing genre was the illustrated play text, performance text, or song book associated with popular drama, which supplies one of the key images discussed in this chapter. Thus many popular comic images produced in the 1830s belong to a transitional phase in which genteel single plate caricature is giving way to the more diverse, more demotic images which fill the pages of *Punch*, the fiction-based monthlies, and the serialized production of the mass-circulation novel. Transitions of this kind will be apparent even in the few images studied here. Two of the images are large single plate lithographs (the lithograph itself being a new technical development of the 1820s which changed considerably the nature of the images being produced from those made by metal engraving or etching).[6] Others are more conventional engraved single plate caricatures, although even in Seymour and Heath's engraved plates the emergence of a multi-image narrative element is discernible. But the final, key image is a technically unsophisticated wood engraving which was printed in a song book which was itself a graphic and verbal companion to popular play texts. By the early 1840s, the single unsupported graphic satirical image was often being absorbed into a wide variety of texts and narratives. Accordingly, this chapter offers an opportunity to look at single plate, allusive, emblematic caricature just at the moment when it was becoming part of less genteel, more verbal, largely narrative early Victorian modes of discourse.

2

Within the confines of a brief chapter, it would be impossible to do justice to the allusiveness and complexity with which 'the march of intellect' was formulated as graphic satire. What follows, however, is an account of some of the dominant satirical images of the period, that is images of street trades, especially those which depict dustmen.

The imagistic connection between dustmen and coal-heavers and the 'march of mind' was, by the time of the launch of *The Penny Magazine* in 1832, well established as a visual trope. The specific association between dust and intellect emerges in the late 1820s from a range of more general humorous graphic representations of the march of intellect. One of George Cruikshank's multi-image single plate etchings, dating from May 1828, defines a number of the figures through which the 'march' was represented. The print was called *The Age of Intellect* and is reproduced in R. A. Vogler's *Graphic Works of George Cruikshank* (New York, 1979, 47). Here the 'march' and the 'pursuit of knowledge' are given bathetic, even mocking, figuration as respectively childish and infantile. In one image on the same plate, '*useful* knowledge' does no more than tell people where to dine cheaply, ignoring the possibility of nourishment for the mind. The plate's central image shows another child-savant lecturing her astonished grandmother, who is trying to recite 'Cock Robin' from a child's picture book, literally how to suck eggs. In both this small image, and the previous one, interestingly direct pre-figurations of Dickens's attacks on utilitarian notions of education are to be found. The pseudo-scientific grandiloquence of the child in explaining (and mystifying) a procedure known to every rural schoolchild is a direct antecedent of Bitzer's book-learnt quasi-knowledge. The '*useful* knowledge' sandwich board man in the bottom right-hand image of

Cruikshank's plate announces the pleasures of the ham and beef shop in Pudding Lane with a placard which asks the public to 'NB Mock Turtle'.

Cruikshank's depiction of 'the march of intellect' in this print depends on a perception of its 'childishness'. He uses an image of the precocity of children, – calling to mind another of Dickens's coinages, the 'infant phenomenon' – to represent social change, so analysing the 'march' as a relatively comic violation of inherited notions of wisdom, age, and understanding.

But there were other contexts in which useful learning might be represented. In one of the prints Dorothy George discusses in *From Hogarth to Cruikshank*, a complex *March of Intellect* of 1829 which 'combines the fantasies of invention with low life luxury' (George 1987, 179) one small-scale but centrally placed visual detail – that of a dustman in full rig eating a pineapple – gives an early example of the emergent visual narrative which combines dirt, disruption, and 'the march of intellect'. This narrative was to become explicit in an undated Henry Heath print called *The March of Intellect* which depicts an evidently poor dustman at leisure wrapped up in a book called 'Introduction to the Pleasures of Science' (George 1987, 179). George concludes 'the jest in such prints was to connect the remarkable movements for self-education and the emergence of a new working-class élite with low life and poverty, which, by tradition, were subjects of comedy' (George 1987, 181), thus making a broad and significant connection between the representation of mechanical invention and and 'the march of intellect' (George 1987, 177–9).

Evidence of the traditionally derisive representation of dustmen, even before they are comically associated with the rise of working-class literacy, is provided by their recurrent presence in popular caricature scrapbooks and miscellanies. Henry Heath's 1840 *The Caricaturist's Scrap Book*, for example, which gathers together his miscellaneous work over many years, represents dustmen in many humorous contexts – drunk in wheelbarrows (*Omnium Gatherum No. 4*); engaged in street repartee (*Scenes in London No. 12*); digging and sifting on a huge dust heap (*Scenes in London No. 29*); irrupting dirtily into a respectable shop (*Scenes in London No. 41*); and reproaching a butcher for cutting meat too fat (*Sayings and Doings No. 6*). Seymour's collected *Sketches* show dustmen enjoying a similarly persistent public and comic presence in topical caricature.[7] But the most interesting image which effectively summarizes this tradition of derision is to be found in one of Heath's short series of *Nautical Dictionary* etchings. *Capsize* shows an enraged dustman overturning a puppet booth, provoked, as the image makes clear, by the continued use of a puppet dustman as a stereotypical figure of fun in the Punch and Judy show. In the background another dustman explains to an amused bystander: 'vy they are always a taking us off'.

Clearly, then, the new caricature formulations of the late 1820s which depict the 'educated' dustman as a central protagonist in the 'march of intellect' draw on, and coexist with, a well-established tradition of ridicule. The caricatures of the late 1820s and early 1830s show an interesting transition from the idea of the dustman (or, often, the coal-heaver) as a buffoon and victim of ridicule to the notion of the dustman as a cultural warrior. This transition is well exemplified in a print published by Fairburn in September 1828 and called *March of Intellect among the Black-Diamond Carriers* (Figure 10). Here, the agency of transformation is not the customary SDUK but rather a genteel popular song which, modulated through the caricature's sturdy if whimsical fantasies, allows the robust central figure to sprout butterfly wings. In the typically even-handed caricature tradition, the satire here is directed as much at the sentimentality and over-refinement of popular parlour songs as at the unlikeliness of a coal-heaver being metamorphosed into an exotic butterfly. This crudely rendered and crowded caricature offers a refreshingly blunt juxtaposition in which the brutally naturalistic figures of the three coal-heavers stand beside such fantastic elements as the 'good fairy' and huge butterfly wings. The genteel presence of the ornate harp at the centre of the image puts both the coal heavers and the fairies into perspective. A woman passer-by and a servant on the cart (who asks rhetorically 'what do you think of the black-diamond fraternity now?') further anchor the print in the contemporary world. The humour here is derived from incongruity, the transforming interruption of the ludicrously genteel into the world of famously uncultured and traditionally comic coalmen. *The March of Interlect, or a Dustman and Family of the 19th. Century* (Figure 11), an undated print engraved and published by J. L. Marks, offers a similarly comic account of the cultural tensions between gentility, aspiration, and the patently ludicrous aping of aristocratic habits by dustmen and their families. I have written at greater length elsewhere on cultural inversions as an aspect of caricature in the 1830s[8] – the point here is to establish the ways in which traditional representations of dustmen as uncultured proletarians are overtaken by a more specialized use of the dustman as a stereotypical representative of the most startling, if potentially ludicrous cultural aspirations to be found among working men and women.

Summarizing the argument so far, a range of readings of the cultural significance of 'the march of intellect' as a social phenomenon became

Figure 10 Anon., 'March of Intellect among the Black-Diamond Carriers'. J. Fairburn, 6 September 1828.

Figure 11 J.L. Marks, 'The March of Interlect, or a Dustman and Family of the 19th Century'. Engraving. J. L. Marks n.d.

available during the late 1820s. Some prints linked the 'march of intellect' to a wider, sceptically amused sense of 'the wonders of the age'. Others read it dismissively as childish precociousness. All the prints agreed in using images of reversal, incongruity, and topsy-turvyness as their central mode of both structure and analysis. Alongside, and increasingly merging with these wider themes, was a well-developed tradition of semi-affectionate ridicule which focused on dustmen and coal-heavers as the most irredeemably dirty, philistine, and intrusive members of the urban proletariat. By 1830, as the SDUK began to provide a central rallying point and ideological focus for 'the march of intellect', the dustman and coal-heaver had emerged as key visual stereotypes, instantly recognizable by their garb and posture, for testing out the political and social claims of self-education and self-improvement among even the most foetid levels of the urban working classes. It is at this moment that a new element enters the caricature repertoire – the mass-circulation, improving, useful-knowledge periodical. With the introduction of a precise cultural medium – cheap, illustrated, weekly-issue serials – for the dissemination of 'useful knowledge' in the 1830s, the aspirations of dustman and coal-heaver towards education, if not towards gentility, become less accessible to the semi-affectionate ridicule traditionally meted out to dustmen, and instead require some more strenuously ideological and political form of discussion. One caricature already discussed – Heath's *March of Intellect* – has already suggested an increasingly politicized context for debate of 'the march of intellect' in caricature. Heath's dustman, as we have seen, is reading in his 'leisure' a book called, strenuously, 'An Introduction to the Pleasures of Science'. Unexceptional enough, except that the sub-title of the book – 'Dedicated to the Majesty of the People' – explicitly links this representation of a comically and incongruously ambitious dustman to the political history of the common people. The remaining caricatures studied in this chapter provide a more detailed account of the politics of cultural debate on 'useful knowledge'. These caricatures bring this debate into sharp focus by concentrating on mass-circulation periodicals, especially *The Penny Magazine*, as a representative symptom of the 'age of intellect'.

3

It is interesting that the two most weakly drawn caricatures studied here, and the two most heavily dependent on labels, explanatory captions, and verbal explication rather than on any visual jokes or aesthetic pleasures, are also the two which offer the most violent denunciation of *The Penny Magazine* and its serial predecessors. If 1832 marks a series of emblematically new beginnings in periodical history (*The Penny Magazine, The Saturday Magazine, Chambers' Edinburgh Journal, Figaro in London*) it also seems to mark conclusively the waning powers of the single plate caricature tradition in sustaining the energy and dynamism of eighteenth-century social and political caricature. Both *Penny Patent Knowledge Mill* and *The March of Literature or The Rival Mag's* (Figure 12) [hereafter cited as *The Rival Mag's*] are large lithographs. While lithography, introduced as a graphic medium in the second decade of the nineteenth century, offered the topographical artist a new and potentially useful range of tonal and colour effects, the weak lines and sketchy tonality as well as the waxy crayon effects characteristic of the medium could easily lead to a loss of visual incisiveness. On the whole you needed to be a Daumier or a Gavarni, who used the medium in a much more painterly way, to make lithography work as an expressive medium for caricature. Interestingly, then, neither of these two images can be understood or enjoyed merely in terms of an immediately impressive or

Figure 12 Anon. [Robert Seymour?] 'The March of Literature or the Rival Mag's'. Lithograph. Thomas McClean, 25 October 1832.

pleasurable graphic image. The verbalness of Victorian print culture is already apparent in these images from the early 1830s.

Patent Penny Knowledge Mill (*BMC* 17267), a large lithograph by Robert Seymour, is totally dependent on words.[9] Indeed, it is hard to describe this image without merely writing out its verbal content as a piece of consecutive prose. The central image is that of a hand-driven mill, somewhere between a giant pepper mill and a crude threshing magazine, with ingredients (wood-block illustrations, Whig philosophy, and liberal theology) being mixed into a 'froth' which is miraculously transformed by the mechanics of the mill into 'pennies' on one side and 'twaddle' on the other. The image is thus compounded out of two elements – one to do with 'grinding' (and so alluding to Blakean educational mills and grindstones as well as industrial milling) and one to do with 'threshing' whereby the harvest of pennies is disaggregated from the chaff of intellectual 'twaddle'. By means of this miraculous caricature machine, a pseudo-industrial process is graphically applied to the manufacture of *values* and *ideology*, thus implicitly compromising value systems by the cash nexus. Grinding out useful knowledge can only be accomplished by grinding the faces of the poor in their own poverty. So the basic thrust of the caricature is to use the metaphor of an industrial process which encompasses both grinding and winnowing at the same time to represent and satirize the forced production and dissemination of 'really useful knowledge' by the Society for the Diffusion of Useful Knowledge both in its pre-*Penny Magazine* period in the 1830s and after *The Penny Magazine* had given critics a particular focus for their attack.

Unusually for caricatures of the 'march of intellect', *Penny Patent Knowledge Mill* represents historical and social change through a depiction of cultural entrepreneurs and (by means of verbal tags) their ideologies and assumptions. Those actually on the 'march', the artisan and proletarian reading public, are excluded from this print, except for the hapless John Bull. The discourse proposed by the print remains sophisticated, concerned with debates within ruling class ideology rather than with any overt derision aimed at the cultural ambitions of the working classes. The attack in this print on the SDUK and its serial publications for their coerciveness, their blatant interest in profit, and their dangerous flirtation with liberalism is fundamentally a critique of Whig ideology rather than of the working-class reader. *Penny Patent Knowledge Mill* has little to say on the value (or absurdity) of popular aspirations to knowledge, however clear its denunciation of the ideologies and entrepreneurs who sponsored those aspirations.

4

The Rival Mag's (Figure 12), while accepting the critique of the 'useful knowledge' movement to be found in *Penny Patent Knowledge Mill*, additionally depicts the proletariat (or, as the print sees it, the dangerous urban rabble) who buy, read, and presumably value *The Penny Magazine*. I have written elsewhere a detailed account of *The Rival Mag's* which offers extensive comments on the eighteenth-century vocabulary of visual emblems, drawn from caricature, that are employed to deride working-class aspirations (Maidment 1996, chapter 3). In comparison with *Penny Patent Knowledge Mill, The Rival Mag's* comes from a different source of political anxiety in so far as it depicts *The Penny Magazine* in relation not to its ideologically motivated, allegedly hypocritical entrepreneurs but rather in relation to its readers, a depraved, syphilitic, potentially disruptive, and politicized urban mob. In *The Rival Mag's*, too, *The Penny Magazine* is described not as a powerfully unique, dominating socio-economic force, but rather as an emblem or representative of a wider 'penny' movement in mass-circulation journalism and literature which the print interprets as a dangerous travesty (a literal and metaphorical 'cheapening') of 'valuable' literature. The central focus of the print is on *readers* as representing the lowest, dirtiest, least cultivated and thus potentially most dangerous elements among the proletariat. But even this denunciation is taken a stage further in the representation of the *Saturday's* readers as *bestial* – the night-soil man's horse, its eyes aglow with its avidity for useful knowledge, is reading the sweep's boy's journal over his shoulder. *The Rival Mag's* is structured not just by traditional caricature tropes of reversal or travesty – the unexpected, but ludicrous, appearance of knowledge among the traditionally unlettered – but also by notions of *irresponsibility*. The night-soil man's absorption in *The Penny Magazine* is represented as a neglect of his proper responsibilities, and hence of his authentic 'place'. When the additional emblematic content of the 'revolutionary' procession is added to the depiction of neglect, then the overwhelmingly reactionary effect of the intersecting visual elements in the print can be felt. In a world characterized by a level of social inversion and reversal which permits the pursuit of knowledge by the hitherto ignorant and unlettered, traditional notions of duty and social responsibility will be neglected, thus allowing the subversive threat of the dirty urban proletariat to manifest itself culturally as well as politically.

The Rival Mag's foregrounds not the unscrupulous and ideologically motivated entrepreneurs of penny culture, but rather the dangers and implausibilities of redeeming the proletariat into knowledge – or at least the kind of knowledge inculcated by *The Penny Magazine*. Accordingly, *The Rival Mag's* forms a second savagely reactionary analysis of 'the march of intellect'. The late caricature mode (a large single plate lithograph), the weak drawing, the high levels of dependence on verbal explanations, the iconography of the politicized procession of urban grotesques all reflect the feeble attenuation of eighteenth-century caricature vigour in the Regency and early Victorian eras. This decline in the visual energy of the caricature mode is not, however, accompanied by any easing of eighteenth-century willingness to represent extreme or outspoken political opinions.

5

The 1830s, however, do show a continuation of eighteenth-century caricature precedents, with their more accommodating and liberal views of proletarian activity. These caricature precedents allow 'the march of intellect' to be regarded primarily as humorous rather than threatening, constructed out of pretentiousness, folly, and a ludicrous wish to ape one's betters rather than any higher political ambition. *The Literary Dustman* (Figure 13), a wood-engraved image from a popular theatrical song book of the 1830s (James 1967, 149) shows the emergence of the educated (or more precisely the *educating*) dustman as a heroic if comic stereotype. Indeed the tone of the entire print might be characterized as one of mocking celebration of proletarian and artisan cultural aspirations. The complexity of this apparently straightforward comic image is largely dependent on the iconographic force of the painting hanging on the wall behind the lounging dustman, who is smoking a cigar and reading *The Penny Magazine* while supposedly superintending his son's music lesson – an interesting leisure combination of self-indulgence and cultural ambitiousness. The painting, paired with a mocking royal portrait, shows two stage figures 'African Sal' and 'Dusty Bob' who originally appeared in Pierce Egan's extremely successful picaresque novel, *Life in London* (1821) (Maidment 1996, chapter 3). The success of the novel was partly ascribed to the illustrations by George and Robert Cruikshank, but it was the large number of stage adaptations of the novel, especially those produced by W. T. Moncrieff, that developed the novel's minor characters from picaresque 'types' or stooges into immediately recognizable popular stereotypes. Dusty Bob, a coal-heaver in the fictional original *Life in London*, metamorphosed on the stage into a 'comic dustman', a stereotype still powerful enough at the beginning

Figure 13 Anon., 'The Literary Dustman'. Wood engraving from W. T. Moncrieff's *Songbook* n.d.

of the next century to produce Doolittle in Shaw's *Pygmalion*. This complex visual allusion, which I have traced in detail elsewhere, maintains a relatively genial eighteenth-century tradition of ridicule of street characters, who were regarded more as 'entertainment' than as potentially civilized human beings. In an early Victorian society which was both seeking to delineate, describe, and analyse street culture and becoming increasingly anxious and even fearful of the public aspects of street life (not least waste disposal), it is important to acknowledge the continuation of less threatened comic traditions of representation. Indeed, the streets are held at bay by this interior scene, which delineates the follies and indulgences of the newly wealthy in their attempts at aping genteel culture as much as it alludes to the brawny and disruptive street presence of the physically as well as culturally challenging figure of the dustman.

Here, then, the increasingly politicized and socially anxious cultural stereotyping of *The Penny Magazine* reading, newly literate dustman is returned to an older tradition of comic topsy-turvyness where absurdity out manoeuvres social threat. Even though the accompanying song from Moncrieff's stage versions of *Life in London* attacks the 'taxes on knowledge' and follows the aspirations of the educated dustman as far as the

Houses of Parliament (James 1967, 149–50), this wood engraving appropriates *The Penny Magazine*, a recent cultural phenomenon, back into longstanding caricature tropes for representing the lowest urban trades, tropes which were constructed equally out of ridicule and delight, rather than out of the fear and anxiety which characterized *Penny Patent Knowledge Mill* and *The Rival Mag's*.

6

This discussion has limited itself to two intersecting elements within the repertoire of visual motifs and tropes that were available to the caricaturist. One element was a social fact, a ubiquitous symptom of the 'march of mind' – the mass-circulation serial publication or magazine aimed at artisan readers. And of all symptoms of 'the march of intellect' none could have been more symptomatic than the publications of the Society for the Diffusion of Useful Knowledge, especially *The Penny Magazine*. The other was less a social fact than a convenient stereotype – the depiction of the urban proletariat through a 'representative' figure drawn from the lowliest, dirtiest, least skilled, least organized city trades, especially dustmen, coal-heavers, and night-soil men. At the level of traditional caricature stereotypes, there were, as we have seen, obvious reasons for picking dustmen as representative urban proletarians in the early nineteenth century. Their distinctive, picturesque, ragged working clothes (especially the famous leather fan-tail hats), their obvious presence on the public streets (evinced by the din of their bells as well as by their appearance), and long traditions of mockery all combined to offer a strong appeal to the caricaturist. But the presence of dustmen was threatening as well as picturesque, a reminder that their physical strength, inarticulacy, lack of skill, and position outside trade organizations all rendered dustmen a potential danger to social order. Yet beneath this central ambiguity between comic grotesquery and social threat lay a deeper source of psycho-social anxiety – the fear of contamination. Soiling all they touched, dustmen were a constant reminder of the detritus and ordure produced by the city. However much Victorian society, like our own, sought to conceal the collection of waste from social consciousness, the physical presence of dustmen on the streets served as a constant reminder of the filth and waste produced by a sophisticated industrial society.

The caricatures discussed here emblematically represent periodicals and SDUK cheap serial publications as part of a wider debate on the proper forms through which the cultural progress of the working classes should take place. In most instances, the representation of serial publications like *The Penny Magazine* in these images also depends to an extent which cannot be accidental on the representation of the 'lowest' trades within the urban working proletariat (dustmen, coal-heavers, and night-soil men) as readers of the periodical press. While this representational intersection between *The Penny Magazine* and its dustmen readers is essentially satirical and drawn within a well-established tradition of caricature denunciation of all things proletarian, some degree of interpretative complexity is identifiable. If the main argument of these assembled caricatures is that the world is threateningly and dangerously turned upside down when dustmen seek to improve themselves by reading *The Penny Magazine* or other improving texts, more accommodating and progressive readings of social change are visible. Some of these images, as this chapter suggests, are little more than a compendium of splenetic reactionary assumptions. But elsewhere a more genial acceptance of the inevitability of a self-educating proletariat, more comic in its pretensions than threatening in its new-found knowledge, is coming into being.

Notes

1. The range of secondary sources which describe these major social and cultural changes are obviously legion. This chapters has drawn mainly on accounts of the relationships between education, class, and the development of literary, graphic, and serial genres aimed at self-educating artisan readers. These accounts include: R. D. Altick *The English Common Reader* (Chicago, 1957); R. K. Webb *The British Working Class Reader* (London, 1955); L. James *Fiction for the Working Man* (Oxford, 1963); ed. L. James *Print and the People* (London, 1976), V. Neuburg *Popular Literature* (London, 1977); D. Vincent *Bread, Knowledge, and Freedom* (London, 1981); D. Vincent *Literacy and Popular Culture 1750–1914* (Cambridge, 1989). More precise secondary sources for visual material and for the Society for the Diffusion of Useful Knowledge are given below.
2. The starting place for this study of caricature representations of the 'march of intellect' was the characteristically suggestive few pages on the subject in Dorothy George's indispensable *From Hogarth to Cruikshank – Social Change in Graphic Satire*. Two other key discussions which stress the importance of graphic discourses in establishing 'useful knowledge' as a cultural construction are Celina Fox's *Graphic Journalism in England in the 1830s and 1840s* and Patricia Anderson's *The Printed Image and the Transformation of Popular Culture 1790–1860*. Both authors provide discussions of the issues addressed in this chapter. The primary source for graphic material is M. D. George, *British Museum Catalogue of Political and Personal Satires* (British Museum XI vols 1870–1954). All references to prints described in these volumes will be made in the following format: *BMC* followed by the print number. Among many other 'March of

Intellect' prints see *BMC* 15604, 15622, 15775, 15779, 15922, 15941, 16027, 16496, 16894, 17282, 17309, 17357.

3. An excellent new account of the methods, interests and representational codes of caricature is provided by Diana Donald's *The Age of Caricature* (New Haven, CT, 1996). Peter Wagner's *Reading Iconotexts* (London, 1995) is a more heavily theorized study of the formation of meanings in eighteenth-century graphic discourses. More precise information on physiognomy and gesture is offered by J. Wechsler, *A Human Comedy: Physiognomy and Caricature in 19th. Century Paris* (London, 1982) and M. Cowling, *The Artist as Anthropologist* (Cambridge, 1989).
4. It is possible to find a number of books and periodicals from this period where sophisticated caricature modes of representation appear in wood-engraved vignettes alongside images drawn from vernacular broadside and chapbook traditions. The three volumes of *The Universal Songster* (Jones & Co. 1825–1827), illustrated by George and Robert Cruikshank, show the increasing elision between polite and vernacular visual vocabularies. The wood-engraved illustrations to periodicals like *The Penny Novelist* (1832ff) show similar variety in representational method. Cruikshank's illustrations to William Hone's various books and pamphlets perhaps provide the most obvious site where vernacular and sophisticated meet – see M. Wood, *Radical Satire and Print Culture 1790–1822* (Oxford, 1994).
5. There is no easily accessible single study which describes the late survival of an eighteenth-century caricature tradition on into the Regency and early Victorian period. Ronald Paulson's 'The Tradition of Comic Illustration from Hogarth to Cruikshank' in R. L. Patten (ed.), *George Cruikshank: A Revaluation* (Princeton, CA, 1992) offers a good brief survey of literary illustration. Simon Houfe's *The Dictionary of British Book Illustrators and Caricaturists 1800–1914* (Woodbridge rev ed. 1981) gives a useful narrative summary of major developments and also individual entries for some of the key transitional artists working in and through the transition from single plate caricature to humorous illustration – artists like Henry Heath, Robert Seymour, and John Doyle. George Cruikshank has a specialist literature of his own, and R. L. Patten's *George Cruikshank's Life, Times, and Art* vol. 1 (New Brunswick, N.J, 1992) describes the context of engraving in the 1820s and 1830s in considerable detail. See also Louis James, 'Cruikshank and Early Victorian Caricature' in *History Workshop Journal* vol. 6 (Autumn 1978), 107–120. G. M. Trevelyan, *The Seven Years of William IV* (London, 1952) gives a detailed account of the political weight of John Doyle's work. D. Kunzle's *The History of the Comic Strip – The Nineteenth Century* (Berkeley, CA, 1990) is a major source for discussion of the visual culture of this period.
6. For the early development of lithography see: M. Twyman, *Lithography 1800–1850* (Oxford, 1970); F. H. Man, *Homage to Senefelder* (London, 1971); and ed. D. Porzio *Lithography: 200 Years of Art, History, and Technique* (London, 1982).
7. I have discussed Seymour's dustmen in considerable detail in my book *Reading Popular Prints 1790–1870*, especially chapter 3. This chapter extends the argument of that chapter to a wider range of material. Seymour's *Sketches* are of particular interest because they were reprinted by Henry Bohn in 1841 with a continuous narrative text supplied by 'Alfred Crowquill' [actually Henry Forrester], thus offering a key example of the way in which a series of separate

visual images on similar themes might be re-formulated as a continuous narrative. I think there is a major research topic to be tackled here in describing the drive towards sustained narrative visible in the re-gathering, re-grouping, and representing of caricature images in this period. The point at which pages of separate graphic images begin to re-formulate themselves (or begin to be re-formulated by their viewers) into linear narratives is worthy of some detailed discussion. Kunzle's book on the comic strip currently offers the only sustained critical account of this process, and he is constrained by his own focus on the comic strip as a genre to omit many related texts.

8. These social inversions have been extensively theorized as an aspect of 'transgression' and the 'carnivalesque'. The opening, and tone-setting, illustration to P. Stallybrass and A. White's *The Politics and Poetics of Transgression* (London, 1986) for example is a German lithograph dating from *c.*1830 which shows 'The Topsy-turvy World' in which horses drive carts pulled by men, cripples give money to well-dressed beggars, and children rock their mothers in their cradles. If contemporary examples are useful in extending this trope, the recent success of the film 'Babe', based on the novel *The Sheep-Pig* by Dick King-Smith, offers a characteristic example. In both film and book a pig takes over the role and responsibilities of a sheep-dog, to triumphant effect. The notion of carnivalesque reversal is useful here, though it is important not to over-emphasize the notion of 'play' in discussing the 'march of intellect'. Further interesting commentary on these issues is provided by a less obviously Bakhtinian commentator, Deborah Epstein Nord. In her fascinating recent book, *Walking the Victorian Streets: Women, Representation, and the City* (Ithaca, NY, 1995), Nord defines a change between eighteenth-century traditions of representing lowlife subjects as a 'subject of ridicule not compassion' and early nineteenth-century practice of dealing with street culture with 'detached amusement' and a 'scientific' impulse to catalogue and sort. What Nord sees here, using mainly written texts as her evidence, is an increasing anxiety in the early Victorian period as to the social threats which underlay the 'carnival' on the streets (24). Certainly Nord's view confirms my own sense that there are competing representational possibilities available here – ridicule, detached amusement, scientific curiosity, compassion, threat. Whether it is possible to identify these possibilities as a coherent sequence of response seems more questionable.
9. Both Fox and Anderson allude to this print. It also forms the illustration to Anderson's useful entry on Charles Knight in the *Dictionary of Literary Biography 106 – British Literary Publishing Houses 1820–1880* (Detroit, London, VA, 1991), 164–70. For further caricatures which specifically refer to the Society for the Diffusion of Useful Knowledge or *The Penny Magazine* see *BMC* 15178, 15535, 15622, 17123, 17247, 17258, 17282, 17285. This chapter has been forced to exclude much interesting material in order to address wider arguments. C. J. Grant's attack on *The Penny Magazine* (*BMC* 17285), drawn on behalf of the views held by radicals who defended an unstamped, topical, and explicitly political literature, deserves particular attention. However, whilst Grant's complex image does blacken *The Penny* with a sweep's boy, there are unfortunately no dustmen to be seen.

Works cited

Anderson, P. *The Printed Image and the Transformation of Popular Culture 1750–1914*. Oxford, 1991.

Fox, Celina. *Graphic Journalism in England in the 1830s and 1840s*. New York, 1988.

George, M. D. *British Museum Catalogue of Political and Personal Satires*. London, vol. X 1820–1827 (1952); vol. XI, 1828–1832 (1954). [abbreviated as *BMC*]

George, M. D. *From Hogarth to Cruikshank – Social Change in Graphic Satire*. London, rev ed. 1987.

James, Louis. *Print and the People*. London, 1976.

Maidment, B. E. *Reading Popular Prints 1790–1870*. Manchester, 1996.

Vincent, D. *Literacy and Popular Culture 1750–1914*. Cambridge, 1989.

8
'Women in Conference': Reading the Correspondence Columns in *Woman* 1890–1910

Lynne Warren

To date it has been extremely difficult to establish anything very tangible about the *real* readers of nineteenth-century popular periodicals (see Ellegård, 1957). Archival resources, such as company records or subscriber lists are scarce, if not non-existent, and few biographies or memoirs mention such ephemeral reading matter as women's weeklies. The publication of names and addresses of readers participating in *Woman's* weekly competitions offers a unique, if limited, research tool. Exhaustive work on several hundred names has to date resulted in the following information: names and addresses would appear to be genuine; certain names appear regularly and over extended periods of time. *Woman's* claim to be read by the middle- and upper-class woman seems to be upheld during the 1890s, although by the 1900s there is evidence of a shift away from the more affluent areas of London, and the emergence of many 'care of' addresses implies a downward swing in social class of the readership. This information, however, represents only one section of the readership and so it must be treated with caution.

Text-based analyses can offer a wide scope for study, but they too can only ever be partial accounts. Because of the difficulties in obtaining information the concept of the implied reader is often substituted for the elusive figure of the real reader – the reader who actually paid her penny and read the magazine. The implied reader is a useful critical tool, but it offers only one way into a particularly complex problem. Implied readers are not, and can never be, anything more than approximations of real people. At the same time it is true that the notion of the 'real' reader is also problematical in the sense that all readers 'construct' themselves in the process of reading. While I would admit that the task of presenting a 'real' reading would be both illusory and misleading, I would argue that the 'real' reader should not be ignored.

In order to go some way towards a reading which includes the figure of the 'real' reader, I offer a synthesis of text and reader-based analyses: exploring the complex relationship between reader and text that arises through the negotiation of identity in the correspondence pages; examining how the power to define the self (particularly in terms of gender roles) was constantly shifting between reader and magazine, and considering the strategies each adopted in order to make their voice heard.

Correspondence in *Woman* can be divided into two distinct types – those featuring the 'voice' of the reader, with or without editorial comment, and those consisting solely of the editorial 'voice'. Reader participation in *Woman* offers a useful site for examining the ways in which readers constructed identities within the margins of the text – whether as progressive 'woman' or conservative 'lady'. Such constructions clearly work on a number of levels, which will be discussed in more detail in the course of the chapter; however, it is worth pausing to identify the two main ways in which reader contributions can be said to help construct a separate self from that implied within the editorial copy. First, readers engaging with other readers' contributions will be prompted either to identify with, or challenge, the 'voices' of these readers (or those of the editorial respondents). This process feeds into the individual reader's construction of her self in relation to the textual community of the magazine (her 'textual identity').

Second, when the reader actually writes in herself, and when she receives a reply (whether publicly or privately), then this process of identification reaches a different level, a change which also applies to other readers who may actually know the correspondent. Thus although such readers may still regard the correspondence columns as a point of departure or of confirmation, they are able to transcend the more abstract reader–text relationship and enter into a process of exchange with more tangible results. Such readers may be looking for reassurance or may wish to register their independence from the dominant values of the magazine; either way, by participating in the creation of meanings the process of identification becomes more dynamic and highly charged.

The appearance of the reader in the text serves both to specify and consolidate the readership, in much the same way as the encoded discourse of, for example, the front cover or the material format (James 1982, 351–2. Beetham 1990, 19–32). This is not an unproblematic process. The success of a magazine was in large part due to its formulation of a textual or 'corporate' identity – inextricably linked to its successful appeal to its targeted audience, conveyed through the clues provided by title, cover design, editorial content and so on. With the proliferation of titles in the latter part of the century, the women's press needed to imagine and appeal to increasingly specific audiences. The construction of the magazine's 'corporate' identity and the imagined community of readers were thus interdependent. However, such communities, once set up, are not so easily manipulated. The multiplicity of voices interacting within the text (asking for advice, competing, commenting on editorial text and on previous contributions) can be interpreted in many different, often conflicting ways. Editorial comment is one way in which the magazine sought to shut down the proliferation of meanings by showing the reader how to interpret them; but ultimately, the power of interpretation rested with the reader.

Readers of *Woman* who wished to correspond or enter any of the competitions were required to attach a coupon clipped from the inside wrapper of the relevant issue. Failure to do so invariably attracted comments from the editorial staff. As a sales technique, the coupon system ensured that the full benefits of participation in *Woman* (access to advice and the possibility of winning a variety of prizes) could only be obtained by readers who bought the magazine rather than borrowing or being given it. As many periodicals at this time were passed on from reader to reader, the coupon system was at this level an attempt to increase sales figures. However, restricting participation to those coupon holders who could prove their purchase effectively limited the range of readers who figured within the text and it must be remembered that any analysis of readership based upon the contributing readers will not necessarily be representative of the overall readership.

As well as encouraging sales, *Woman's* coupon system also functioned as an attempt at audience building, feeding into the notion of the magazine as a community. The coupon was, then, a physical symbol of belonging to the community of *Woman* readers and correspondents, and not only provided proof of the reader's licence for participation but also invested her, as a legitimate member of the 'community', with full access to all the privileges offered. The very act of purchase contributed to the reader's sense of belonging, based upon the notion of reader and magazine having a mutual interest in the construction and maintenance of both text and identity. This definition of a community reinforced by the 'right' to participate, however, should not be regarded as representative of the entire community. As I have pointed out, magazines were not read only by the individuals who actually purchased them. Popular texts do tend to get passed around – periodicals in particular are noted for their large and informal circulation – and it is unlikely that *Woman* would prove to be any different. This means that there is a potentially large portion of *Woman's* readership which is not represented in the text, either in the correspondence columns or the published competition responses. Clearly this implies that within the readership itself there was a division – between those imbued with full participatory rights and those who had no opportunity to intervene directly in the text. What is impossible to resolve, though itself an interesting question, is how this affected the reader's construction of a self in relation to the text – whether the opportunity to intervene actively within the printed pages of the magazine made readers more or less likely to identify with the community represented by the magazine and the readers' voices which figured there. It seems feasible to suggest that participating readers may have been more likely to see the magazine as having closer links with their life and sense of self (even where they may have been challenging assumptions made there) than those whose experience of the magazine was founded on a rather more unevenly balanced relationship.

The correspondence column was a regular feature of the magazine until it ceased publication, although its precise format underwent certain changes which can be directly related to changes in editorial control. Between 1890 and 1900, under the management of Fitzroy Gardner and Arnold Bennett, correspondence consisted of advice columns divided into specific areas of interest, corresponding to aspects of the feminine sphere as conventionally defined. These columns, devoted to Health, Dress, Furniture, and Domestic & Sundries, consisted of editorial responses only. The typeface is significantly smaller and the layout denser than that of the editorial material; and this, coupled with the location of the columns among the advertisements at the back of the magazine, serves to mark them off as separate from the main body of the text. Readers are identified by pseudonym, a convention of most correspondence throughout *Woman's* publication. Pseudonyms provided an anonymity for readers which, apart from exceptional cases, was enthusiastically promoted by the magazine.

Following Ida Meller's assumption of editorial control in October 1900,[1] the correspondence columns became integrated into the main body of the text. This marked a radical departure since for ten years the correspondence had followed the same format – surviving the change in editorial control from Gardner to Bennett. The original column's succinct titles are replaced by those which borrow from the register of the editorial copy: 'Round the House', 'Toilet Talk', 'What Shall I Wear?' and 'Home Work. And How to Make it Pay'. In terms of actual subject matter there is little change; the columns are still conducted through replies only, although the space allocated is roughly halved. Integrating the columns within the main text also has the effect of 'rescuing' them from their previously isolated position. Instead of turning to the back pages to read these columns the reader finds them slotted in amongst articles on fashion, cookery, and the like. Obviously the juxtaposition of correspondence and editorial copy imbues the correspondence with different connotations from those which might be apparent to the reader when the columns are surrounded by advertising. The association of consumerism with correspondence, when columns are embedded among advertisements for precisely those products that are recommended as solutions to the reader's problems, becomes much less explicit when those columns are included in the main body of the text.

It might also be argued that readers receiving a reply in the relocated columns are encouraged to regard themselves as more closely involved in the production of the text than when they appear as an appendage to the rest of the magazine. Clearly this is not entirely unproblematic – the contraction of distance between reader and text within the framework of the correspondence column also encroaches upon the sense of independence from the dominant discourses of the magazine. Readers' correspondence, while being given a surface authority – granted by the acknowledgement of their 'right' to be regarded as equally important as feature articles – become at the same time diminished. This is reflected in the decreased space occupied, and in the guarded manner of presentation. When readers' letters are moved into the body of the text with no dissimilarity between typefaces to mark their separate status, they can no longer be allowed the same freedom of presentation. The alteration in titles works to subsume the correspondence within the parameters of the magazine itself, erasing any explicit sense of disparity between the two.

The use of correspondence columns to engender a sense of community has a somewhat ambiguous status during the 1890s. As has already been pointed out, these columns, far from being an overtly privileged writing form, seem to be marginalized – included at the back of the magazine amongst the adverts and isolated from the main body of the text with a typeface significantly smaller and more densely set than the rest of the text. The position and appearance, then, is not immediately appealing, and serves to convey a sense of separation from the rest of the editorial material. One reason for this may be that the letter, which is highly personalized, is a site over which the magazine has least editorial control. Potentially, the letter proffers a multiplicity of discourses and as such may be seen as a threat to the sense of unity aimed at by the magazine. By declining to publish individual letters the periodical could exercise some degree of control over the meanings and identities evolved from them. Michel de Certeau's assertion that contemporary culture hierarchizes the activities of reading and writing, so that 'To write is to produce the text; to read is to receive it from someone else without putting one's own mark on it, without remaking it' (de Certeau, 1988, 169), suggests that allowing the actual voice, the 'mark', of the reader to intrude into the text becomes a highly charged process. While the magazine was essentially heteroglossic, with a multiplicity of voices, the greater part of these were presented as belonging to the magazine – contributions that had been either written or commissioned by the editorial staff. Writing in this instance, then, might be regarded as having been economically sanctioned. By paying for the material the magazine exerts its authority over these voices through a commercial exchange in which it becomes the legitimate owner – the contract is purely economic with none of the sort of obligations one might expect to find when contributions are of a different, more communal nature. With this other sort of exchange, when readers are complicit in creating the text from the point of view of community, then they might reasonably expect to find themselves on a more equal footing with the editorial team. Clearly this sort of relationship would be unacceptable to a commercial organ whose affiliations are not just to the readership. As an avowedly commercial venture, the magazine needed to maintain a 'corporate identity' which relied upon controlling, to a greater extent, the form and content of the text. With the publication of readers' letters a space is created which, through its economic independence, is outside the unifying drive of the magazine. Controlling the opportunities for correspondence – by fragmentation and by 'silencing' the readers' voice – allows the magazine to present these columns as having a nominally independent status while preserving its own dominant position.

While most replies make the initial enquiry explicit, others leave one mystified as to what exactly is being referred to, such as the reply to 'Nancy' – 'I cannot suggest anything better than our WOMAN motto, "Forward! But not too fast"' (21 March 1894, 22). By making no attempt to indicate what they are responding to, the correspondence columns force the reader to fill in the gaps. While this filling in is performed to a greater or lesser degree with all the replies, the publication of such manifestly cryptic answers not only allows this to occur, but implicitly encourages such activity. This raises the question of whether such open-ended responses provide a point of departure for readers to develop their own narratives and personal fantasies. Certainly it would appear that the pleasures of these columns were quite distinct from other forms of writing in the magazine. Not only are the columns highly self-referential –

directing readers to past answers and assuming an awareness of the various threads of discourse running throughout them, but they also contain letters from readers commenting on replies to other readers in past issues. 'Petite-Yvonne' clearly followed the correspondence however inattentively. In reply to her comments on a response printed four weeks earlier she is told that: 'You must have mis-read my reply to "Little Blossom," as I did not recommend her a depilatory, on the contrary I advised her to use nothing of the kind' (30 May 1894, 24). The sort of interaction, in which the reader not only follows the 'conversations' but is sufficiently motivated to comment on another reader's problem (and the magazine's response to it), suggests that the correspondence column functioned significantly to create a sense of belonging. That such a sense existed is supported by answers such as the one given to 'Corrie':

> I hope you may see this. A lady having noticed my answer to you a fortnight since, has very kindly written offering to give you any information you may require on the subject of embarking as a laundress... If you will send me your name and address I will put you into communication with her (9 April 1891, 22).

Answers such as these reveal the extent to which readers often transcended the textual community to come into actual contact with each other. The exchange between readers within the limits of the text becomes more highly charged with the possibility of it also occurring in reality – whether this was ever very much the case is largely irrelevant, as long as it was seen to be a possibility. We may contemplate how far this served to reinforce a discourse crossing class boundaries – were readers encouraged to regard themselves as equal? Or is the case of the reply to 'Corrie' simply another way in which the upper and upper-middle classes extended patronage to those in the lower ranks? Clearly this is not a question which may be easily resolved.

Readers often sought to personalize their relationship within the column – both to the magazine itself and, as the replies to other readers suggest, to the readership as a community. It would be wrong to suggest that a reader's personal sense of self is endangered through misrecognition or misrepresentation within the magazine. Individual personal identity derives from any number of circulating discourses and texts. However, I would suggest that the urge to personalize is based, in part at least, upon a tacit acknowledgement of the reader's relative powerlessness in constructing a self within the correspondence column. In order to correspond the reader was forced to conform to the various conditions laid out by the magazine: use of a pseudonym; attachment of the relevant coupon; use of separate pieces of paper for each enquiry; confining letters to those areas prescribed by *Woman*. If we are to regard the reader's letter as a site in which subjectivity resides, then such conditions and limitations clearly affect the reader's 'vocabulary of self' (Gergen 1989, 73). In the case of *Woman's* correspondence, the magazine justifies the superiority of its voice through its textual conventions. The readers, in order to correspond, have to master these, internalizing at the moment of writing a subjectivity shaped by the magazine's conditions of participation. The urge to personalize correspondence, in the face of the many limitations imposed by the magazine, may be a reaction to this loss of mastery over the representation of the self. By introducing a personal element (whether to thank the editor of that particular column for past advice, to ask for recognition, or simply to congratulate them on their good work) the correspondent effectively breaks the frame of the correspondence and hopefully ensures an acknowledgement which transcends the superior relation of the advice-giver.

Numerous replies begin with reassurances that the columnist remembers the correspondent, which suggests some attempt at establishing a relationship between reader and magazine. This relationship, however, is manifestly uncertain – the need or desire to be remembered suggests a degree of anxiety on the part of the reader. The construction of identity through the reader's relationship with the magazine and with the textual community it represents may lead to her investing it with a significance which requires reaffirmation. Margaret Beetham suggests that: 'For many – perhaps most – readers the desire to be confirmed in the generally accepted or dominant discourse may be a more powerful need than the dream of a different future or the desire to construct alternatives' (Beetham 1990, 30). If a reader's sense of confirmation in the dominant discourse of *Woman*, or of belonging to the community represented within the magazine, rests on being remembered it follows that without acknowledgement or recognition her identity, as a member of the textual community, comes into question. That this relationship was often reciprocal is suggested by a reply to 'Penelope-Anne': 'I remember you well, and am so glad that all the things pleased you. I am always glad to hear that my advice has given satisfaction, it encourages me more than my correspondents have any idea of' (30 May 1894, 22). That readers were actively involved in constructing identities for the editorial staff is a further indicator of the interpretative practices over which the magazine itself had little control. Readers were quite clearly caught up in imagining or fantasizing about the 'real world' of the editorial office.

Such a practice suggests that this world was one that had some resonance for the reader, and one in which some personal investment was made. The earlier discussion of how readers personalized their discourse with the magazine is further evidence of an interpretative process that is, to an extent, suggestive of an emotional involvement. The preface to a response to 'An Admirer of *Woman*' demonstrates quite clearly that readers were in fact interpreting the text in ways which transcended the cues provided by the magazine, such as the pseudonym given to the conductor of the column dealing with health. Although this writer had been given the obviously female name of 'Medica' at least one reader clearly had her doubts about his / her gender, provoking this reply:

> I wonder what made you imagine I am a man. Had I been I am quite sure I should not have felt the sympathy for you that I did on learning that your skin which, until recently, has always been so beautiful, is showing a decided tendency to fall into lines. Truly a horrible discovery for a woman to make (18 July 1894, 24).

Resisting the construction placed upon her by a reader, 'Medica' is not merely reinforcing an editorial definition of what it is to be 'feminine', but is actually in the same position as the readership who are also being constructed by the text, and must also make the decision to accept or reject that definition. It may be argued that in accepting this, 'Medica' (or any of the other editorial voices) is also, in effect, appropriating the readers' fantasy in order to further his / her own textual identity – creating a stronger sense of this 'real world' s/he supposedly inhabits. The construction of identity would, then, appear to be a mutually ongoing process. Although the representation of the editorial team as an intimate and 'real' circle of women is clearly artificial, made up as it was of those writing under pseudonyms and under their own names, some of whom were actually men, there is a drive to represent themselves as 'real' and, moreover, as real women. Attempting to construct a 'real world' of the editorial office prompts responses such as the self-referential and intimate reply given to 'Pilgrim' who finds herself drawn into an exchange between two of the editorial staff: 'How strange that you should write to me on the subject of a small Continental outfit. "Lady Vain" and I were only discussing the question the other day, with the result of which (*sic*) you will find in Snuggery Small Talk' (4 July 1894, 18). These 'writers' however, are not real people: they do not have 'real' names: even when presented as ordinary women interested in the same sort of things as their readers they apparently address each other by their pseudonyms.

What is clearly at work here is an attempt to create a 'world' in which writers for the magazine could quite reasonably be chatting about clothing, the result of which is then reproduced in the text.

Although the construction of community and identities on the part of both reader and editorial staff is a mutual process, there is also ample evidence to suggest that the magazine sought to diminish the power of the reader. Within the parameters of the issue the reader is relatively powerless against the restrictions and admonishments imposed by the magazine – they have little or no recourse within the text to correct or challenge what is written. Readers are criticized for poor handwriting, spelling, and grammatical errors as well as for their choice of pseudonym. The editorial voice is presented as naturally superior, privileged over that of the reader who is constantly returned to the magazine for information, advice, and reassurance. Certainly, those columns which omitted the readers' voice made the process of asserting individual subjectivities more problematic for the readership, although they did not prevent such incidents. As the textual creation of identity within the correspondence column hinges on the ability to both write and receive a reply, we might expect that limiting the range of enquiry would function to limit the formation of a textual identity. However, there is evidence within the text of alternative subjectivities being constructed by the readers, notwithstanding the boundaries imposed by the format of the magazine. Thus we see answers such as that offered to 'Self-Help': 'I do not know of any such penny weekly paper as you require; it seems to me that this paper would exactly meet the requirements of your friend if it gave politics, which I am most thankful to say it does not do' (30 May 1894, 22). In the same column in which 'Self-Help' is chastised, a correspondent is taken to task over her choice of 'Tom' as a pseudonym rather than something more feminine (30 May 1894, 24).

After 1899 the correspondence columns explicitly figured the reader's voice. Although I have suggested that the absence of a voice has implications for the creation of an identity within the boundaries of the textual community represented by the magazine, I would not suggest that the appearance of this voice was unproblematic.

The major textual characteristic which distinguishes the correspondence figuring the reader's voice from that in which this voice is absent is the narrative structure. With the publication of both letter and editorial response there is a much more explicit sense of a narrative than that which can be deduced from the response-only columns. I have already discussed how the publication of often cryptic replies to specific readers in the 1890s, while offering a source for personal fantasy, made the columns difficult to follow and yet there is ample evidence that readers not only followed them but were inspired to write in and comment on the advice tendered to other readers. The degree of dedication that this sort of involvement requires is significantly different when the correspondence is neatly presented as following a clear and definite (as well as more impersonal) structure. What is also significant is that, paradoxically, meanings begin to close down with the publication of both reading and editorial voice. While the reader of correspondence in the 1890s had only the editorial voice to provide clues to the content of the original letter, the content and meaning of that letter itself was much more in flux. Lacking the full text, readers are forced to fill in the gaps, in the process of which they are able to construct meanings which are, obviously, of greater relevance to their own situation; once the original letter is published the gaps are not so obvious.

Publication of 'original' letters, however, did not guarantee an 'authentic' voice for the reader. As with correspondence in the periodical press even today, letters underwent selection and editing, with text often being significantly altered in the process. There is considerable evidence within the pages of correspondence of editorial intervention in the original text. For example, in 'Answers' in 1901, the answer to V. B.'s query, 'Is it necessary when a hospital nurse is in the house to include her as one of the family?' is concluded with the statement that 'Yes, you will be responsible for the laundry' (29 Jan. 1902, 30). While a reply to Bella, whose published question simply asks, 'I want you, if you will, to give me some hints on catering on £3 a week. There are three of us and two servants', asserts that, 'from what you tell me of your means I am inclined to think that you use too much meat' (15 Jan. 1902, 26). These answers clearly point to other, unprinted, elements of the original letter. Additionally, it would seem probable from the brevity of, and uniform language employed in, the letters that a significant amount of editorial shaping was going on. Whether this would be apparent to the original reader, whose reading practices were unlikely to replicate those of the critically informed researcher, is a point which is impossible fully to resolve. I would suggest, however, referring back to an argument put forward in the opening section of this chapter, that such correspondence being presented as the authentic product of the readership was intended to be read as such. Even if readers had been sceptical about the origins of the actual language employed, its appearance on the printed page acted as a means of shaping their responses to it.

What also marks off the correspondence that actually figures the 'voice' of the reader is not only that it is generally allotted a page of its own, rather than constituting a column either appended to a 'main' feature or ensconced among the advertisements in the back pages, but that it is also presented as being conducted or directed by the readers. 'Our Social Club' was the first to do this. Instead of providing readily-formulated categories within which readers were able to respond, this page not only asked the readers to contribute subjects for discussion but also provided a space in which far more disparate topics were discussed. The editorial introduction points to the hidden activity of readers who, despite the lack of any general correspondence column, had nevertheless written in:

> The receipt of numerous letters from readers of WOMAN referring to various topics that are of general interest has induced me to start a new feature this week, and I introduce Our Social Club, which will, I hope, form an interesting medium through which the ideas of our readers may be exchanged (10 Oct. 1900, 20).

This certainly suggests the existence of a tension between the way the magazine had previously conceived the needs and interests of its readers and what that readership had, in fact, desired. That readers had persisted, even ten years into the publication of *Woman*, in writing letters for which there was no legitimate space in the text is highly significant, pointing to a level of engagement that transcends the framework of community which is editorially encoded. Although the magazine had set itself up as representing a textual community of readers it had limited the opportunities for intervention within the text to those defined by itself as acceptable. That readers persisted in corresponding on subjects outside those areas that were explicitly permitted suggests that they were involved in negotiating a relationship with the magazine based upon their own needs and desires rather than on those allowed them within the parameters of the text.

What I hope to have demonstrated is that the negotiation of identities, both on the part of the reader and within *Woman*, was a dynamic process, requiring an understanding of the various interpretative practices utilised by both sides. Neither the voice of the reader nor the editorial voice can claim an unassailable ascendancy. Although there is ample evidence of the unifying drive of the magazine to present itself as a coherent whole, and its audience as a homogeneous group, it is also clear that this process was not unchallenged. What this points to is that the construction of the textual community, the 'world' in which identities are constructed and challenged, is a two-sided process, albeit an unequal one, in which magazine and readers are involved in a struggle to assert their own meanings, and in which it is not a simple matter of the magazine imposing a way of reading on a passive readership. In fact, readers and editorial staff seem, at times, to be complicit in the creation of a discursive world in which it seems that a constant process of misrecognition is occurring – with both sides active in the imagination, imposition and appropriation of identities.

Note

1. Ida Meller (1864–1934), was taken on by Bennett as his assistant editor in 1897, and became *Woman's* first female editor; her editorship lasted less than twelve months, ending 11 September 1901.

Works cited

Beetham, Margaret. 'Towards a Theory of the Periodical as a Publishing Genre' in Laurel Brake, Aled Jones, and Lionel Madden (ed), *Investigating Victorian Journalism*. London, 1990, 19–32.

de Certeau, Michel. *The Practice of Everyday Life*, trans. Steven Randall. Berkeley, CA, 1988.

Ellegård, Alvar. *'The Readership of the Periodical Press in Mid-Victorian Britain', Göteborgs Universitets Arskrifft* LXIII. Göteborg, 1957.

Gergen, Kenneth J. 'Warranting Voice and the Elaboration of the Self' in John Shotter and Kenneth J. Gergen (ed), *Texts of Identity*. London, 1989.

James, Louis. 'The Trouble with Betsy: Periodicals and the Common Reader in Mid-Nineteenth Century England' in Joanne Shattock and Michael Wolff (eds), *The Victorian Press: Samplings and Soundings*. Leicester, 1982, 349–66.

Part III

Writers / Authors / Journalists

9

Dickens as Serial Author: A Case of Multiple Identities

Robert L. Patten

Within a print culture, many forms of communication may not be constructed for or produced in book format. Private letters, state documents, manuscript poetry circulated within a coterie as John Donne's poetry was (Wollman 1993, 85–97), and dramatic scripts are some instances of writing not designed for book publication. Another non-book publication format is the periodical. The fictions issued in nineteenth-century magazines and newspapers provide complex and interesting cases where privileging the volume edition over the periodical issue distorts the nature of the fiction and the history of the book. The cases are important exceptions to the primacy of book publication throughout the century, for, as the eminent literary historian George Saintsbury has said, 'there is no single feature of the English literary history of the nineteenth century... which is so distinctive and characteristic as the development in it of periodical literature' (Saintsbury 1913, 166).

What is a book? According to the *Oxford English Dictionary* (s.v. 'book', sense 3 *gen.*), a book is a 'printed treatise or series of treatises, occupying several sheets of paper or other substance fastened together so as to compose a material whole'. Books are sheets of paper on which something self-contained and internally coherent has been printed; all the printed material relates in some way to the book's title and 'treatise'. The contents of books are fixed in space – 'fastened together so as to compose a material whole' – and they are also fixed in time, since the Renaissance, to a date at which copyright protection commences. Finally, many books are defined by being authored, or edited, by one, or more than one, person. Indeed, marketing a book by the author's name has been a commercial strategy for centuries: Stephen King's and Barbara Cartland's names are more important than the titles of their newest productions.

When, around 1700, magazines were invented in England, the designers started out by creating a character, a persona whose observations constituted the 'printed treatise' of the publication. Thus Richard Steele produced Isaac Bickerstaff, a principal correspondent for *The Tatler* (1709–11), and Steele with his fellow editor Joseph Addison developed the persona of 'Mr Spectator' for *The Spectator* (1711–12, 1714). In 1731, the first periodical with the word 'magazine' in the title appeared; its persona was also announced in the title, *The Gentleman's Magazine*. A century later, the proprietors and editors of *Punch* created Mr Punch, who never quite qualified as a gentleman. Magazines aim to develop their own distinctive personality; editors, designers, writers, and artists all direct their efforts toward that end. For instance, *The New Yorker* has a distinctive personality, complete with a character, Eustace Tilley, who at the end of February every year examines a butterfly through his monocle. This magazine prints only certain kinds of prose and poetry and cartoons, only certain kinds of 'Talk of the Town'. Magazines, then, like books, often market themselves as coherent texts 'fastened together so as to compose a material whole' and authored by an individual. But magazines are in other ways very different from books.

To illustrate one aspect of the stake our society has in preferring the book, consider this entry in a recent bookseller's catalogue: 'AN EXCEPTIONALLY FINE COPY. REMARKABLY CLEAN. *DICKENS* (*Charles*) Oliver Twist,...3 volumes, FIRST EDITION. First issue,...choicely bound for Hatchards,...£1500' (Traylen 1993, item 185). This notice is misleading in fundamental ways. The 'material whole' for sale for fifteen hundred pounds is not the first issue or first edition of most of the text of the novel now known as *Oliver Twist*. It does not print the title by which the story was first identified; it is not a book whose author was named Charles Dickens; and it is not even the 'material whole' the book was when it first appeared. This three-volume edition, authored by 'Boz', was originally published by Richard Bentley in three volumes bound in cloth covers on 9 November 1838, six months *before Oliver Twist* finished its serialized run in the pages of *Bentley's Miscellany*. This bookseller's copy has been rebound in half red morocco with gilt tops and gilt letters: the cloth covers have been stripped off, leather covers substituted, the edges of the pages trimmed, and the tops of the pages gilded to protect them from dust. So this is not 'an exceptionally fine copy' of the first three-volume book edition of the novel but rather a copy that has been made finer by an expensive bookbinder (unnamed) and bookdealer (Hatchard's, on Piccadilly near London's Piccadilly Circus). This copy sells for fifteen hundred pounds because it is handsomely bound and because it contains early state title pages, a suppressed plate, and three leaves of publishers' advertisements that have nothing to do with the text of the 'treatise', that is, *Oliver Twist*.

By contrast, battered copies of the monthly numbers or half-yearly volumes of *Bentley's Miscellany* containing the serialized version of *Oliver Twist* can be found on bookstalls for a few pounds. And for a little over half of the fifteen hundred pounds that would buy a single novel, one can purchase 28 volumes of the *Cornhill Magazine*, 1860–1873, whose first editor was William Makepeace Thackeray, Dickens's great rival as journalist and novelist. Within those 28 volumes are hundreds of stories and articles and the first issue of four serial fictions by Anthony Trollope, two by Thackeray, and one each by Elizabeth Gaskell, Wilkie Collins, and George Eliot (Traylen 1994, item 203). For even less money, twelve hundred dollars, an antiquarian bookseller recently offered 10 half-yearly volumes of *Bentley's Miscellany*, 1837–41, containing, as the dealer correctly says, '*the true first version of Oliver Twist*' (Windle, list 23, item 96). In addition, these 10 volumes contain a half-dozen other significant novels, scores of poems and articles, and over a hundred first-issue etchings by George Cruikshank.

But dealers, collectors, museums, libraries, bibliographers, textual scholars, and the general public have been conditioned to think of the book as the substantial, first publication of most serials, from Dickens's *Pickwick Papers* to Thomas Hardy's *Tess of the d'Urbervilles* and Joseph Conrad's *Lord Jim*. Books have authors who are solely responsible for their contents – Charles Dickens, for example, even though those 'early state' title pages of the 'exceptionally fine copy' of *Oliver Twist* identify the author not as Charles Dickens but as what was then a well-known pseudonym, 'Boz'. And books get passed down through the generations, while periodicals are tossed out by families or broken up for the prints by dealers or reduced to microfilm, -form, or -fiche by librarians. The book is semi-permanent and simply displaces the format in which the text first appears. This is true for virtually every magazine serial, not just fiction, but also history, theology, biography, philosophy, criticism, and other genres that, as Saintsbury reminds us, were often in the nineteenth century first published in instalments in periodicals (Saintsbury 1913, 166).

Over the last twenty years we have learned a lot about what this preference for the book produces in the way of copytext distortions. But I'm interested here, not in the textual controversies surrounding serialized, machine-printed works, but in the distortions that our preference for the book version has imposed on our very conception of periodical fiction. Defining the text as an identifiably authored and self-contained book obscures the ways in which fictions were affected by the periodicals in which they first appeared. Four aspects of book fiction that are significantly altered by magazine serialization are the convention of single authorship, the genre of fiction itself, the apparent self-containedness of the 'material whole', and the timing and effect of the story's reception.

That magazines are collaborations among proprietors, editors, writers, and sometimes illustrators should not be surprising, since the word 'magazine', from the Arabic *makhazin* meaning storehouses (related to the French *magasin*, department store), denotes a warehouse storing many things. Collaboration among those running a magazine – the editorial 'we' is not always only a convention – is the norm even for the production of periodical fiction. Some of the most amusing, and possibly accurate, accounts of proprietors' constructing their publications are contained in Anthony Trollope's *The Way We Live Now* and George Gissing's *New Grub Street*. But though we have biographies of publishers and publishing houses, and narratives of publishers' interventions in the works of such famous authors as Sir Walter Scott and Dickens, little work has been done on the theory or general practice of proprietal intervention in periodical texts. However, recently some scholars have studied the kinds of effects editors may have on the medley of materials they assemble and publish. For instance, Mark Parker has shown how John Scott, editor of the *London Review* in the 1820s, transmitted Romantic ideology by manipulating the style, tone, subject, and placement of Charles Lamb's and William Hazlitt's essays and by framing them with surrounding essays in each issue to highlight their political implications (Parker 1991, 693–711). Laurel Brake has demonstrated the influence of the differing editorial practices at the *Westminster Review* and the *Fortnightly Review* on the separately published essays that make up Walter Pater's *Renaissance* (Brake 1995, *passim*).

To assess proprietal and editorial interventions in *Oliver Twist*, we need first to recall that the proprietor, Richard Bentley, had started up his magazine as a rival to his former partner Henry Colburn's successful *New Monthly Magazine*. Bentley, determined to outdo the competition, aimed at capturing the market for humour: the *Wit's Miscellany*, he intended to call it at first. Then he changed to a name that links his own to that of the magazine's persona: *Bentley's Miscellany*. (One wag said the title was ominous – Miss Sell Any.) Humour, not politics, was, as the 'Prologue' announced, the keynote of the venture (Maginn 1837, 2–6). For contributors to this miscellany of papers melded together into

Bentley's, Richard Bentley bought up the team who were making a decided hit with their humorous pen and pencil sketches of lower-middle-class urban life: Charles Dickens and George Cruikshank. Their success was in the sketch, something either visual or verbal that was quickly executed and quickly absorbed, and that recorded the evanescent and fugitive detail of modern life.

The text of *Oliver Twist*, a long novel that is seldom humorous and that from the beginning attacks political measures, should be understood at the outset as a contribution that contrasts to the ostensible purpose and tone of *Bentley's Miscellany* and to *Sketches by Boz*, magazine and newspaper scenes and tales written by Dickens and subsequently, when collected into volumes, illustrated by Cruikshank. The magazine's purpose and tone, in turn, are defined against the competition, archly described in the 'Prologue':

> We do not envy the fame or glory of other monthly publications. Let them all have their room...One may revel in the unmastered fun and the soul-touching feeling of Wilson, the humour of Hamilton ...and the Tory brilliancy of half a hundred contributors zealous in the cause of Conservatism...Elsewhere, what can be better than Marryat, Peter Simple, Jacob Faithful, Midshipman Easy, or whatever other title pleases his ear.... In short, to all our periodical contemporaries we wish every happiness and success.... We wish that we could catch them all.
>
> (I (Jan. 1837) 4–5)

But, the 'Prologue' continues, using the editorial first-person plural although the text was composed by a hired contributor, 'Our path is single and distinct. In the first place, we have nothing to do with politics'.[1]

Out of the context of a magazine promising 'wit' and eschewing 'politics', *Oliver Twist* loses some of its shock value.[2] And unless we keep in mind the opening announcement that this will be a witty non-political miscellany, we may miss some of the underlying tension that surfaced in Richard Bentley's quarrels with Dickens about his contributions to the *Miscellany*, for Dickens was never so keen on eschewing politics as Bentley was. Yet only one scholar, Kathryn Chittick, has studied *Bentley's Miscellany* as an entity in its own right and indicated how the proprietor, through his choice of writers and selection of materials, promulgated his conception of a wit's miscellany.[3]

Moreover, proprietor and editor further shaped the contents of the journal by attempting to buy up the writers then making *Fraser's Magazine* so popular. They succeeded in hiring many of the contributors to *Fraser's*, including its outstanding editor William Maginn, who wrote much if not all of the 'Prologue'. And then they signalled that coup in the 'Prologue', hailing 'the wit of our contributor Theodore Hook', 'the fascinating prose or delicious verse of our fairest of *collaborateuses* Miss Landon', and 'the polyglot facetiae of our own Father Prout'. The contents of any issue of *Bentley's*, including its fiction, should be understood in the context of what other journals were publishing and what other writers and illustrators were available. To some extent authors and artists had their own styles, but to some extent too they were melded collectively into a house style, the polyphonic 'miscellany' unifying into a 'material whole'.[4] The best serial fiction in or out of magazines, such as Thackeray's *Vanity Fair* (published in 20 numbers over 19 months), exploits that verbal and visual polyphony.

Dickens both edited and supplied text for *Bentley's* under the pseudonym 'Boz', the author of *Sketches by Boz*. At the conclusion of his first contribution, about the mayor of the provincial town of 'Mudfog', Boz explains that 'this is the first time we have published any of our gleanings from this particular source', that is, the Mudfog papers. He also suggests that 'at some future period, we may venture to open the chronicles of Mudfog' (Boz 1837, 63). This fiction of a manuscript source resembles that preserved through the first numbers of Dickens's concurrent monthly-parts serial, *The Pickwick Papers*. *Oliver Twist* commences in the February number of *Bentley's*, under the title of *Oliver Twist, or, the Parish Boy's Progress*, by 'Boz'. As the first sentence of the story makes clear, this is a continuation of the Mudfog papers: Oliver is born in the Mudfog workhouse. Thus naming Dickens as the author of *Oliver Twist*, as the first volume publication and the bookdealers' catalogues do, distorts the textual record and obscures the ways in which the narratorial voice, Boz, is constructed out of those voices that precede and surround it.[5]

That distortion was compounded when Dickens's authorized biographer, John Forster, made *Oliver Twist* into the first of three versions of Dickens's own childhood hardships, as if the story had its only origin in Dickens's own life. In fact there were other claimants for the originator of the story. Dickens's 'coadjutor' on *Sketches by Boz*, George Cruikshank, may indeed, as he later said, have suggested to Dickens that they collaborate on a story about the rise of 'a boy from a most humble position up to a high and respectable one'.[6] For fifteen years, Cruikshank had wanted to illustrate a contemporary version of Hogarth's narrative series of 12 pictures, *Industry and Idleness*. Hogarth's pictures track the parallel but contrasted fortunes of two apprentices: one, Francis Goodchild, through industry rises to 'a high and respectable' position as Lord Mayor of London, while the other, Tom Idle, through idleness falls into bad company and is eventually sentenced by his former fellow apprentice Goodchild and hanged on Tyburn Hill. A successful recent exhibition of Hogarth's work had made such a project more viable.

If Dickens did adopt Cruikshank's suggestion, which chimed with some of his own ideas for what he told Richard Bentley was 'a capital notion for myself, and one which will bring Cruikshank out' (Dickens 1965, 224), then that resulting story has wider contexts than Dickens's own life. But those wider contexts and those other contributors were deliberately effaced. Charles Dickens constructed himself over the decades as a separable and distinct, unique author. The promulgation of the figure of the writer Dickens became an immensely powerful merchandising ploy, selling not only his books but also political programmes, charitable causes, wines and teas, and thousands of manufactured products, from printed handkerchiefs to china, deriving from Dickens's work. Dickens's biographers, following John Forster's lead, reinforced the notoriety and singularity of Charles Dickens, author, rather than 'Boz', the editor and collaborator. Severing Dickens's periodical fictions from their magazine contexts allowed them to be reinterpreted as self-referential biographical disclosures. Largely a marketing strategy that made Dickens a household name, a fortune, and a model of a genius generating text out of trauma, this elevation of the book over the serial and its consequences persuaded even George Cruikshank's biographer to say in 1882, 'In all the range of Dickens's work, there is nothing more essentially his own than "Oliver Twist"' (Jerrold 1882, 223–4). But writers closer to the initial context within *Bentley's*, knowing both Dickens and Cruikshank, were of a different opinion. Samuel Warren, one of the most successful authors of the day, said in 1842 that the illustrations influenced Dickens, and 'thus the writer follows the caricaturist, instead of the caricaturist following the writer' (Warren 1842, 785).

Another writer with a claim on Oliver's paternity is the novelist William Harrison Ainsworth, who was from the 1830s to the end of the century a very popular writer, especially of fiction for boys. Dickens's closest friend, Ainsworth had put Boz in touch with the publishers, artists, and journalists who helped to make Dickens's early reputation. By 1836 Ainsworth had embarked on a series of novels exploring the lives of thieves. The first, *Rookwood*, about the famous highwayman Dick

Turpin, had been reissued with illustrations by Cruikshank, who had also been interested for seventeen years in collaborating with some writer on an illustrated life of a thief. The second, *Jack Sheppard*, about a romantic young burglar, overlapped *Oliver Twist* in *Bentley's Miscellany*. Ainsworth collected a lot of written and pictorial material about eighteenth-century miscreants, including James Sykes, a companion of Jack Sheppard and a possible inspiration for Bill Sikes. Dickens and Ainsworth were together frequently, discussing a slew of present and future projects. Yet Ainsworth's contribution to *Oliver Twist* usually goes unnoticed.

The point of all this is that in the magazine context *Oliver Twist*'s authorship is a multiple thing – the proprietor Richard Bentley; the editor Charles Dickens; other journals such as *The New Monthly* and *Fraser's Magazine*; other contributors to *Bentley's* such as the stable of *Fraser's* writers and those already being published by Bentley; the pseudonymous author 'Boz', known for his 'sketches' of metropolitan scenes, characters, and tales; the illustrator George Cruikshank; pictorial progresses by Hogarth and his successors; and a close friend (and future editor of the *Miscellany*) Ainsworth – all influenced the subject, style, tone, and content of the serialized fiction. Serialization deconstructs the single author as sole creator, and does so as part of a larger collaborative project within which the serial is framed. Conversely, prioritizing the book as a single-authored material object over the multiply-authored serial deconstructs the periodical, erasing all the extratextual influences, turning the voice from polyphonic to monophonic, and connecting the text to one author's imagination rather than to the culture of a magazine and of a historical moment.

Serialization also tends to deconstruct the genre of fiction. *Oliver Twist*, beginning as a Mudfog paper, starts as a further instance of the articles and tales Dickens had written for *The Monthly Magazine* and the *Morning* and *Evening Chronicle*; these in turn were partly based on sketches of the city by other journalists and graphic artists. Like the *Old Curiosity Shop*, *Oliver Twist* begins more like a short story than an extended fiction. Like *Pickwick*, the Mudfog papers claim affinity with parliamentary reports, memoirs, and posthumous papers. In *Bentley's Miscellany*, the instalments of *Oliver Twist* are interspersed with other serials and with essays and poems about everything from angling in the days of Izaak Walton to an ode upon the birthday of Princess Victoria. It was not unusual, in these settings, for periodical stories to incorporate poems, criticism, philosophy, history, travelogues, or even musical scores.

Most notably, many magazine fictions articulate an affinity with drama. *Oliver Twist* is confessedly constructed according to melodramatic principles: 'It is the custom on the stage', Boz the narrator tells us, 'in all good, murderous melodramas, to present the tragic and the comic scenes in as regular alternation as the layers of red and white in a side of streaky, well-cured bacon' (Dickens 1837, 437). The story is situated among many articles about the theatre: the opening number of *Bentley's Miscellany* starts with an essay by Theodore Hook on George Colman the Younger. The novel employs a great deal of theatrical characterization and dialogue, such as Sikes's stagy 'Wolves tear your throats!' (Dickens 1838, 313). Like other serials, *Oliver Twist* was on the boards before it was in boards, even before it was completely written: the first dramatization, by Gilbert À Beckett, appeared after only half the instalments had been composed. Cruikshank increasingly directed the illustrations toward their theatrical realization (Hill 1980, 429–59); and for Dickens the prospects, and retrospects, of dramatizing the story significantly affected the text's design. Knowing, as he did about half the way through composition and publication, that the actor-manager Frederick Yates wanted to stage the story before it was completed in print, Dickens may have enhanced the role of Fagin in the concluding chapters in anticipation of the dramatization. Certainly Dickens wanted his serial readers to respond viscerally, as if in a theatre.[7] Further, *Bentley's Miscellany* is itself refigured as a stage rather than as a magazine or sheets of printed paper bound up as a 'material whole': the 'Editor's Address' on the completion of the first half-year is couched as the manager's stage address at the end of a theatrical season.

Nor is this the end of the generic ambiguities of serial fictions. As in *Oliver Twist*'s case, the graphic tradition often played a prominent role; this story is subtitled a 'progress', a term denoting a sort of life-contour, physical and moral, often told through a series of pictures. And midway through the text's composition Dickens renegotiated his author's contracts with Richard Bentley, who was not only the proprietor of the magazine but also the stipulated publisher of Dickens's next two novels. That renegotiation changed the serialized fiction *Oliver Twist* from a work fulfilling a contractual monthly contribution of 16 pages for the *Miscellany* into a novel to be released in three volumes half a year before the serial dragged to a close in the magazine. So even in format *Oliver Twist* was neither conceived of nor produced as a novel for much of its initial appearance. Its affinities with theatre, pictures, sketches, journalism, short stories, and parliamentary proceedings are at least as influential in its initial stages as the generic boundaries of the novel. Serial fictions interweave with other genres in immediate, material ways that their later repackaging into self-contained novels conceals.

My third concern is the way serial fictions, not being self-contained, incorporate the times. Eminent scholars have identified many topics of the day to which serials make reference.[8] The topicalities of periodical fiction tend to be journalistic, metropolitan, topographical, sensational, and narratable. For instance, the gradual spread of a mysterious form of equine hoof-and-mouth disease might throw the Scottish highlands into a panic, but though topical it probably would not become the subject of a successful serial fiction: it would not make the daily newspapers, would not appeal to city dwellers, could not be located in a familiar neighbourhood or landmark building, would not accelerate the pulses of readers seeking sensation, and could not be cast as an enthralling narrative. Murder, on the other hand, fulfils all the requirements, and by the 1850s it was a leading subject of periodical fiction.

Magazine serials also circulate references of a more general character than local, topical, and timely ones. Two categories of fundamental referentiality are ideology and aesthetics. (Each of these kinds of reference involves the other, but for present purposes I have separated them.) By ideology, I mean those clusters of ideas, abstract or pragmatic, that issue in the mythologies and expressions of an era.[9] Scholars have studied the political ideology of selected nineteenth-century British journals, from the Whig quarterlies in the first decades of the century to John Stuart Mill's Utilitarian *London Magazine* at mid-century and the imperialism fostered by *Blackwood's Edinburgh Magazine* in the last decade. But little has been done about more diffuse ideological concerns. In the case of *Oliver Twist*, Richard Stein has demonstrated that the story participates in widespread writing about orphanhood and lost identities surfacing at the beginning of Victoria's reign (Stein 1987). A population that was becoming more mobile across spaces and classes had vested its proofs of identity in detachable and forgeable adjuncts such as birth and marriage certificates, names, and other tokens. These may fail to confirm identity, legitimacy, property, class, status, rights, and name, as they failed for Oliver and his mother.

There are other anxieties of the late Regency and early Victorian period that ally with this fear of alienable identity. Cruikshank and the *Fraser's* crowd, especially Ainsworth, were, as we have seen, interested in the lives of thieves. 'Crime', Martin Wiener recently noted, 'was a central metaphor of disorder and loss of control in all spheres of life' (Wiener 1990, 11). This interest in crime related to perennial debates about nature versus nurture, that is, whether bad was innate like a bad seed or produced by a bad environment. A related issue vexed Dickens for decades: did parental neglect foster in the child a determination to succeed in respectable society (Charles Dickens) or did it lead to criminality (the Artful Dodger) and/or death (Oliver's orphan friend Dick)? The early Victorian cultural fascination with criminality seemed to find particular expression in the dramas of incarceration and excarceration. Many critics have noted the cramped, confined world of Dickens and Cruikshank's collaborations – the sense of structures, persons, environments, laws, and fortune closing down, squeezing, deforming the individual.[10] Such paradigms do seem to reflect the cultural emphasis on policing the self that bourgeois industrial capitalism promulgated.

Conversely, and less often noticed, there is in the 1830s an equal interest in escape, excarceration, in how individuals might defy those pressures that could identify and fix them. Nowhere is that escapism more celebrated than in *Jack Sheppard*. This was the serial Ainsworth wrote that began to appear at the front of each issue of *Bentley's Miscellany* after *Oliver Twist* had been published in three volumes but while it was still running in instalments in the periodical. The letterpress by Ainsworth and the illustrations by Cruikshank celebrate the eighteenth-century prison escapist Jack Sheppard, who, though an idle and foolish apprentice, is unfairly bedevilled by the law and repeatedly foils its attempts to imprison him. In the text and pictures and later in the theatre, his famous escape from Newgate prison stages his eel-like ability to slip out of shackles and cells. They also dramatize the site of the old Newgate prison, as Fagin's time in the condemned cell, like Dickens's earlier sketch 'A Visit to Newgate' in *Sketches by Boz*, brings to life the modern prison.

Had Dickens continued as editor of *Bentley's Miscellany*, and had Cruikshank continued as illustrator, their next collaborative serial fiction, following *Jack Sheppard* in the magazine, would have been *Barnaby Rudge*, a story about the burning of the old Newgate prison. The original title Dickens assigned to that work was *Gabriel Vardon, the Locksmith of London*, a title that again plays on the theme of incarceration and excarceration. And so, within the pages of *Bentley's Miscellany*, all illustrated by Cruikshank, would have appeared three serial fictions about crime, dealing in sequence with the modern Newgate (*Oliver Twist*), the eighteenth-century prison (*Jack Sheppard*), and its destruction during the Gordon Riots (*Gabriel Vardon*). These unrealized plans for a triple serial on prisons illustrate one way in which magazine publication shows deep ideological currents that separate book publication obscures.[11]

Victorian periodicals were also much occupied with aesthetic issues. The idea of the aesthetic comprises a great range of topics, from the typographic design of the journal and the accuracy, beauty, and instructiveness of its 'illuminations' to debates about the moral (or immoral) functions of literature and the appropriate conduct of readers. As Kelly Mays shows, by mid-century periodicals regularly instructed their customers in the proper – that is, moral – way to read. Critics denounced desultory, casual, random, escapist, and sensationalistic reading practices; they condemn short attention spans, skimming, and flitting from topic to topic; and they criticized mothers for reading while minding their children because the children would grow up to associate reading with surveillance (Mays 1995, 165–94).

At the time of *Oliver Twist*, William Maginn was producing for *Bentley's Miscellany* a series of articles on character in Shakespeare that are, in Chittick's judgement, 'of impressive acuteness' (Chittick 1990, 96). His was another way of directing reading, writing, social, and punitive practices, as well as of commenting on theatricality – all themes touched on in this chapter. One of Maginn's papers, about the villain Iago in *Othello*, appears at a crucial spot in the journal: between the first chapters of Ainsworth's *Jack Sheppard*, about an eighteenth-century boy thief, and the last chapters of Dickens's *Oliver Twist*, about nineteenth-century boy thieves (Maginn 1839, 43–50). In this paper Maginn implicitly argues against Samuel Taylor Coleridge's notion that Iago's actions can only be explained as the products of 'motiveless malignity'. All Shakespeare's characters, he says, 'are real men and women, not mere abstractions'. Their virtues and vices are compounded 'and are not unfrequently blended'. Maginn maintains that wickedness is prompted by the same impulses that stimulate persons 'to the noblest actions – ambition, love of adventure, passion, necessity'. Thus, in an article printed between the opening of *Jack Sheppard* and the chapters of *Oliver Twist*, recounting Sikes's last day and Monks's capture, Maginn ventures to say that 'the creatures of Boz's fancy, Fagin or Sikes, did not appear in every circle as the unmitigated scoundrels we see them in Oliver Twist'. It is due to the exigencies of the story, Maginn reasons, that readers see only one side of their characters. 'But, after all, the Jew [Fagin] only carries the commercial, and the housebreaker [Sikes] the military principle to an extent which society cannot tolerate' (Maginn 1839, 43).

Maginn uncovers an aesthetic that appeals more to empathy and capitalistic entrepreneurship than to social controls. The feeling that covers the globe with shipping, he says, also animates the breasts of thieves and highwaymen: 'Robber, soldier, thief, merchant, are all equally men' (Maginn 1839, 43). (No wonder archconservative Mary Russell Mitford hated the idea of making lawbreakers attractive and police venal.) In conclusion, Maginn calls for yet another inscription of the Hogarthian 'progress' contrasting industry and idleness: 'It would not be an unamusing task to analyze the career of two persons starting under similar circumstances, and placed in situations not in essence materially different, one ending at the debtors' door of Newgate amid hootings and execrations, and the other borne to his final resting-place in Westminster Abbey, graced by all the pomps that heraldry can bestow' (Maginn 1839, 43–4).

Thus Maginn's serial criticism of Shakespeare's characters accomplishes five things. One, it establishes an aesthetic principle – successfully drawn characters are compounded of good and evil. Two, it derives reading practices from that principle – readers should recognize and empathize with motives that lead to bad actions as well as to good ones. Three, it declares a moral order – 'robber, ... merchant, all are equally men'. Four, it relates dramatic criticism to the magazine's current serial fictions – Fagin and Bill Sikes are also 'equally men', appearing much more lovable and redeemable offstage, out of the plot. And five, Maginn's essay sets an agenda for future writings – another fiction contrasting the industrious and idle apprentice. In such ways this and the many other articles interspersed with poetry and serialized fiction collectively articulate the aesthetic presumptions and reading practices that inform the journal and affect its contributions.

With regard to the last topic, reception, the difference is obvious: novels issued in book form as 'material wholes' on a single publication date get reviewed after the text has been written and printed; serial fictions, which appear periodically in small sections over months or years, tend to get reviewed as each instalment is published. Thus amateur and professional readers of serial fiction are encouraged to speculate about the story and the characters, to project the future, and to offer the writer advice. Moreover, magazine fictions are, as we have seen in the case of *Oliver Twist*, frequently dramatized before completion. Readers often bombarded the producers of the periodical or the fiction with suggestions for changes, some of which might be adopted, others resisted, so that the narrative, by a kind of back-formation, could revise its own previous trajectory in the course of subsequent instalments. (A good example of this back-formation is Dickens's making amends in *David Copperfield* to the outraged original of Miss Mowcher by changing the character in the middle of serialization from a procuress into a nemesis: another instance of Maginn's thesis that outside the plot a villain may be exemplary.) Thus readers of serial fiction could exert influence on the narrative as it was being composed, something readers of books cannot usually do.

There is another way that serial fictions can be affected by serial responses. When a magazine serial becomes popular, it gets copied, imitated, pirated, plagiarized, often before the story has been completed in manuscript, much less in print. Such imitations and anticipations rob the original producers of the story of some of their revenue and some of their options, both for the story and for merchandising the product. Hence Dickens and others were deeply concerned, when *Oliver Twist* was being published, about passing legislation to strengthen copyright. As Mark Rose reminds us in his excellent book on copyright, 'plagiarism' is from the Latin word for 'kidnappe' – a plagiarist appropriates the offspring of a creator's body, the text, just as a kidnapper appropriates the offspring of a parent's body, the child (Rose 1993). Dickens resented the fact that his stories were taken over by others – his publishers, dramatists, rival authors, foreign publishing houses. He objected to losing money, to having his story mangled by incompetents, to finding his climaxes anticipated, and to enriching others by his imagination and industry. Dickens's resentment, which he shared with many authors and publishers, found its way into his story of an orphan boy whose identity is stolen at birth by a legion of kidnappers, who take Oliver's story and make it their own. Oliver's half-brother Monks tries to ruin Oliver's mother's reputation by rewriting it and to invoke those clauses in the paternal will that would disinherit Oliver by rewriting him as a criminal. Nancy stages a public scene in which Oliver is her naughty brother, and more privately she tries to make him her son, a reinscription of his identity and relationship to Nancy that drives Sikes mad with jealousy. Giles and Brittles, the Maylies' servants, and the surgeon who attends the wounded boy aver that Oliver is a thief. Even Mr Brownlow constructs many stories about Oliver, sometimes agreeing with Grimwig and Bumble, before he puts together the true narrative of Oliver's past. The story of *Oliver Twist*, therefore, is not just about the loss of personal and familial identity and legitimacy and inheritance. It also absorbs Dickens's and his age's preoccupation with losing control over one's own story, having it kidnapped – plagiarized – by others. The very popularity of serials made them vulnerable to commercial highwaymen who hijacked them for their own profit.

We have looked at four areas of difference between novels designed to be produced as books and serialized fiction. Serials tend to be multiply-authored, being constructed by the producers, editors, and other contributors to the magazine in which they appear. Serials have permeable generic boundaries, incorporating a miscellany of literary genres and appealing to readers in ways more dynamic and theatrical than contemplative. Serial fictions are topical both specifically and more generally; annotated editions for modern classroom use often pick up particular references but neglect the general ideological and aesthetic resonances. And serials interact, during the course of production, with a wide variety of readers whose responses impact successive instalments. When magazine fictions are repackaged as books these points are often obscured or denied or cancelled. The result, in case after case of which *Oliver Twist* is only one example, is a significant distortion of the history of the book.

Notes

1. Chittick (1990, 102) says that Dickens may have written the second part of the 'Prologue', although the table of contents to *Bentley's Miscellany*, vol. 1, identifies Dr Maginn as the sole author. Maginn was unquestionably the author of the passages just quoted.
2. In the introduction to his *Companion to 'Oliver Twist'* (1992, 5), Paroissien establishes the novel's principal ideological contexts and argues that 'a determination to respond to contemporary issues shaped Dickens's writing plans'.
3. Chittick 1990, 92–113. This chapter, in a book on all of Dickens's writings in the 1830s, is necessarily brief, but it is the best account of the magazine and of Bentley's proprietorship that we have.
4. Dickens's correspondence with Bentley through the first nine months of editing the *Miscellany* indicates some of the decisions he and his publisher made about which pieces to accept or reject, what to rewrite, and how to arrange the contents of numbers (Dickens 1965, vol. 1).
5. For illuminating discussions of the shaping of narratorial voice in Dickens, see Jaffe 1991; and for Dickens's unsuccessful struggle to establish a coherent persona for his weekly miscellany *Master Humphrey's Clock*, see Mundhenk 1992.
6. See Cruikshank's version of the origin of the novel in his letter to *The Times*, 30 December 1871.
7. In different ways, the following authors have analysed aspects of Dickens's theatricality: Axton 1966; Eigner 1989; Garis 1965; Hughes and Lund 1991; Meisel 1983; D. A. Miller 1988; and J. Hillis Miller 1959.
8. See Altick 1991; McMaster 1990; and numerous articles by K. J. Fielding on Dickens's topicality published in, among other journals, the *Dickensian*.
9. Cf. *OED*, s.v. 'ideology', 2: 'The study of the way in which ideas are expressed in language'. Williams (1966, s.v. 'ideology') distinguishes between Napoleonic and Marxist usage, where ideology is abstract or false theory not based on experience and materiality, and a more neutral usage meaning the system of ideas appropriate to a particular class.
10. See Chesterton 1906; J. Hillis Miller 1971; and Stein 1987.
11. For more about the connections among these three 'Newgate' novels, see my *George Cruikshank's Life, Times, and Art*, vol. 2.

Works cited

Altick, Richard. *The Presence of the Present: Topics of the Day in the Victorian Novel*. Columbus, OH, 1991.

Axton, William. *Circle of Fire: Dickens' Vision and Style and the Popular Victorian Theater*. Lexington, KY, 1966.

Boz. 'Public Life of Mr. Tulrumble, Once Mayor of Mudfog', *Bentley's Miscellany* 1 (Jan. 1837), 49–63.

Brake, Laurel. 'The "Wicked *Westminster*", the *Fortnightly*, and Walter Pater's *Renaissance*', in *Literature in the Marketplace: Nineteenth-Century British Publishing and Reading Practices*, ed. John O. Jordan and Robert L. Patten. Cambridge and New York, 1995, 289–305.

Chesterton, G. K. *Charles Dickens*. New York, 1906.

Chittick, Kathryn. *Dickens and the 1830s*. Cambridge, 1990.

Dickens, Charles. *Oliver Twist. Bentley's Miscellany* 1–5 (Feb. 1837–April 1839).

——*Letters*, ed. Madeline House and Graham Storey. vol. 1. Oxford, 1965.

Eigner, Edwin M. *The Dickens Pantomime*. Berkeley, CA, 1989.

Garis, Robert. *The Dickens Theatre, A Reassessment of the Novels*. Oxford, 1965.

The Gentleman's Magazine.

Hill, Jonathan E. 'Cruikshank, Ainsworth, and Tableau Illustration', *Victorian Studies* 23.4 (Summer 1980), 429–59.

Hughes, Linda K., and Michael Lund. *The Victorian Serial*. Charlottesville, VA, 1991.

Jaffe, Audrey. *Vanishing Points: Dickens, Narrative, and the Subject of Omniscience*. Berkeley, CA, 1991.

Jerrold, Blanchard. *The Life of George Cruikshank*. vol. 1. London, 1882.

Jordan, John O., and Robert L. Patten. *Literature in the Marketplace: Nineteenth-Century British Publishing and Reading Practices*. Cambridge and New York, 1995.

McMaster, Rowland. *Thackeray's Cultural Frame of Reference: Allusion in the Newcomes*. Basingstoke, 1990.

Maginn, Dr [William]. 'Prologue', *Bentley's Miscellany* 1 (January 1837), 2–6.

——'Shakespeare Papers. – No. VIII. Iago', *Bentley's Miscellany* 5 (Jan. 1839), 43–50.

Mays, Kelly J. 'The Disease of Reading and Victorian Periodicals', in Jordan and Patten 165–94.

Meisel, Martin. *Realizations: Narrative, Pictorial, and Theatrical Arts in Nineteenth-Century England*. Princeton, NJ, 1983.

Miller, D. A. *The Novel and the Police*. Berkeley, CA, 1988.

Miller, J. Hillis. *Charles Dickens: The World of His Novels*. Cambridge, MA, 1959.

——'The Fiction of Realism: *Sketches by Boz, Oliver Twist*, and Cruikshank's Illustrations'. In *Charles Dickens and George Cruikshank*. Los Angeles, CA, 1971.

Mundhenk, Rosemary. 'Creative Ambivalence in Dickens's *Master Humphrey's Clock*', *SEL* 32.4 (Autumn 1992), 645–61.

Parker, Mark. 'The Institutionalization of a Burkean-Coleridgean Literary Culture', *SEL* 31.4 (Autumn 1991), 693–711.

Paroissien, David. Introduction to *Companion to 'Oliver Twist'*. Edinburgh, 1992.

Patten, Robert L. *George Cruikshank's Life, Times and Art*. vol. 2. New Brunswick, NJ and Cambridge, 1996.

Rose, Mark. *Authors and Owners: The Invention of Copyright*. Cambridge, MA, 1993.
Saintsbury, George. *A History of Nineteenth Century Literature (1780–1895)*, 166. 1896. Reprint, New York, 1913. Quoted by Brake in Jordan and Patten.
The Spectator
Stein, Richard L. *Victoria's Year: English Literature and Culture, 1837–1838*. New York, 1987.
Traylen, Charles W. Catalogue 113 (1993), item 185.
——Catalogue 114 (1994), item 203.
Warren, Samuel [Q. Q. Q., pseud.]. 'Dickens's American Notes for General Circulation', *Blackwood's Edinburgh Magazine* 52.326 (Dec. 1842), 783–801.
Wiener, Martin J. *Reconstructing the Criminal: Culture, Law, and Policy in England, 1830–1914*. Cambridge, 1990.
Williams, Raymond. *Keywords: A Vocabulary of Culture and Society*. New York, 1966.
Windle, John. List 23, item 96.
Wollman, Richard B. 'The "Press and the Fire": Print and Manuscript Culture in Donne's Circle', *SEL* 33.1 (Winter 1993), 85–97.

10
Authorship, Gender and Power in Victorian Culture: Harriet Martineau and the Periodical Press

Alexis Easley

The contributions of Victorian women writers to periodical journalism have received little attention in recent criticism. Yet most major women writers of the Victorian period began their careers as contributors to the periodical press and continued to publish in the reviews throughout their lives. Due to the convention of anonymity associated with most Victorian periodicals, journalism was a relatively accessible medium of discourse for women writers. In addition to publishing their fiction and poetry in major periodicals such as *Blackwood's Magazine*, women writers also published political essays, travelogues and book reviews. Masked behind the universalized 'editorial we', women writers were able to express themselves publicly on the pressing social and political issues of their day – without risking fame, which for a Victorian woman always translated into something more like infamy.

Though the periodical press provided women writers with the opportunity to write on issues outside the female sphere, it did not allow them to escape gender definitions entirely. Even though the narrative voices encoded in most Victorian periodicals were anonymous, they were not correspondingly genderless. The assumption of male narration and male readership was almost universal in the most prestigious Victorian periodicals. Thus, women journalists often wrote in 'drag', referring to themselves and their readers using masculine gender markers. By alternating between their anonymous – often 'masculine' – voices in the periodical press, and their socially constructed 'feminine' voices as authors of signed books, women writers expanded the range of possible audiences and subject matter for their work.

The purpose of this chapter is to investigate how one woman writer – Harriet Martineau – was able to pursue a successful career as both the author of signed books and the writer of anonymous periodical essays.

This analysis will demonstrate how Martineau was able to negotiate the hazards and pleasures of literary fame, alternately revealing and concealing her gendered identity as a means of gaining cultural power. Further, it will demonstrate how Martineau was able to resist and reconstruct cultural stereotypes of the 'female author' by expanding the number of possible literary mediums, topics, and narrative voices open to women writers. Through the multiplicity of her own narrative voice – at once masculine and feminine, journalistic and literary, public and private – Martineau provided an important model for later women writers such as George Eliot and Margaret Oliphant.

In 1821, Harriet Martineau began her career in journalism by publishing her first anonymous essay in *The Monthly Repository*, a small-circulation periodical associated with the Unitarian church. Intent on establishing herself as a writer, she submitted an essay in the form of a letter to the editor entitled 'Female Writers of Practical Divinity', signed only with the letter 'V'. To her surprise, the essay appeared in the next issue of the journal. It is easy to sense her delight in her description of the moment when she revealed her literary identity to her brother. Thomas Martineau approached Harriet with the latest edition of *The Monthly Repository*, exclaiming over an essay written by 'a new hand', not knowing it was his sister's creation. Gleefully listening to his unknowing praise of her work, Martineau responded to her brother, saying that the article 'did not seem any thing particular' (Martineau 1877, I 92). She records the following exchange:

> 'Then', said he, 'you were not listening. I will read it again. There now!' As he still got nothing out of me, he turned round upon me, as we sat side by side on the sofa, with 'Harriet, what is the matter with you? I never knew you so slow to praise any thing before'. I replied, in utter confusion, – 'I never could baffle any body. The truth is, that paper is mine'.
>
> (Martineau 1877, I 92)

This passage records the pleasure of knowing – but only slowly revealing – authorial gender and identity. Over the next eleven years, Martineau was able to evade the publicity associated with being identified as a 'female author' by publishing anonymously in *The Monthly Repository*, writing on subjects as diverse as slavery, religion, and political economy.

Once Martineau began publishing signed books, it became increasingly difficult for her to evade socially constructed stereotypes of the 'female author'. This negative publicity began in earnest after the publication of her first major signed work, *Illustrations of Political Economy* (1832–4). To publish a book so early in her literary career, let alone one that carried her own signature, was indeed a brave step. The result was instant notoriety – and infamy. By daring to publish 'serious' literature under her own name – in this case, a series of didactic stories based on the precepts of political economy – Harriet Martineau came to be defined as a feminine transgressor, a public identity that subjected her to a great number of personal attacks in the press. Though almost all reviewers praised Martineau's genius, they, also tended to focus on her emotionality and inexperience – what they defined as her 'femininity'. As one critic for *The Edinburgh Review* suggested, 'A young lady can scarcely possess the experimental knowledge of mankind, without which a confident imagination must occasionally run wild in the paradise of its own conceptions' (Empson 1833, 10). The political implications of Martineau's *Illustrations of Political Economy* were seen as being beyond the scope of a young woman, whose proper purview was the domestic sphere. As *Fraser's Magazine* put it, 'it was indeed a wonder that such themes should occupy the pen of any lady, old or young, without exciting a disgust nearly approaching to horror' (Maginn 1833, 576).

Much of this negative publicity was instigated by a review article published in *The Quarterly Review* in 1833. Focusing on Martineau's treatment of Malthusian theories of population growth in *Illustrations of Political Economy*, this article questioned the propriety of a woman – let alone an unmarried woman – who dared to discuss the sexual themes implied in Malthus's theories. *The Quarterly Review* writes:

> But no; – such a character is nothing to a *female Malthusian. A woman* who thinks child-bearing a *crime against society!* An *unmarried woman* who declaims against *marriage!!* A *young woman* who deprecates charity and a provision for the *poor!!!*
>
> (Scrope *et al.* 1833, 151)

By focusing on Martineau's character – highlighting her social status as a young, unmarried female – this review attempted to criticize not only the Malthusian principles behind Martineau's work, but also her right to write about issues outside her proper sphere of influence. Though Martineau claimed to have learned a great deal about the importance of courage in authorship from reading this review, she also acknowledged that it had damaging effects on her career. 'For ten years', she writes, 'there was seldom a number which had not some indecent jest about me, – some insulting introduction of my name' (Martineau 1877, I 158).

During this time, Martineau's books were subjected to various forms of social censorship. She heard stories of parents who out of fear of impropriety kept her books out of the reach of their children. Likewise, Martineau heard tales of unnamed 'literary ladies' who claimed that in her place, they 'would have gone into the mountains or to the antipodes, and never have shown their faces again' (Martineau 1877, I 156).

The negative publicity associated with Martineau's early attempts at book publication caused her to reflect on the role of the author in Victorian society. In her essay, 'Literary Lionism', published in *The London and Westminster Review* in 1839, she expressed concern about the tendency of society to 'lionize' its writers – to admire their public personalities more than their work. Such lionism, she claimed, would have a damaging effect on intellectual production because it encouraged egotism and mediocrity, rather than moral sympathy and literary excellence. Literary lionism was especially hazardous for the female writer, who might arrive at a private dinner party only to be

> seized upon at the door by the hostess, and carried about to lord, lady, philosopher, gossip, and dandy, each being assured that she cannot be spared to each for more than ten seconds. She sees a 'lion' placed in the centre of each of the first two rooms she passes through, . . . and it flashes on her that she is to be the centre of attraction in a third apartment.
>
> (Martineau 1839, 267)

Such attention not only placed women writers in the uncomfortable position of being public performers but also exposed their work and lives to the scrutiny of the general public that reserved its 'hardest treatment' for female authors (Martineau 1839, 268).

As an example of the ideal conditions of authorship, Martineau turned to the medieval period, where 'literature was cultivated only in the seclusion of monasteries' and where the 'prospect of influence and applause was too remote to actuate a life of literary toil' (Martineau 1839, 262–3). In order to recreate these ideal conditions of authorship:

> The author has to do with those two things precisely which are common to the whole race – with living and thinking . . . The very first necessity of his vocation is to live as others live, in order to see and feel as others see and feel, and to sympathize in human thought.

> In proportion as this sympathy is impaired, will his views be partial, his understanding, both of men and books, be imperfect, and his power be weakened accordingly.
>
> (Martineau 1839, 272)

In this way, Martineau defined authorship as a sympathetic, rather than egotistical process. By living as others live, authors are able to absorb the thoughts and feelings of their fellow humans, rather than reveling in their own narrow perspectives.

Within Victorian society, this kind of sympathetic distance was difficult to achieve, especially for women writers, who had a hard time retreating into the anonymity of the private, domestic life after being lionized (and ridiculed) by the reading public. It was only through the anonymity of the periodical press that Martineau was able to evade the limitations imposed by socially constructed definitions of the 'female author'. In her *Autobiography,* Martineau recounts tales of these successful evasions, communicating her pleasure at not revealing her authorial identity to her masculine readers. For instance, when her first anonymous leaders appeared in *The Daily News* in 1852, Martineau was delighted to discover that her work created

> such a noise that I found that there would be no little amusement in my new work, if I found I could do it. It was attributed to almost every possible writer but the real one. This 'hit' sent me forth cheerily: and I immediately promised to do a 'leader' per week.
>
> (Martineau 1877, II 82)

This sense of amusement in concealing her gendered identity from the reading public pervades many women writers' narratives surrounding their attempts at anonymous and pseudonymous authorship. By possessing the secret knowledge of their own identities, women writers were able to gain an ironic distance on the commentary of critics and their tendency to gender the authorial role in Victorian society.

Even though the periodical press provided Martineau with the means of evading the socially constructed identity of the 'female author', it did not allow her to evade gender altogether. This was because the 'editorial we' of many journalistic publications was generally assumed to be masculine, and Martineau often wrote for them as if she were male, gendering her narrative 'we' with masculine descriptors. For instance, in an article entitled 'Female Industry', published in *The Edinburgh Review* in 1859, Martineau writes:

> We must have a release from the ragged edges, loose buttons, galling shirt-collars, and unravelled seams and corners which have come up as the quality of needlewomen has gone down. Let our wives undertake the case of the remnant of the poor sempstresses, – the last, we hope, of their sort.
>
> (Martineau 1859, 328)

Here, masculinity is clearly associated with the editorial 'we', and femininity is objectified as 'our wives' and the 'poor sempstress', who makes faulty masculine clothing.

It is interesting to contrast a passage like this one with another work published under Martineau's own name. For instance, in *Household Education,* a work of non-fiction serialized in *The People's Journal* in 1847, Martineau writes:

> I remember being fond of a book in my childhood which yet revolted me in one part. It told of the children of a great family in France, who heard of the poverty of a woman about to lie in, and who brought and made clothes for herself and her infant.... I used to blush with indignation over this story; indignation on the poor woman's account, that her pauperism was so exposed.
>
> (Martineau 1847, 37)

Here, Martineau's 'we' is replaced by a confessional 'I' that seems to attach directly to the authorial name published at the head of the article. Such a narratological position suits her rhetorical situation as a contributor to a reform-minded periodical designed to address both male and female readers. By moving back and forth from an anonymous 'masculinized' narrative voice to a personalized narrative voice, Martineau was able to develop flexibility as a writer. Depending on her subject matter, readers, and editorial guidelines, Martineau was able to manipulate her own identity to suit her rhetorical purpose.

This narrative flexibility was only possible if Martineau was able to control who possessed knowledge of her authorial identity in any given rhetorical situation. When writing for 'masculine' journalistic publications such as *The London Daily News, The Edinburgh Review,* and *The Westminster Review,* Martineau was always carefully protective of her anonymity and that of her sources. The result is a discourse of secrecy that emerges in her correspondence with fellow writers. For example, in 1858 Florence Nightingale approached Martineau about writing a series of articles for *The Daily News* based on some 'confidential' reports she had written on the sanitary conditions of the army. Because of her politically sensitive position as a consultant to Sidney Herbert, Nightingale was especially concerned that her identity as the source of the articles be kept secret. Martineau wrote back to Nightingale, reassuring her that

> In whatever I did, in the war-time & after it, in relation to yourself, my object was to have you entirely *let alone*; – in regard to theological opinions, offers of praise, assurances of fame, descriptions & criticisms of your management, – (& even in my own mind, subscriptions of money for your objects.) What I could do I did to keep the crowd off you, & leave you air & space & liberty. The thing was impossible, of course; but it was right to try.
>
> (Martineau 1990, 166).

Such a discourse over privacy is characteristic of much of the Nightingale–Martineau correspondence, which spans thirteen years. The content of these letters, often marked 'confidential' or 'private', reads today like a who's-who of mid-Victorian politics. Working cooperatively behind the scenes on issues as diverse as sanitary reform and women's education, Martineau and Nightingale engaged in a discourse of secrecy that characterizes so much of women's professional correspondence relating to journalistic publication during the Victorian period.

However, even with the most careful attention to the preservation of their anonymity and that of their sources, women writers were still often 'found out' by their critics. Because Martineau's identity as a woman writer was so controversial and well-known, it was increasingly difficult for her to adopt the anonymous cloak of the periodical press. This break in journalistic cover was something that concerned Florence Nightingale as she continued in her covert publishing ventures with Martineau. In correspondence relating to a later collaborative project – an article entitled 'Mutual Relations of the Sick and Well' – Nightingale warns Martineau:

> Do not write anything which, you would not wish to have known, is by you. The article will be remarked, questions will be asked, and I never knew anything that people *wished* to know (of this kind) that did *not* – at last – 'leak out'. If a Review Article does not fall dead, – and depend upon it this will not – people always ask, whose is it? and people always find out. Ultimately *everybody* will know that *you* have written it.
>
> (Nightingale 1860, ff. 61–6)

Indeed, try as she might, it was almost impossible for a literary lion like Harriet Martineau to avoid being identified as a 'female author'.

As her fame grew, her literary disguises became increasingly transparent to the public gaze. And her gendered identity, rather than her work, became more and more the subject of public scrutiny. In the last years of her life, she made several attempts to reconstruct her public image. One of the most interesting of these reconstructions was Martineau's self-written obituary, published in *The Daily News* in 1876. Written in the third person, this obituary points out the high and low points of her own life and character, remarking, 'But none of her novels or tales have, or ever had, in the eyes of good judges or in her own, any character of permanence' (Martineau 1877, II 565). Amusingly enough, she punctuates such self-effacing commentary with plugs for her *Autobiography*, which was scheduled to be released to the public shortly after her death. She writes, 'Harriet Martineau's forthcoming Autobiography will of course tell the story of the struggle she passed though to get her work published in any manner and on any terms' (Martineau 1877, II 564). Ironically, Martineau attempts to perform the ultimate act of self-effacement by writing her own obituary in an anonymous journalistic voice, and at the same time she attempts to perform the ultimate act of self-advertisement by puffing her own book.

As it turned out, the anonymity Martineau was trying to maintain in writing this obituary was soon lost. After her death, the editors of *The Daily News* retitled the obituary 'An Autobiographic Memoir' and printed it with Martineau's by-line. In this way, Martineau's attempt at self-reconstruction had the effect of turning her into public spectacle – the woman writer who would vainly write her own obituary and puff her own book while posing as an anonymous journalist. It was just these kinds of negative stereotype that Martineau had been attempting to dismantle throughout her literary career.

In another attempt to reconstruct her public image, Martineau recalled all of the letters she had written to her many correspondents and asked that those that were not returned be destroyed. Forbidding her literary executors from publishing any of her letters posthumously, Martineau offered in their place her two-volume *Autobiography* that, published posthumously, would provide the essential details of her life and character. In her *Autobiography*, she carefully details her development as a writer, her artistic craft, her meetings with famous people, and her literary achievements. She also retaliates against the many attacks lodged against her work and character during her career. For example, in retelling the story of Croker and Lockhart's damaging review of

Illustrations of Political Economy, Martineau writes, 'The compassion that I felt on this occasion for the low-minded and foul-mouthed creatures who could use their education and position as gentlemen to "destroy" a woman whom they knew to be innocent of even comprehending their imputations, was very painful' (Martineau 1877, I 156). With perhaps more righteous glee than 'pain', Martineau uses her *Autobiography* to have 'the last word' in her long-running disputes with those who would use her gender in their *ad hominem* attacks against her work.

Another means she used to reconstruct her authorial identity was to describe her attitude toward literary fame. In addition to including a revised version of her early essay, 'Literary Lionism', in the *Autobiography*, she also reflects on her own artistic motivations:

> Authorship has never been with me a matter of choice. I have not done it for amusement, or for money, or for fame, or for any reason but because I could not help it. Things were pressing to be said; and there was more or less evidence that I was the person to say them.
>
> (Martineau 1877, I 143)

What first appears to be an expression of 'feminine' modesty can perhaps be read as a statement of Martineau's resistance to the definitions of the gendered author so prevalent in her society. For Martineau, the author's role was to reflect and shape the discourses circulating within her culture as a whole, rather than actively to express ideas that originate within the authorial persona.

Such a view of the authorial role provides insight into Martineau's views on female authorship expressed in the *Autobiography*. While Martineau had for many years championed the right of women to pursue professions, here she criticizes the woman writer who draws attention to her own personal life in her written work, who '[violates] all good taste by her obtrusiveness in society, and oppressing every body about her by her epicurean selfishness every day, while raising in print an eloquent cry on behalf of the oppressed'. (Martineau 1877, I 302) Such selfish literary acts, Martineau claimed, had the effect of drawing attention to the gendered identity of the writer, instead of furthering ideas and actions that would enable women to escape confining definitions of the female author. Only by attempting to write from an ungendered, depersonalized perspective could women come to be treated equally in all of the professions.

Yet by writing her own 'portrait of the artist', Martineau was in many ways guilty of the same form of self-advertisement and literary egotism that she criticizes in her *Autobiography*. By flaunting her achievements and focusing on the development of her own identity as a writer, she in many ways reinscribes the same stereotypes of female authorship that she was trying to evade. That is, through self-definition, Martineau singularized her authorial identity in the same way her critics had attempted to pin her down to a particular feminine stereotype. Of course, this is not to say that Martineau's *Autobiography* didn't also chart new territory for women in its representation of a woman professional writer. Indeed, it is remarkable in its frank depictions of the hazards and pleasures of being a powerful woman writer and cultural icon. However, at the same time that it presented these new images of the woman writer, it reinforced the stereotype of the vain, self-promoting 'female author'.

Only as an anonymous periodical writer had Martineau been able to achieve – at least partially or temporarily – the kind of humility and self-effacement that she had long held as her artistic ideal. The periodical press provided an opportunity for Martineau to begin her literary career without exposing her gendered identity to public view. After she began publishing signed books, the periodical press enabled her to reclaim at will the power associated with anonymity – to write on issues of social and political concern in a male or female narrative voice and to an audience composed of men and/or women.

However, at the same time that the periodical press liberated the expression of women writers like Martineau, it also constrained their careers in some significant ways. With their specifically defined audiences, periodicals reinforced the Victorian notion of separate spheres – that there were 'masculine' periodicals written by and for men, 'mixed gender' periodicals written by and for both men and women, and 'feminine' periodicals, written by and for women. Of course, these strict gender bifurcations masked a great deal of gender diversity behind the single-gendered, yet anonymous, 'editorial we' of most periodicals. But for women, the fact remained that they could only speak out freely on issues outside the female sphere if they were 'dressed' as men, that is, if they wrote in a masculine voice.

In addition to reinforcing the notion of separate spheres, the periodical press also played a key role in formulating the constrictive definitions of the 'female author' that women were attempting to evade. Reviews of women's books created negative stereotypes associated with female authorship during the Victorian period. Thus, paradoxically, while the periodical press played a key role in constructing gender stereotypes associated with male and female authorship, it also provided the anonymous medium through which authorial gender and identity could be disrupted and resisted. It is only by gaining an understanding of how women writers dealt with the opportunities and constraints associated with periodical publication that we can begin to understand the complex gender issues and narrative structures in women's non-fiction prose.

Through negotiating the fame associated with signed publication and the obscurity associated with anonymous publication, women writers created careers for themselves during a historical time period when there were few professional opportunities for women. Further investigation of their experiences with the periodical press can also help us to understand the sometimes contradictory views on authorship expressed by women writers such as Harriet Martineau. By defining the woman author in multiple terms – as both visible and invisible, self-promoting and self-effacing, intentional and passive, masculine and feminine – Martineau expanded the possibilities for women writers, helping to complicate Victorian notions of the 'female author' and 'feminine' writing.

Works cited

[Empson, William.] 'Mrs. Marcet-Miss Martineau', *Edinburgh Review* 115 (April 1833), 1–39.

[Maginn, William.] 'Miss Harriet Martineau', *Fraser's Magazine* 5 (Nov. 1833), 576.

[Martineau, Harriet.] 'Literary Lionism', *London and Westminster Review* 32 (April 1839), 261–81.

Martineau, Harriet. 'Household Education (No. 9)', *The People's Journal* 4 (July 1847), 36–7.

[——]'Female Industry', *Edinburgh Review* 109 (April 1859), 293–336.

—— *Autobiography*. 2 vols. Boston, MA, 1877.

—— *Selected Letters*, ed. Valerie Sanders. Oxford, 1990.

Nightingale, Florence. Letter to Harriet Martineau, 8 Feb. 1860. Florence Nightingale Papers, British Library vol. L, MSS. 45, 788, ff. 61–6.

[Scrope, G. Poulett, John Croker and John Lockhart.] 'Miss Martineau's Monthly Novels', *Quarterly Review* 49 (April, July 1833), 136–52.

11
Work for Women: Margaret Oliphant's Journalism[1]

Joanne Shattock

> [Mrs. Lewes] thinks people who write regularly for the Press are almost sure to be spoiled by it. There is so much dishonesty, people's work being praised because they belong to the confederacy.
>
> (George Eliot 1978, VIII 466)

That comment, recorded in 1869 by George Eliot's friend Emily Davies, serves to reinforce one model of the professional woman writer in the nineteenth century, a model best represented by Eliot herself, in which journalism was an apprenticeship, abandoned when financial security was achieved and 'serious' writing could begin. Eliot's was a masculine success story, some would argue, both in the single-mindedness with which the writing was pursued, and the critical acclaim and reputation which ensued.

For Margaret Oliphant, George Eliot became the writer against whom she measured herself from the early days with Blackwood, when it was made clear that the latter was the firm's most valued asset. Oliphant was touchingly honest about her mixed feelings for her celebrated contemporary. She resumed the writing of her fragmented autobiography in February 1885 after an interval of over twenty years, immediately after reviewing J. W. Cross's biography of Eliot for the *Edinburgh Review*. 'I have been tempted to begin writing by George Eliot's life – with that curious kind of self-compassion which one cannot get clear of. I wonder if I am a little envious of her', she recorded candidly. Later she continued:

> How I have been handicapped in life! Should I have done better if I had been kept, like her, in a mental greenhouse and taken care of? . . . I was, after all, only following my instincts, it being in reality easier

> to me to keep on with a flowing sail, to keep my household and make a number of people comfortable, at the cost of incessant work, and an occasional great crisis of anxiety than to live the self-restrained life which the greater artist imposes on himself.... No one even will mention me in the same breath with George Eliot. And that is just.
>
> (Oliphant 1974, 5–7)

Elisabeth Jay, in her biography, *Mrs. Oliphant: 'A Fiction to Herself'*, argues that a more appropriate model for Oliphant would be contemporary novelists like Fay Weldon, Anita Brookner, Penelope Fitzgerald, and Martin Amis, writers both male and female who successfully combine reviewing with creative writing, and who do not regard journalism, in Jay's words, as 'the financial penalty for being a minor artist' (Jay 1995, 4). She notes that Oliphant made her first submission to become a *Blackwood's* reviewer on the strength of her success as a novelist and short story writer – the reverse of the Eliot model. For Oliphant, Jay argues, reviewing fuelled her creativity rather than exhausting it. She was a voracious reader of the work of other writers due to the demands of her reviewing, and her own work was as a result enriched.

It could also be argued that for many nineteenth-century professional women writers Eliot was not necessarily the paradigm they sought to emulate. There were others among Oliphant's contemporaries whose writing careers paralleled hers, women who worked in a variety of genres in which fiction and poetry were not automatically privileged, energetic women whose work she knew. Harriet Martineau in many ways was much better placed to become the model for the Victorian woman journalist. Other possibilities included Anna Jameson, Geraldine Jewsbury, Mary Howitt and Eliza Lynn Linton. In this chapter I want to explore the issue of gender in the nineteenth-century periodical press by focusing on Margaret Oliphant as an example of a professional woman journalist.

Oliphant was undoubtedly a writer condemned by her industry. Virginia Woolf in *Three Guineas* (Woolf 1938, 166) invited her readers to 'deplore the fact that Mrs. Oliphant sold her brain, her very admirable brain, prostituted her culture and enslaved her intellectual liberty in order that she might earn her living and educate her children'. The charge is that because she had to support her family she wrote too much too quickly, that the biographies and literary histories, and, as Woolf described them, the 'innumerable faded articles, reviews, sketches of one kind and another which she contributed to literary papers' exhausted her creativity and prevented the writing of a few good novels.

For her contemporaries it was her prolixity which was always commented upon. John Skelton, in an article in *Blackwood's* in January 1883, celebrating her *Literary History of England in the end of the Eighteenth and beginning of the Nineteenth Century* (1882) noted her 'unwearied and facile pen [which] has been constantly at work now for more than thirty years' ([Skelton] 1883, 73). Skelton was writing primarily of her fiction, and he compared her not unfavourably with Scott, adding: 'I do not suppose that Mrs. Oliphant is one of the writers who consciously entertain or profess, what is called in the jargon of the day, "high views of the literary *calling*" but it may be said of her that she has never written a page which she would wish unwritten' ([Skelton] 1883, 76). Faint praise indeed.

Interestingly it was usually her non-fictional writing, in other words her journalism, which drew adverse comments. Edith Simcox described a conversation in 1878 with George Eliot, when the talk was 'of translations, ignorance in print, and the unprincipledness of even good people like Mrs. Oliphant who write of that whereof they know nothing' (Eliot 1978, IX. 228). This was not a comment about a novelist who had produced some seventy odd titles at this time. It was a comment on a too prolific journalist. Henry James, writing of her sympathetically in an obituary, reflected that no one had practised criticism

> more in the hit-or-miss fashion and on happy-go-lucky lines than Mrs. Oliphant.... She wrought in 'Blackwood' for years anonymously, and profusely, no writer of the day found a *porte-voix* [megaphone] nearer to hand, or used it with an easier personal latitude and comfort. I should almost suppose in fact that no woman had ever, for half a century, had her personal 'say' so publicly and irresponsibly.
>
> (James 1914, 358)

James's comment is interestingly double-edged. He notes the prolixity of her reviewing, but he also notes her influence. If we wanted further testimony to the latter we could recall Hardy's irritated description of her review of *Jude the Obscure* ('The Anti-Marriage League') as 'the screaming of a poor woman in Blackwood's' (Hardy 1912).[2] Jay argues that Oliphant's reputation suffered at the end of the century at the hands of a 'male clubland taking its revenge for the long years of George Eliot's supremacy' (Jay 1995, 245). It is too easy, I think to see Eliot as the dominant force in women's writing from the mid-century onward. It was not necessarily so perceived by her contemporaries. Oliphant's sheer ubiquity as a writer and reviewer was sufficient to encourage detractors.

Her reviewing was phenomenal in its bulk and considerable in its impact. In her writing life of over forty years there were few of her contemporaries, male or female, not to mention writers of the past, whose work did not come under her scrutiny. In this she was the precursor not only of contemporary women writer/reviewers, but also of Virginia Woolf and Alice Meynell, two early modernist women writers for whom journalism was a persistent strand throughout their writing lives.

Oliphant was in no doubt that she was entering a masculine world. She recognized that to make a living by writing for the periodical press it was the male-dominated press she needed to infiltrate. It was through her mother's intervention that she secured an introduction to George Moir of *Blackwood's*, a distant cousin. It never occurred to her or to her mother (as Jay points out) that she should attempt to get work from any of the women's or children's magazines or the more popular family-oriented periodicals which were coming into existence. Writing to John Blackwood in 1855 at the beginning of their long relationship she expressed some concern about her material: 'I am sometimes doubtful whether in your most manly and masculine of magazines a womanish story-teller like myself may not become wearisome' (Oliphant 1974, 160). In response to her self-doubt she evolved a neutral or ungendered voice for her reviewing which was on the whole more successful than George Eliot's comparable attempts to blend her voice with those of her male colleagues in the *Westminster Review.*

The opening paragraphs of her reviews were often self-consciously masculine or tended to expansive pronouncements, as if she was deliberately writing in an unfamiliar mode, after which her voice dropped into a more relaxed tone:

> It is a dangerous thing to have your life written when you are dead and helpless, and can do nothing to protest against the judgment.... But if biography is thus dangerous, there is a still more fatal art, more radical in its operation, and infinitely more murderous, against which nothing can defend the predestined victim. This terrible instrument of self-murder is called autobiography.
>
> ([Oliphant] 1877, 472)

That was the opening of her review of Harriet Martineau's *Autobiography*, uncharacteristically abrasive and somewhat stilted, after which she resumed her normal tone. Other openings affected the magisterial or the near platitudinous, as in 'Greatness is always comparative: there are few things so hard to adjust as the sliding-scale of fame' ([Oliphant] 1855, 554), or 'Civilization, like every other condition of humanity, has its dark as well as its bright side' ([Oliphant] 1858, 139), the beginning of an article on the 1857 Matrimonial Causes Act which also, uncharacteristically, invoked the reader: ('Let not so serious an introduction damp your courage, oh reader just and kind!' [Oliphant] 1858, 140). Occasionally she identified herself as a woman, as in her 1866 article on Mill's proposals for female suffrage which began: 'The present writer has the disadvantage of being a woman. It is a dreadful confession to put at the beginning of a page; and yet it is not an unmitigated misfortune' ([Oliphant] 1866, 367). She declared herself too in a somewhat uncomfortable discussion of the emphasis on sexuality in Grant Allen's *The Woman who Did*: ('It is painful to me, a woman, to refer to it', [Oliphant] 1896, 144). For most of her reviewing, though, Oliphant's voice was ungendered, carefully modulated, confident, not strident.

She began her reviewing career in 1855 when the controversy over anonymity versus signature was about to break. She remained wedded to the old system to the end, arguing in her 1882 *Literary History of England*:

> We believe that criticism is always most free, both for praise or blame, when it is anonymous, and that the verdict of an important publication, whether it be review as in those days, or newspaper as in our own, is more telling, as well as more dignified, than that of an individual, whose opinion, in nine cases out often, becomes of inferior importance to us the moment we are acquainted with his name.
> [Skelton] 1883, 90)

'Anonymity is a great institution – I think I shall go in for that henceforward in everything but novels', she told Blackwood in 1870 (Jay 1995, 244) and she stuck to her decision. Her reviews in *Macmillan's* in the 1880s were initialled and the few in the *Contemporary Review* in the same period signed, but otherwise she remained anonymous throughout her forty years of reviewing.[3] It was to prove a valuable key to negotiating the gender barrier in the world of journalism.

Most of her colleagues in the reviewing fraternity were men. A letter to Blackwood in 1873 in response to criticism of one of her reviews revealed her sense of the masculine university-dominated scene in which she found herself:

> But your correspondent's opinion, even had I had the advantage of knowing it before I wrote my paper, is not mine, and I can only review

> books at all on the condition that I express my own feelings in respect to them. It is of course in your power to bid me refrain from reviewing any particular book or any books at all, but I can only say what I think and not what other people think, whatever the universities may say. I read every word of both books mentioned to my pain and sorrow.... The tremendous applause which has greeted this performance is a good specimen of the sort of thing which I am anxious to struggle against – the fictitious reputation got up by men who happen to be 'remembered at the Universities,' and who have many connections among literary men.
>
> (Oliphant 1974, 240–1)

For the most part she kept her gender and her own circumstances out of her reviewing, even in a number of pugnacious articles on women's issues which ran from the 1850s through to 1896,[4] in which she forcefully espoused a conservative position, but rarely gave away any personal details. Only in her article on Mill's proposals for extending the suffrage to female householders, among whom she was numbered, did she reveal something of herself:

> The class for which he proposes to legislate is not the most interesting section of woman kind. It is the class of female householders, lone women who pay their own rent and taxes, and have their own affairs to manage, and 'nobody to look after them,' according to the vernacular.... It is We, gentlemen, with whom you will have to do; we who have withered on the stalk, or taken many a buffet from the world; who are respectable, but no longer charming; whose hair is growing grey.
>
> ([Oliphant] 1866, 369–70)

'This petition', she continued, 'was forwarded to Ourselves (if, indeed, a woman's pen may venture upon that sublime pronoun) for our signature' ([Oliphant] 1866, 368). 'We are even, some of us', she hinted coyly, 'admitted to the honour of inscribing our opinions in the pages of Maga' ([Oliphant] 1866, 367).

Late in her career, perhaps with tongue in cheek, she embarked upon a series of review articles for *Blackwood's* under the heading 'The Old Saloon' (January 1887 – June 1892), the title reminiscent of the magazine's early days and suggestive of a masculine clubland in which she took no part. In the opening article she invoked the shades of her male predecessors, John Wilson, Lockhart, James Hogg and others, in the room which had been the focal point of the magazine's intellectual and social life: 'Here where we sit our predecessors have sat for three parts of a century, discoursing wit and wisdom such as no longer meets the common ear' ([Oliphant] 1887, 126). Contributions still come to the magazine, she noted, 'from many a manly pen' as far as the British Empire extends, adding, 'she has her ladies too, but shall we own it? perhaps loves them less' ([Oliphant] 1887, 127).[5]

She wrote occasionally for the *Edinburgh* and the *Fortnightly*, and for the *Contemporary Review*. Two of her novels were serialized in *Longman's Magazine*, and others in *Macmillan's* and in the *Cornhill* to all of which she contributed occasional articles. In addition she wrote for the *Spectator* and for the *St. James's Gazette*. But *Blackwood's* gave her a central platform from which to speak throughout her career. John Blackwood was very much her mentor, performing a function similar to that played by W. J. Fox in Harriet Martineau's early reviewing days with *The Monthly Repository*. Unlike George Eliot she had no G. H. Lewes to look after her business affairs, and she conducted all her negotiations, sometimes too diffidently, herself. But despite her enterprise and her independence the ultimate journalistic prize, as Jay points out, eluded her. Her ambition was to equal her male colleagues, Trollope, Dickens, Thackeray and Leslie Stephen, as the editor of a serious journal, a position which would have brought with it financial security as well as status. Feminine enterprise like Mrs Henry Wood's *Argosy*, Mary Elizabeth Braddon's *Belgravia*, or Mrs Gatty's *Aunt Judy's Magazine* would not have sufficed. But it was not to be. She edited a minor book series for Blackwood but the editorship of a periodical equivalent in status to the magazine did not come her way

The spectrum of subjects she tackled could in no way be regarded as gendered. She wrote on fiction, poetry, religion, art, travel, history and biography. Politics she eschewed, apart from the current debates about women, but she reviewed political memoirs. She did not write much on philosophy, but it occasionally formed part of her articles on religion and other cognate subjects. She was sufficiently fluent in French to translate Montalembert and she reviewed French literature competently. Not for nothing did she refer to herself as Blackwood's 'general utility woman'.

By Elisabeth Jay's reckoning she was even-handed in her reviews of male and female writers. Undoubtedly her reviews of her female contemporaries and predecessors stand out from her journalism in the way in which Virginia Woolf's do. Oliphant was perceptive about personal circumstances and the problems of writing for a living, and particularly attuned to the merits as well as the failings of women writers. She was generous and sympathetic to Mary Russell Mitford (*Blackwood's* 75, June 1854), for example. She presented a convincing appreciation of the impact women made on the cultural scene even when they wrote comparatively little, in a review of recent biographies of Hester Piozzi and Mary Delany ([Oliphant] 1862). She wrote intelligently of three female autobiographers, Alice Thornton, the Duchess of Newcastle and Madame Roland, in a series of seven articles on autobiographies (*Blackwood's* 129, Jan. 1881–133, April 1883). She showed a shrewd sense of the strengths as well as the conscious limitations of Jane Austen in a review of J. E. Austen-Leigh's *Memoir*, an article whose critical judgements, particularly on individual novels, stand up well today. Austen, she noted:

> confines herself to a class – the class of which she has herself the most perfect knowledge – striking out with an extraordinary conscientiousness which one does not know whether to call self-will or self-denial, everything above and everything below. Lady Catherine de Burgh and the housekeeper at Pemberley – conventional types of the heaven above and the abyss below – are the only breaks which Miss Austen ever permits herself upon the level of her squirearchy.... If it were not that the class to which she thus confines herself was the one most intimately and thoroughly known to her, we should be disposed to consider it, as we have said, a piece of self-denial on Miss Austen's part to relinquish all stronger lights and shadows; but perhaps it is better to say that she was conscientious in her determination to describe only what she knew, and that nature aided principle in this singular limitation.
>
> ([Oliphant] 1870, 293–4)

Austen's juvenilia give an early indication of her temperament and 'that amused, indifferent, keen-sighted, impartial inspection of the world as a thing apart from herself, and demanding no excess of sympathy, which is characteristic of all the work of her life' ([Oliphant] 1870, 295).

When it came to her assessment of her female contemporaries, some of whom were her friends and others her rivals she was shrewd, generous, knowledgeable and devoid of spite. She leapt to the defence of Jane Carlyle (Oliphant 1883) the victim, as she saw it, of Froude's predatory biographical speculations, a subject which became for her almost an obsession. Her review of the *Memoirs* of Anna Jameson and of Fanny Kemble's *Records of a Girlhood* saluted two working women who earned her respect for combining professional careers with the vicissitudes and demands of their domestic circumstances.

> They were in full tide of their lives, if not beginning to wane, when the agitations of recent times were but beginning; which did not hinder them, however, from stepping into the busy current of active life when necessity made it desirable so to do – finding work that suited them, and doing it, as well as if all England got up in church on Sunday and said, 'I believe that women ought to be allowed to work'. ([Oliphant] 1879, 207)

Her review of Harriet Martineau's *Autobiography* ([Oliphant] 1877) was sympathetic but measured in its praise. She disliked Martineau's air of self-importance, the 'self applauses' of the work as she described them, her inflated sense of her influence on her times, her ill-tempered comments on her contemporaries, and her ungenerous treatment of her mother. Martineau's autobiography constructed a masculine success story, in Oliphant's view, the kind of autobiography she could never have contemplated.

Her review of J. W. Cross's *George Eliot's Life* ([Oliphant] 1885) was an intelligent assessment of Eliot as Cross presented her as well as an enthusiastic response to her early novels, the latter a position typical of most posthumous assessments of Eliot. She took issue with Cross's chosen mode of presenting his material, that of letting Eliot tell her own stories through her letters, comparing it with Froude's biography of Carlyle, and adding, 'Carlyle has been made out of his own mouth to prove himself a snarling Diogenes . . . and George Eliot by the same fine process has been made to prove herself a dull woman' ([Oliphant] 1885, 517). 'Is this the woman who wrote Adam Bede?' she asks disbelievingly ([Oliphant] 1885, 519) after failing to find letters which demonstrated Eliot's humour. Contemporary and posthumous assessments of the *Life* have agreed with her.[6]

She quoted at length the well-known passage from Eliot's Journal which Cross included, 'How I came to write Fiction' and commented perceptively: 'The reader will make the acquaintance of a remarkable character . . . the strongest of all female writers, he will find in her what is almost the conventional type of a woman – a creature all conjugal love and dependence, who is sure of nothing until her god has vouched for it, not even her own powers' ([Oliphant] 1885, 551).

Two themes connect these reviews of works by nineteenth-century women writers, their emphasis on women's education and on women's employment. Both ideas were very much in vogue, and both, Oliphant contended, had existed long before the present movements for their improvement. Mary Russell Mitford and Jane Austen were excellent examples of how women had managed to become well-educated long before ladies' colleges had ever been thought of, very much, she might have added, like herself:

> They were both well educated, according to the requirements of their day, though the chances are that neither could have passed her examination for entrance into any lady's college, or had the remotest chance with the University Inspectors; and it is not unconsolatory to find, by the illumination which a little lamp of genius here and there thus throws upon the face of the country, that women full of cultivation and refinement have existed for generations before ladies' colleges were thought of, notwithstanding the universal condemnation bestowed upon our old-fashioned canons of feminine instruction.
>
> ([Oliphant] 1870, 290)

Only occasionally in the Austen / Mitford article is there the slightest hint of envy that the section of society into which those women had been born was inestimably advantaged:

> They were well born and well connected, with a modest position which not even poverty could seriously affect, and the habit from their childhood of meeting people of some distinction and eminence, and of feeling themselves possessed of so much share in the bigger business of the world as is given by the fact of having friends and relations playing a real part in it. No educational process is more effectual than this simple fact, and Jane Austen and Mary Mitford were both within its influence.
>
> ([Oliphant] 1870, 290)

Jay argues that Oliphant had no chips on her shoulder as regards her own social background, and that she used her own career to celebrate the social mobility which the world of letters afforded. Occasionally in her reviewing, there are hints to the contrary.

Oliphant saw the significance of the current agitation for paid employment for women but was scornful of the notion that this was a novel concept, or that legislation could effect any useful changes:

> The present generation considers itself to have invented the idea that women have a right to the toils and rewards of labour, not withstanding the long array of facts staring them in the face from the beginning of history, by which it is apparent that whenever it has been necessary, women *have* toiled, have earned money, have got their living and the living of those dependent upon them in total indifference to all theory.
>
> ([Oliphant] 1879, 206)

Throughout history, women have worked as necessity demanded, demonstrating talents and ability in a range of activities, and will continue to do so, she argued: 'With all respect for the eloquent advocates of work for women, a capable woman is just as likely to make a livelihood for herself if she wants it, and get a good return for her pains, as a man is' ([Oliphant] 1866, 367). And she might have added, 'Like me'.

The writers she admired, writers of either sex, were those who worked for their living, who did not allow themselves the luxury of theorizing about 'art' and its demands but for whom art and life and work were interwoven. In the introduction to her edition of the *Autobiography* (Oliphant 1990, vii), Elisabeth Jay writes perceptively of the 'intimate and complex relationship between the writing and the need that generated it' which ran throughout Oliphant's writing life. In the *Autobiography* Oliphant bristled with resentment against the images of a life dedicated to 'art', the literary 'success story' as projected by both Eliot's biography and Trollope's autobiography:

> I have never had any theory on the subject. I have written because it gave me pleasure, because it came natural to me, because it was like talking or breathing, besides the big fact that it was necessary for me to work for my children.... They are my work, which I like in the doing, which is my natural way of occupying myself, though they are never so good as I meant them to be. And when I have said that, I have said all that is in me to say.
>
> (Oliphant 1974, 4–5)

Oliphant's conservative views on the woman question are well known – her opposition to the vote ('So far as the designs of God may be judged from His works, He did not intend us either for ploughing or voting', [Oliphant] 1866, 376) and to divorce ('the man and the woman united in the first of all primitive bonds, the union upon which the world and the race depend, *are* one person', [Oliphant] 1869, 581). In her journalism she was an eloquent exponent of the doctrine of separate spheres, a view all the more disappointing to twentieth-century readers who have come to regard her as an early modern professional woman.

For her, journalism, work which could be undertaken in the home in a way that dressmaking or sewing could also be undertaken by women with talents in that direction, provided the ultimate solution for women who needed or sought financial independence but who at the same time refused to compromise their domestic responsibilities and roles. Nor did she regard journalism as an inferior form of work. Jay is right I think in arguing that for Oliphant journalism was not an apprenticeship which was to be shed as soon as success in a higher sphere looked distinctly possible. Nor was it, as it was for Woolf, one suspects, the routine but not unpleasurable chore which paid the bills. For Oliphant writing biographies, literary histories, and 'the innumerable... articles, reviews, sketches of one kind and another' was inseparable from her total writing life, a life which itself was inseparable from the domestic circumstances which demanded it. She did not place her various kinds of writing in a hierarchy. 'Margaret Oliphant: Journalist' was a title of which she would have been proud.

Notes

1. An earlier version of this article was published in Barbara Garlick and Margaret Harris, eds., *Victorian Journalism: Exotic and Domestic*, St. Lucia, Queensland, 1998. I am grateful to the editors for giving permission to republish this article with some amendments.
2. Oliphant's review was perceptive, particularly on the characters of Jude and Sue, but emotionally charged in its comments on the sexual explicitness of some of the scenes, and critical too of what she regarded as a popular propensity to devalue marriage. Hardy's opinion of Mrs. Oliphant was not high, even before the review of *Jude*. 'Mrs. Oliphant always admires what public opinion has decided that it is right to admire, & patiently repeats the old estimates, & quotes the old stories', he noted of her book on *The Makers of Florence*, 1876. (*The Literary Notes of Thomas Hardy*, ed. Lennart A. Bjork, vol. I Text, Gothenburg Studies in English 29, Göteborg, 1974). Of the review of *Jude* he wrote to Grant Allen (whose *The Woman Who Did* was reviewed in the same article): 'Talk of shamelessness: that a woman who purely for money's sake has for the last 30 years flooded the magazines & starved out scores of better workers, should try to write down rival novelists whose books sell better than her own, caps all the shamelessness of Arabella, to my mind'. (*The Collected Letters of Thomas Hardy*. ed. Richard L. Purdy and Michael Millgate, vol. II, Oxford, 1980, 106). I am grateful to Shanta Dutta for drawing these comments to my attention.
3. Her article 'The Anti-Marriage League', *Blackwood's* 159 (Jan. 1896) was initialled.

4. See *The Wellesley Index to Victorian Periodicals*, vol. 5 Toronto, 1989, for a list of Oliphant's contributions to *Blackwood's, Edinburgh Review, Fraser's Magazine* and *the Contemporary Review* in which these articles appeared.
5. 'The reader, however, must consider these words as a parenthesis, not intended in this connection', she added [Oliphant] 1887, 127).
6. See the introduction to David Carroll ed., *George Eliot: The Critical Heritage* London, 1971.

Works cited

Eliot, George. *The George Eliot Letters*, ed. Gordon S. Haight, 9 vols. New Haven, CT, and London, 1954–78.

Hardy, Thomas. 'Postscript to the Preface, April 1912', *Jude the Obscure*. London, 1912.

James, Henry. 'London Notes, August 1897', *Notes on Novelists*. London, 1914.

Jay, Elisabeth. *Mrs. Oliphant: 'A Fiction to Herself'*. Oxford, 1995.

Oliphant, M. *Autobiography and Letters of Mrs. Oliphant*, ed. Mrs. Harry Coghill, introd. by Q. D. Leavis. Leicester, 1974.

—— *The Autobiography of Margaret Oliphant*, ed. Elisabeth Jay. Oxford, 1990.

[Oliphant, M.]. 'Modern Novelists', *Blackwood's Magazine* 77 (May 1855), 554–68.

—— 'The Condition of Women', *Blackwood's Magazine* 83 (Feb. 1858), 567–86.

—— 'The Lives of Two Ladies', *Blackwood's Magazine* 91 (April 1862), 401–23.

—— 'The Great Unrepresented', *Blackwood's Magazine* 100 (Sept. 1866), 367–79.

—— 'Miss Austen and Miss Mitford', *Blackwood's Magazine* 107 (March 1870), 290–313.

—— 'Harriet Martineau', *Blackwood's Magazine* 121 (April 1877), 472–96.

—— 'Two Ladies', *Blackwood's Magazine* 125 (Feb. 1879), 206–24.

—— 'The Old Saloon', *Blackwood's Magazine* 141 (Jan. 1887), 126–53.

O., M. 'The Anti-Marriage League', *Blackwood's Magazine* 159 (Jan. 1896), 135–49.

Oliphant, M. 'Mrs. Carlyle', *Contemporary Review* 44 (May 1883), 609–28.

[Oliphant, M.]. 'Mill's *The Subjection of Women*', *Edinburgh Review* 130 (Oct. 1869), 572–602.

—— 'Life of George Eliot', *Edinburgh Review* 161 (April 1885), 541–53.

[Skelton, John]. 'A Little Chat about Mrs. Oliphant', *Blackwood's Magazine* 133 (Jan. 1883), 73–91.

Woolf, Virginia. *Three Guineas*. London, 1938.

12
Israel Zangwill's Early Journalism and the Formation of an Anglo-Jewish Literary Identity

Meri-Jane Rochelson

One hundred years ago, Israel Zangwill was probably the best-known Jewish writer in the English-speaking world. With reforming interests broad enough to include pacifism and feminism, he eventually became the leading English spokesman for Zionism and was a tenacious if often controversial Jewish activist. Zangwill's most enduring legacy is his fiction about Jewish life; works such as *Children of the Ghetto* (1892), *Dreamers of the Ghetto* (1898), *They That Walk in Darkness* (1899), and *Ghetto Comedies* (1907) established him as both an interpreter of the Jews and their conscience in an era of conflict, change, and assimilation. But before he gained his international reputation, Zangwill established himself as a literary personality simultaneously in the Anglo-Jewish and general interest[1] British press, creating the unique persona that he would maintain throughout his career. A look at Zangwill's earliest work in periodicals reveals how he used editorial positions in two newspapers – a Jewish news weekly published on Friday and a general interest humour paper that appeared on Saturday, the Jewish Sabbath – to develop an identity both Jewish and English and to build the base of what might be termed a crossover readership for his writings on Jewish and non-Jewish subjects. The combination of boldness and compromise in these efforts reflects a proud but uneasy status and self-image that Zangwill shared with many English Jews at the end of the nineteenth century.

Throughout his career Zangwill was a prolific journalist, and his writing in the periodical press helped solidify his position as a significant figure in the larger English literary milieu. As was common in the late Victorian period, most of his short fiction (including fiction on Jewish subjects) made its original appearance in periodicals as diverse as *Cosmopolis, Cosmopolitan*, the *Outlook, Atlantic*, and the *Illustrated London News*.

His novel *The Master* (1895) was serialized in *Harper's*. However, journalism for Zangwill was not simply a way to publish literary writing destined for volume form. In the 1890s, he enjoyed his status as a journalist as much as the rewards of periodical publication. Between 1892 and 1895 he became known as a 'New Humour' writer when his comic stories appeared in Jerome K. Jerome's *Idler* and he contributed to its roundtable feature, 'The Idler's Club'. His columns of reviews and literary and social commentary appeared in the *Pall Mall Magazine* and the New York *Critic* from 1893 to 1896 (the *Critic* column beginning its run in 1894), and literary gossip about him graced the pages of the *Bookman*. When the *New Review* ran a symposium on sex education in 1894 ('The Tree of Knowledge') Zangwill was a contributor. He became, in short, the quintessential 1890s literary man.

Apart from his first published story, 'Professor Grimmer', which won the prize in a *Society* competition when its author was seventeen, most of Zangwill's early journalism was in Jewish publications. Between 1883 and 1885 he edited and wrote for *Purim*, a humour annual (Winehouse 1970, 12–13).[2] From 1888 to 1890 three of his stories appeared in *The Jewish Calendar, Manual, and Diary*. Zangwill's 1889 article on English Judaism in the newly formed intellectual periodical *Jewish Quarterly Review* led him to the commission to write *Children of the Ghetto*, the novel that solidified his reputation. And from 1888 to 1891 he wrote a column called 'Morour and Charouseth' (loosely translated 'bitter and sweet') in the *Jewish Standard*, a paper sub-titled 'the organ of Orthodoxy'.

Zangwill used the pseudonym 'Marshallik' in 'Morour and Charouseth,' and in his first column explained that a *marshallik* is the Jewish court jester: 'No Jewish institution is complete without its fool, and some are blessed with a good many. But *The Jewish Standard* is to have one only' (Marshallik 1888b, 6). The jab at communal complacency was to be characteristic of Marshallik's reign. Zangwill's column supplied a satiric view of current controversies in the Jewish community, puffed forthcoming books by Jewish writers (including Zangwill), and presented a humorous Jewish angle on national political and cultural events. By October Zangwill was the *Standard*'s sub-editor (possibly even its editor; for two different accounts see Winehouse 1970, 49 and Udelson 1990, 71). He wrote many of the *Standard*'s leaders, though he himself was not orthodox, and contributed stories and poems. Zangwill's last 'Morour and Charouseth' column is dated 27 February 1891, just a few months before the paper itself ceased publication.

For most of the last year of Marshallik's run, Zangwill also edited the humorous penny weekly *Puck*, later titled *Ariel, or the London Puck*. This paper, for the most part, showed little evidence of its editor's Jewish identity, and as the periodical developed during Zangwill's tenure it increasingly adopted the look of *Punch, or the London Charivari*. However, Ariel and Marshallik had a great deal in common. How exactly Zangwill came to edit *Puck* is unclear, although his 1888 satiric novel *The Premier and the Painter* – a 'prince and pauper' story with a political theme not specifically Jewish – had widened his reputation. It has been suggested that his work on the *Jewish Standard*, noticed by Jerome K. Jerome, led to this new journalistic commission (Udelson 1990, 72; Wohlgelernter 1964, 22). *Puck/Ariel* lasted about two years under Zangwill's stewardship. I have not been able to locate circulation figures, but *Willing's British and Irish Press Guide* shows its duration to have been fairly respectable in relation to that of the many other comic papers that proliferated at the end of the century. And while it was increasingly modelled on *Punch*, *Ariel* compares favourably with most of the other humour papers, as well. Its content, both satiric and serious, is entirely original (albeit at times derivative), distinguishing it from such papers as *Snap-Shots* that reproduced cartoons from the American press. Moreover, Zangwill went far beyond even James Henderson – his Fleet Street neighbour who published *Snap-Shots, Scraps, Funny Folks, Comic Cuts*, and others – in using *Ariel* to establish an editorial identity.

When Zangwill took over the editorship of *Puck* in March 1890, it was an eight-page newspaper with very little editorial content. Its pictorial emphasis was signalled by large colour front-page portraits of contemporary celebrities and three additional pages of colour illustrations and advertisements. The few columns that appeared, covering stage, sport, and so on, were signed by 'Oberon', 'Peaseblossom', and other pseudonyms drawn from *A Midsummer Night's Dream*. Zangwill did away with most of these character signatures, while retaining Robin Goodfellow as the author of 'Puck's Panorama', a regular page-length column on current affairs. The stage column was replaced by the full-page 'Puck at the Play', which seems to have been written by Zangwill himself, as were other ongoing columns such as 'Puck's Paper-knife' (literary) and 'Puck's Platform' (editorials/leaders). Recurrent sports columns and comic pieces were produced by other writers.

To make room for all this new content, and to underscore *Puck*'s new, more literary emphasis, Zangwill increased the size of the paper to twelve pages while decreasing the number of columns per page from four to three, making them wider and easier to read. At the same time, he decreased the number and importance of colour illustrations, eliminating colour cartoons and moving, and eventually discontinuing, the

'Puck's Portraits' series. Full-page political cartoons remained, one per number, in black and white, each accompanied by an explanatory poem (as in *Punch*) written most likely by Zangwill himself. The artist 'Crow' contributed a series of drawings and verse called 'Puck's Theatrical Alphabet'. This regular feature, placed in the middle of 'Puck's Panorama' and continuing in the *Ariel* format, meant in effect an additional full page of theatrical emphasis, in keeping with Zangwill's involvement in theatre circles at the time. Crow, along with Mark Zangwill (the editor's brother) and other artists, contributed smaller cartoons of modern life similar to those found in *Punch*. Crow's cartoons illustrating words better left unspoken, for example, are clearly imitations of du Maurier's series 'Things One Would Rather Have Expressed Differently', published in the same period.

The paper became even more *Punch*-like on 5 July 1890, when it changed its name to *Ariel, or the London Puck*.[3] It adopted *Punch*'s page size and layout, its volume-consecutive page-numbering system, and even the single square bracket separating the date from the other matter at the top of each page. But Zangwill created his own complex of recurring features. To 'Puck's Panorama' and 'Puck at the Play' were added 'Ariel's Album' (biographical sketches accompanied by small satiric portraits), 'Ariel's Arts and Letters Club', and 'Ariel's Arm Chair', which succeeded 'Puck's Platform'. The editorial identity established so carefully in Zangwill's *Puck* incarnation was thus carried forward and reinforced as Puck became Ariel. The name change is itself significant. Zangwill inherited the title *Puck*, and changed it to avoid confusion with a prominent American humour paper by that name. But although Puck and Ariel are both friendly spirits in Shakespeare plays, they are not identical. 'Ariel', however, is also a Hebrew name meaning 'lion of God', and may have been a subtle allusion to the editor's background. The cherubic winged toddler that represented Ariel on the paper's masthead and elsewhere never had quite the distinction of 'Mr. Punch'. Still, in the editorial voice behind *Puck* and *Ariel* Zangwill created a persona at least as compelling as that of his paper's rival.

The 'I. Zangwill' who became well-known to the reading public through *Ariel* and the *Jewish Standard* was an urbane individual of progressive cultural inclinations, an Ibsenite and insider in theatrical circles, an English Jew who (sometimes subtly, sometimes less so) insisted upon enlightening his non-Jewish readers on matters of Jewish concern while reminding his Jewish readership of his status in the larger social and literary milieu. At first glance the signs of the Englishman in Marshallik are easier to spot than those of the Jew in *Ariel*. Commentary on current social and cultural issues including women's rights and colonial politics demonstrated to the *Jewish Standard*'s readers that Zangwill was no narrow sectarian. On 7 March 1890 Marshallik satirized the 'causes' of 1890s reformers:

> Hebrew literature in England is going ahead. The following lectures are promised at the Jews' College Literary Society: 'Water as a Shocker', by Anacreon Jones; 'Tobacco as a Tainter', by Father O'Flamingo; 'Stays as a Squeezer', by Mrs. Lucelaste; 'Skirts as a Divider', by the Countess of Babbleton; 'Meat as a Malady', by Dr. Howlingson; 'Music Halls as a Misfortune', by Mr. McDigall; 'Dancing as Devilry', by Canon Crutches; 'Marriage as a Mania', by Mrs. Nobody Cared.
>
> (Marshallik 1890, 9)[4]

Marshallik's note on the Anglo-German Agreement of late June 1890 demonstrated his awareness of colonial matters. The agreement, which received major coverage in the British press, including lead cartoons on 28 June in *Punch* and *Funny Folks* as well as in *Puck*, involved the cession of Heligoland, a small rocky island in the North Sea, to Germany in exchange for large territories in East Africa. All three cartoons expressed hesitancy over whether Britain had really got the best of the bargain. Zangwill's *Puck* additionally pointed to the racism inherent in the scramble for Africa.[5] Marshallik, however, on 27 June, saw the agreement as a chance to lampoon inter-communal rivalries among his Jewish readers: '"The Jews of Heligoland" is a fine topical title; but my imagination is rather fagged, and I don't feel equal to writing the article. I wonder if there are any Sephardim on the island [Jews of Iberian descent, who were known for their feelings of superiority]; and if so, how they like the prospect of becoming German Jews' (Marshallik 1890a, 10). Marshallik similarly found a Jewish angle in the 'Stanley craze', informing readers of the *Standard* that although 'Stanley has not yet been shown to be a Jew.... the first journalist to interview the explorer in Paris was a Jew named Jacques St. Cere' (Marshallik 1890f, 10). Each of these examples represents a larger number of similar instances.

Finally, Marshallik's literary parodies revealed him to be conversant with historic English forms as well as with current controversies. 'The Ballad of ye Tearfulle Bryde', one of several poems in an ersatz Middle English (Marshallik 1888a, 10), dealt with the breach of promise issue that seemed to consume the British public – an issue that of course received comment in *Ariel* as well. Both periodicals featured parodies of Gilbert and Sullivan among frequent examples of satiric verse, and, indeed, Zangwill's interest and involvement in the theatre served as a bridge between 'English' and 'Jewish' content in *Puck/Ariel* and the *Jewish Standard*. Zangwill as Marshallik emphasized the prevalence of Jews in the London theatrical world: 'When good Americans die, they go to Paris', he wrote. 'On the same principle, I fancy that when good Jews die they go to the play' (Marshallik 1890g, 8).

Zangwill frequently commented in 'Morour and Charouseth' on productions he reviewed in 'Puck at the Play', although his *Puck/Ariel* persona was at times more reticent than Marshallik in introducing Jewish issues. In the case of Henry Arthur Jones's *Judah* (1890), for example, he criticized the play severely in both columns, though only in the *Jewish Standard* did he discuss its protagonist's supposedly Jewish background (Marshallik 1890d, 10; 1890c, 11; Anon. *Puck* 1890a, 11). When the theatrical columns of *Ariel* put forth a consistently pro-Ibsenite stance, Marshallik, too, promoted J. T. Grein's production of *Ghosts* – adding that 'Mr. Grein, the indefatigable organiser, also claims kinship with the Semite, so that Jews are found playing the same part as they did in the Middle Ages, of conductors for the diffusion of ideas from one country to another' (Marshallik 1891a, 10). The 'Ariel's Arm Chair' devoted to Grein did not mention his Jewish background at all (Anon. *Ariel* 1890c, 354).

On another occasion, reviewing Arthur Wing Pinero's *The Cabinet Minister*, Zangwill used both of his platforms to criticize the depiction of a Jewish money-lender. In this case he was more forthright in the persona of Puck, while as Marshallik he appeared peculiarly ambivalent. To his Jewish readers he began almost apologetically, noting that '[u]npleasant as are the features of the portrait so cleverly executed by Mr. Weedon Grossmith [the actor], their general accuracy can scarcely be impugned.... As the "Judaism" of this delightfully humourous creation is nowhere obtruded, it is impossible for Jews to complain that they are being unfairly treated' (Marshallik 1890e, 10). But while Zangwill was careful to avoid accusing Pinero of anti-Semitism, he questioned the realism of the character's development with an argument based on Jewish social habits: 'Joseph Lebanon in his own circles may have been the man Mr. Pinero and Mr. Grossmith represent him, but would he not have adapted himself more to the new surroundings when he was thrown into refined company? The adaptiveness of Jews is one of their most marvellous traits'. (Marshallik 1890e, 10). In *Puck*, the play's realism was similarly questioned, with non-Jewish characters taking their share of criticism. 'Mr. Weedon Grossmith's Joseph Lebanon, money-lender and cad, is a rich comic creation', wrote the reviewer (Anon. *Puck* 1890c, 10). Still, 'one would like to know... Whether Mrs. Gaylustre [the money-lender's aristocratic sister] was likely to have such a brother as Mr. Lebanon? . . . Whether the behaviour of the aristocrats to Mr. Lebanon is not as caddish as his own?' (Anon. *Puck* 1890c, 10).

The 'general assumption that money-lender and Jew are convertible terms' (alluded to in a notice of Sidney Grundy's play 'A Pair of Spectacles' in 'Morour and Charouseth', (Marshallik 1890b, 9) was a particular cause of dismay to Zangwill, who attacked it more directly in a theatrical parody in *Ariel* on 18 October. The paper printed what was said to be a scene from 'The Sixth Commandment: or, Beneficent Murder of his Own Reputation. By Robert Buchanan'. It included the character 'Father Abramoff', who states: 'I am a stage Jew. As I happen to be really a Jew I know how unreal my part is. But no matter. As all the world knows, I was born a money-lender' ('Robert Buchanan' 1890, 244). On 23 and 24 January 1891, Zangwill again took up the 'stage Jew' issue in both publications. 'I wonder how much more "the Jews" will be synonymous with "usurers" ', remarked Marshallik in reference to the play *Joan of Arc* – which, he added, was 'written . . . by a Professor of History at Cambridge, who should be above perpetuating the traditions of prejudice. If anyone wants to do a good turn for the Jews, let him simply publish a complete list of modern money-lenders. It will be found that Jews have anything but a monopoly' (Marshallik 1891b, n.p.). Zangwill briefly mentioned a more positive view presented in another play, *A Note of Hand*, which – as he commented the next day in *Ariel* – 'actually utilises a Jewish money-lender as the benevolent *deus ex machinâ*. This is a welcome change' (Anon. *Ariel* 1891j, 61). Thus theatrical topics provided one way for Zangwill to refute anti-semitism as the editor of *Ariel*.

Zangwill also spoke out (albeit often circumspectly) on current issues concerning the Jews. British readers were well aware of the persecution of Jews in Russia during the 1880s and 1890s. In 1882, an anonymous two-part article in the *Times* (later attributed to the Jewish man of letters, Joseph Jacobs [Efron 1994, 62–3]), had publicized widespread pogroms against Russian Jews, as well as occupational and residential restrictions, emphasizing the complicity of government officials. In the early 1890s the persecutions continued as a cause for public outrage in the mainstream British press. Indeed, it was *Punch* and not *Puck* that between 1890 and 1892 had the larger number of lead cartoons on the persecutions in Russia. Still, *Puck* and *Ariel* published two lead cartoons and three commentaries on the Russian persecutions between April 1890 and January 1892, and a powerful lead cartoon on the persecution of Jews in Corfu in May 1891.

Ariel's editorial comments on Russia (at least one of which was written by Zangwill himself) carefully maintain the authorial persona of an English Christian, as do the rather distanced cartoons and poems. The earliest of the leaders, a paragraph on 'Russia' in 'Puck's Platform' (Anon. *Puck* 1890f, 6), commended Russian university students for demanding that Jews be admitted on an equal basis. While warning against violence, the writer added that 'short of assassination there is no specific, as there is no language, too strong to apply to the internal condition of Russia'. In December 'Ariel's Arm Chair' was devoted to 'Unholy Russia', and here Zangwill's style is evident. 'The people who begot CHRIST has been the CHRIST of peoples', he wrote, using a phrase that would frequently reappear in his writings (Anon. *Ariel* 1890a, 386). It is significant that this column was inspired by a multi-denominational protest meeting held at the Guildhall earlier in the week. Unable to claim a Christian identity, and unwilling in his role as editor to assert a Jewish one, Zangwill could assume authority as part of a multi-cultural community gathered 'a few days before CHRISTMAS to protest against the methods of the nation which claims to possess the only orthodox form of Christianity'. He made sure to add, perhaps optimistically, that 'England has now got rid of her own Judæophobia' (Anon. *Ariel* 1890a, 386).

On June 6 1891, 'Robin Goodfellow' raised the issue of emigration to Palestine as a solution for Russian Jews. Frank about English resistance to increased Jewish immigration at home, he put the question, 'Why should not England arrange with Baron Hirsch and other wealthy Israelites to settle them in Palestine, there to form a State under British protection? I am well aware that the Sultan would want squaring; but England can surely get round him, which is the same thing' ('Robin Goodfellow' 1891c, 362). Altruistic reasons for English involvement ('because England is always in the van, or should be, of any humanitarian movement') were bolstered by appeals to political and military self-interest.

The two cartoons on Russia appeared later in 1891 and 1892, and both were by George Hutchinson, who would go on to illustrate *The King of Schnorrers* and other stories by Zangwill. The first specifically takes up the subject of Jewish resettlement. Captioned 'The Modern Moses', it depicts Baron de Hirsch, the financier of the Jewish nationalist movement, in English businessman's dress carrying a staff labelled £20,000,000. He is followed by two downtrodden bearded men in Russian dress carrying a banner marked 'Jewish Colonization Company Unlimited', and behind them march masses of working-class men, women, and children in a docile yet determined procession (Hutchinson 1891b, Figure 14). The cartoon is a two-page foldout, one of the innovations of *Ariel*'s third and more literary phase. Its title, 'The Modern Moses', may have been influenced by Sir John Tenniel's depiction in *Punch* (Tenniel 1890, [67]) of the Tsar as a new Pharoah (Figure 15).

As the accompanying poem makes clear, however, the Jews following Hirsch are on their way not to Palestine but to New Jersey, as part of one of the more pragmatic resettlement plans offered in that era. Its first two stanzas reinforce the ambivalent stereotyping in the illustration, emphasizing the role of the Jewish financier in bringing about a material redemption ('See the moneyed modern MOSES/Leading files of Hebrew

Figure 14 George Hutchinson 'The Modern Moses', cartoon. *Ariel*, 26 September 1891.

Figure 15 John Tenniel. 'From the Nile to the Neva', cartoon. *Punch*, 9 August 1890.

noses, /From the Russian PHARAOH's hand, /Onward to a Promised land!' and so on). While the poem ends on a note of support, *Ariel* at this time appears to have been more reticent than *Punch* as an organ of Jewish advocacy. Its clinging to shreds of anti-Semitic stereotype may have been a sign of Zangwill's insecurity as a Jew in the public eye.

Figure 16 George Hutchinson. 'Janus: or, a look a head', cartoon. *Ariel,* 5 January 1891.

Indeed, in the first number for 1891, Zangwill had published Hutchinson's foldout illustration of the 'hopes and fears' for the new year; among the 'fears' is the figure of a Jew, fist held tightly around a bag of gold (figure 16). A few months later, in what appears to be a set of comic captions for works by well-known artists, one reads: 'The Jewish Doctor: "I wonder what they want for that Chippendale chair"' (Anon. *Ariel* 1891f, 328). So it is not entirely surprising that *Ariel*'s second cartoon about Russia concerned specifically the Tsar's denial of famine conditions, and that the accompanying poem only briefly and incidentally mentioned anti-Jewish persecution (Hutchinson 1892b, 73, 67). However, Zangwill's Jewishness found means of expression in political commentary both before and after the examples noted above. The editorial cartoon about persecutions of Jews in Corfu, for example, appeared in *Ariel* just a week after the doctor joke (Anon. *Ariel* 1891e, 345; figure 17). It employed a wolf and lamb motif that *Punch* had used in reference to Russia after the Guildhall protest meeting (Anon. *Punch* 1890b, 290). The attached poem in *Ariel* forcefully reminded English readers to feel some responsibility toward Jews on an island they had lately governed. With similar power, an article titled 'Puck Interviews the Jew Baiter' gave satiric justice to Stöcker, a notorious anti-Semite who had recently been profiled in the *Pall Mall Gazette* (Anon. *Puck* 1890g, 4).

Zangwill used 'Ariel's Album' to introduce readers to Baron de Hirsch (Anon. *Ariel* 1891g, 324) and Sir Henry Aaron Isaacs, Lord Mayor of London (Anon. *Ariel* 1890b, 307). While the profiles and illustrations were satiric, they commented directly and favourably on their subjects' Jewish background. The column on Hirsch was an occasion to comment wryly on the boundaries of British toleration: 'In this happy England of ours locks, bolts, and bars fly asunder at the touch of a millionaire'. Still, Zangwill (Anon. *Ariel* 1891g, 324) concluded that '[t]here are many worse things that Baron de Hirsch might do with his millions than devote them to succouring the poor and oppressed of the race to which he belongs' (Anon. *Ariel* 1891g, 324). A leader on the sweating system attacked those who called for an end to (primarily Jewish) immigration in the service of keeping wages up. Asserting that '[i]t is the immigrants from the vast and shadowy regions of the Potential – the thousands who are weekly brought into the world by [English] parents incompetent to provide for their future.... who crowd the labour markets' (Anon. *Puck* 1890b, 2), Zangwill argued against immigration restrictions and at the same time reflected the ambivalence of many liberal-minded established Jews reluctant to condemn a system in which Jews gave other Jews employment (see Black 1988, 39–40).

Figure 17 George Hutchinson. 'The Wolf and the Pascal Lamb or Aesop Up to date', cartoon. *Ariel*, 30 May 1891.

Zangwill occasionally used his editorship of *Puck* and *Ariel* to educate his readers on Jewish culture. An essay on 'Easter Eggs' (Anon. *Puck* 1890e, 9) commented extensively on their origins in the Jewish Passover commemoration, and then went on to discuss the uses of eggs in other Jewish ceremonies. Smaller references, too, reminded readers of Zangwill's Jewishness; in May 1891 a poem, 'In Old Jerusalem' (Anon. *Ariel* 1891i, 309), celebrated the bazaar held to raise funds for the new Hampstead Synagogue – an event that surely did not have major coverage outside the Anglo-Jewish press.[6] In his role as a prominent Jewish author, Zangwill had contributed a story (though, perhaps characteristically, not a 'Jewish' one) to the bazaar's commemorative programme. Books by Joseph Jacobs were promoted in *Ariel*, and 'Mrs. Danby Kaufmann of Bayswater' – a now quite unknown contribution to an Anglo-Jewish literary controversy – was criticized, just as both were promoted or criticized in the *Jewish Standard*.[7] Thus Jewish issues were naturalized in *Puck* and *Ariel* for a presumably non-Jewish audience.

Of course, 'presumably' is an important word here. *Ariel* was advertised in the *Jewish Standard*, and it is reasonable to assume that many Jewish readers purchased the paper precisely because it was Zangwill's. Such readers would have been curious to see whether (and if so, how) Zangwill presented himself as a Jew to those outside the community. They would most likely have been pleased at his references to Jewish affairs, while not expecting him to go overboard in announcing his ethnicity. Thus Zangwill's insertion of Jewish content into *Puck* and *Ariel* may have been intended to gratify his Jewish readership as well as to educate non-Jews in the audience. Indeed, the image of a Jewish editor trying to inspire sympathetic views of Judaism in the general public would in itself have ingratiated Zangwill to his Jewish readership. Thus in his management of *Puck* and *Ariel* altruism and self-interest were mutually reinforcing motivators, as Zangwill ultimately obtained goodwill for his people and for himself.

In the end, if the readers of the *Jewish Standard* saw Israel Zangwill as a jocular and urbane Englishman and 'co-religionist' (even if he did pretend to greater orthodoxy than he observed), the readers of *Ariel* found in him a playful and therefore non-threatening critic in the vanguard of public opinion, and a Jew who stood by his people, while not above a joke at their expense now and then. The combination was effective, and after a few years Zangwill was ready to move on. In June of 1891 both *Ariel* and the *Jewish Standard* carried notices of Zangwill's book of comic fiction *The Bachelors' Club*. By this point, too, the character of *Ariel* was changing. Expanded to sixteen pages, it was increasingly devoted to serializations of comic works (including Zangwill's own *Old Maids' Club*), with the ponderous and abstract cartoons of 'Cynicus' for a while replacing the more topical lead drawings. It had become, perhaps, a more literary paper, but also a much less lively one. Changes in format suggest the paper was under pressure to increase circulation; for whatever reason, it ceased publication abruptly in February 1892.

Zangwill went on to become a frequent contributor to the *Idler*, in which he published a major Jewish work of fiction, *The King of Schnorrers* (1894). Throughout the rest of his career he wrote both 'Jewish' and 'general' works, insisting on the universal relevance of both and infusing both with Jewish values. In an age that valued novelty and the exotic, Zangwill's success may have been aided rather than impeded by his Jewish literary subject matter. However, as recent scholarship has shown, social prejudice against the Jew during the same period often set Jewishness and Englishness in opposition to each other.[8] Zangwill as an author rejected both ghettoization and literary assimilation. By publishing simultaneously in Jewish and general interest papers, and by writing in both on a great variety of subjects, Zangwill insisted on the normality of Jewish experience in English life.

Notes

1. The distinction between the two has often been represented in the terms 'Jewish' and 'secular'. But although these terms are convenient they imply an inaccurate dichotomy, since even Zangwill's writing on Jewish subjects is generally not 'religious' literature. To distinguish simply between 'Jewish' and 'non-Jewish', however, would also be inaccurate, as British literary circles in Zangwill's day included many Jews, and (as my references to *Punch* [below] indicate), Jewish content appeared in publications not specifically Jewish. I have thus attempted to make clear what is in fact a genuine distinction, while using terminology that does not obscure more subtle areas of overlap.
2. Although Winehouse notes that no copies of *Purim* are extant, the British Library holds the volume for 1883. It is edited anonymously.
3. The apposition may have first appeared in a later number; in the one page I have reprinted from 5 July, the page masthead reads only 'Ariel'.
4. A number of these titles are surprising coming from a future feminist, but Zangwill in those early years was rarely above selling out a principle to please a crowd.
5. The caption on *Puck*'s cover cartoon of 28 June 1890 reads 'Exchange is No (?) Robbery'. The poem of the same name connected with the picture ends as follows:

   So rule the Fates; in Life's chess-game,
     White triumphs over black.
   Destruction, like a forest flame,
     Pursues the negro's track;
   And yet, perhaps, not wholly gain
     That he should cease to be.
   His life cannot be wholly pain
     And 'immorality'.

   Through palls of fear and ignorance
     Some gleams of sunlight creep –
   The joys of love, the chase, the dance,
     Wild battle, rest, and sleep.
   Are Europe's realms so rich in joy,
     That we to end these thirst?
   Were it not better to destroy
     Our own barbarians first?

   (Anon. *Puck* 28 June 1890, 2)
6. I am grateful to Mr Harry S. Ward of London for giving me a reproduction of the original bazaar programme, which includes Zangwill's story of the late Victorian journalism scene, 'An Honest Log-Roller'.
7. On Jacobs, see *Ariel* 25 Oct. 1890, 270 and 16 May 1891, 319; also the *Jewish Standard* 5 June 1891, 4. On 'Mrs. Danby Kaufmann...' see *Puck* 19 April 1890, 4; *Jewish Standard* 28 Feb. 1890, 9.
8. On the interrelations of Jewishness and Englishness in nineteenth-century culture, see, for example, the works cited by Cheyette, Ragussis, and Kushner.

Works cited

Anon. 'Blind Authority' [poem], *Ariel* (30 Jan. 1892a), 67.
Anon. 'The Modern Moses' [poem], *Ariel* (26 Sept. 1891a), 195.
Anon. 'Reviews', *Jewish Standard* (5 June 1891d), 4.
Anon. 'The Wolf and the Paschal Lamb or AEsop Up to Date' [illus], *Ariel* (30 May 1891e), 345.
Anon. [artist's signature unclear]. 'Ariel's Academy', *Ariel* (23 May 1891f), 328.
Anon. 'Ariel's Arts and Letters Club', *Ariel* (23 May 1891g), 324.
Anon. 'Ariel's Arts and Letters Club', *Ariel* (16 May 1891h), 319.
Anon. 'In Old Jerusalem' [poem], *Ariel* (16 May 1891i), 309.
Anon. 'Puck at the Play', *Ariel* (24 Jan 1891j), 61.
Anon. 'Ariel's Arm Chair: Unholy Russia', *Ariel* (20 Dec 1890a), 386.
Anon. 'The Russian Wolf and the Hebrew Lamb (After a well-known Picture)' [illus], *Punch* (20 December 1890b), 290.
Anon. 'Ariel's Arm Chair', *Ariel* (6 Dec. 1890c), 354.
Anon. 'Ariel's Album: Sir Henry Aaron Isaacs', *Ariel* (15 Nov. 1890d), 307.
Anon. [Untitled Note], *Ariel* (25 Oct 1890e), 270.
Anon. 'Puck at the Play', *Puck* (31 May 1890a), 11.
Anon. 'Puck's Platform', *Puck* (17 May 1890b), 2.
Anon. 'Puck at the Play', *Puck* (3 May 1890c), 10.
Anon. 'Puck's Paper-knife: The Gentle Reader and the Captious Critic', *Puck* (19 April 1890d), 4.
Anon. 'Easter Eggs', *Puck* (12 April 1890e), 9.
Anon. 'Puck's Platform: Russia', *Puck* (5 April 1890f), 6.
Anon. 'Puck Interviews the Jew Baiter', *Puck* (5 April 1890g), 4.
'Baroness von S.' [Israel Zangwill]. 'Diary of a Meshumad', *The Jewish Calendar, Manual, and Diary 5651*/1890–91, 54–76 [collected in *Ghetto Tragedies*, 1893 and *'They That Walk in Darkness'*, 1899].
——'Satan Mekatrig: A Romance of the London Ghetto', *The Jewish Calendar, Manual, and Diary 5650*/1889–90, 52–76 [collected in *Ghetto Tragedies*, 1893 and *'They That Walk in Darkness'*, 1899].
Bell, J. Freeman [Israel Zangwill]. 'Under Sentence of Marriage', *The Jewish Calendar, Manual, and Diary 5649*/1888–9, 54–79.
Black, Eugene C. *The Social Politics of Anglo-Jewry, 1880–1920*. Oxford, 1988.
Cheyette, Bryan. *Constructions of 'the Jew' in English Literature and Society, 1875–1945*. Cambridge, 1993.
Efron, John M. *Defenders of the Race: Jewish Doctors and Race Science in Fin-de-Siècle Europe*. New Haven, 1994.
Hutchinson, George. 'In Darkest Russia' [illus], Ariel (30 Jan. 1892b), 73.
——'The Modern Moses' [illus] *Ariel* (26 Sept. 1891b), n.p. [foldout].
——'Janus: Or, A Look a Head: The God of January's Hopes and Fears for 1891' [illus], *Ariel* (3 Jan. 1891k), 8–9
Kushner, Tony, ed. *The Jewish Heritage in British History: Englishness and Jewishness*. London, 1992.
'Marshallik' [Israel Zangwill]. 'Morour and Charouseth', the *Jewish Standard* (13 Feb. 1891a), 10; (23 Jan. 1891b), Supplement, n.p.; (27 June 1890a), 10; (30 May 1890b) 9; (23 May 1890c), 11; (16 May 1890d), 10; 2 May 1890e), 10; (25 April

1890f), 10; (18 April 1890g), 8; (7 March 1890h) 9; (28 Feb. 1890i) 9; (8 June 1888a), 10; (11 March 1888b) 6.
Ragussis, Michael. *Figures of Conversion: 'The Jewish Question' and English National Identity*, Durham, North Carolina, 1995.
'Robert Buchanan'. 'The Sixth Commandment: or, Beneficent Murder of his Own Reputation', *Ariel* (18 Oct. 1890), 244
'Robin Goodfellow'. 'Puck's Panorama', *Ariel* (6 June 1891c), 362.
Tenniel, John. 'From the Nile to the Neva' [illus], *Punch* (9 Aug. 1890), 67.
'The Tree of Knowledge' [symposium]. *New Review* 10 (1894) 675–90.
Udelson, Joseph H. *Dreamer of the Ghetto: The Life and Works of Israel Zangwill.* Tuscaloosa, AL, 1990.
Winehouse, Bernard, 'The Literary Career of Israel Zangwill from its Beginnings Until 1898', Unpublished PhD dissertation, University of London, 1970.
Wohlgelernter, Maurice. *Israel Zangwill: A Study.* New York, 1964.
Zangwill, Israel, 'An Honest Log-Roller', *The King of Schnorrers: Grotesques and Fantasies*. London and New York, 1894, 171–5.
——'English Judaism: A Criticism and a Classification', *Jewish Quarterly Review* 1 (1889), 376–407.
——'Professor Grimmer', *Society* (16 Nov. 1881), 10–12; (23 Nov. 1881), 8, 10–11.

Part IV

Negotiating Gender

13

America's First Feminist Magazine: Transforming the Popular to the Political

Amy Beth Aronson

A contribution to America's first feminist magazine, *The Lily* (1849–58), captures a critical moment in the evolution of woman's media image in the nineteenth century, a moment made possible by the magazine itself. The writer, Jane Frohock, proclaims: 'It is woman's womanhood, her instinctive femininity, her highest morality, that society now needs to counteract the excess of masculinity that is everywhere to be found in our unjust and unequal laws' (Frohock, 1856, 150).

The contributor uses the ideological touchstones of domesticity, the popular gender discourse of her day, to pursue feminist politics. The idea of woman's instinctive 'morality' is a cardinal claim, and she moves, phrase by phrase, through another prevailing idea: 'femininity' is seen as the necessary corrective to male dominance of the public, law-making domain. Frohock's sentiments read like textbook domestic discourse except for their direct object; here, 'womanhood' explicitly targets the law. Two conflicting gender discourses – 'feminine' and feminist – cooperate rather than compete. They form a sequence that both propels and sustains the writer's call for action.

The discursive development on which this transformation depends was promoted by *The Lily's* formal qualities and periodical practices, and the moment in press history in which the magazine was produced. From its origins in colonial Philadelphia, the American magazine was a dynamic environment, in which a range of generic contents and contributing readers engaged in dialogue. The magazine functioned in cultural practice as a forum, providing a space for rhetorical and ideological exchange through which new language, new images, new stories can – and did – emerge.

The Lily exemplifies this process, but with a difference. It worked with and within a gender agenda, to evolve new ways of speaking and writing about women. The magazine hosted repeated confrontations with the popular gender discourse of the women's presses of the day – with sentimental-domesticity's image of woman's innate moral purity, virtue, and 'influence' (Welter 1966, 151–74); with the corresponding vision of social organization, 'the woman's sphere'; and with formula of the sentimental-domestic or 'woman's' novel. Issue after issue, such dominant gender discourse was deployed in various formats and debated in the various departments of the magazine, a process of persistent articulation and reflection, repetition and recycling, that encouraged discursive revisions over time. Miscellaneous content, periodical publication, and the serial styles of women's magazine reading and writing all helped such new meanings to emerge. Finally, other conditions of the periodical industry, such as the practice of clipping copy, further assisted in the process of transformation. Together, these periodical pressures and practices challenged the boundaries of popular"femininity,' pushing it into new realms of female empowerment.

Gender politics and media history

The first generation of American feminist magazines was a group of six national players. In addition to *The Lily,* there was the intellectual and policy-oriented *Una* (1853–5), a 16-page monthly edited by Paulina Wright Davis of Providence, Rhode Island. *The Pioneer and Women's Advocate* (1852–3), Anna W. Spencer's four-page, semi-monthly, was also published near Providence. *The Genius of Liberty* (1851–3), an eight-page monthly edited by Elizabeth Aldrich of Cincinnati, Ohio, was the only one of the group to refuse advertising on principle. *The Woman's Advocate* (Jan. 1855–8 or 1860), Anne McDowell's Philadelphia monthly, was written, edited, printed, and sold exclusively by women. And *The Sibyl* (1856–64), edited by Dr Lydia Sayer Hasbrouk of Middletown, New York, began as a dress reform magazine, but, like *The Lily,* soon expanded its scope beyond that single-issue focus.

The Lily, the most successful and longest-lived of the first feminist magazines, was an eight-page monthly edited until 1855 by Amelia Bloomer, mostly in Seneca Falls, New York, the birthplace of American feminism. Launched in January of 1849 carrying the tag line 'Devoted to Temperance and Literature,' the magazine was at first supported by the newly-reorganized Female Temperance Society based in the town. Anna C. Mattison, a local poet, was co-editor at the outset, but retired in April of 1849. By the early 1850s, Bloomer had discovered that the issue of temperance led to a broader debate about women's rights; she was publishing increasingly radical articles by outspoken women writers, such as the first published work of Elizabeth Cady Stanton (in April 1850), and three illustrated editorials endorsing the 'Bloomer costume,' soon labelled in ridicule of the editor (*The Lily*, July 1851, 53; Sept. 1851, 66; Sept. 1851, 69). The content was too much for the Female Temperance Society. As of April 1852, Bloomer was editing and publishing the magazine on her own. Now carrying the motto 'Devoted to the Interests of Women', *The Lily* ran a range of contributions – prose, poetry, short stories, letters, reviews, and various kinds of fragments – concerning women's rights issues, including questions of poverty, women's work, domestic tyranny, dress reform, and even suffrage (Marzolf 1977, 222–3; Russo and Kramarae 1991, 11–12). A subscription still sold for 50 cents per year, and that price remained even after *The Lily* began publishing bi-weekly in 1853 and after it resorted to semi-monthly publication in 1855. By that time, Bloomer had sold the magazine to a Richmond, Indiana, literary lady, Mary B. Birdsall, a former editor of the Ladies' Department of *The Indiana Farmer* (1851–61). Although Bloomer stayed on as Corresponding Editor through December 1856, *The Lily* became somewhat more moderate in its voices and views under Birdsall's leadership during the last three years of its life.

The first feminist magazines were part of a small, but recognizable tradition of reform journalism in the United States (Marzolf 1977, 220–21). By definition, the women editors and writers were unconventional in their professional status and political visibility. Although the editors of the feminist magazines were movement leaders, not professional journalists (Marzolf 1977, 220), several went on to careers in the industry, becoming part of the rising tide of women involved in the American press. Indeed, editors and writers of the feminist magazines worked in dialogue with popular women's magazine editors, often making explicit the politics implied by the presence of women working in the media industries.

Still, many scholars of women's writing, media, and journalism have analysed feminist periodicals mainly in terms of their interplay with the 'masculine', mainstream press of the day. In their collection, *The Radical Women's Press of the 1850s* (1991), for example, journalism historians, Ann Russo and Cheris Karamarae, explain that these feminist magazines 'publicized a broad range of women's issues and protests at a time when many men editors throughout the country were transmitting messages hostile to women's rights' (Russo and Kramarae 1991, 2). Similarly, Martha Solomon, in her edited collection, *A Voice of Their Own: The Woman Suffrage Press, 1840–1910* (1991) and Maurine

Beasley and Sheila Gibbons in theirs, *Taking Their Place: A Documentary History of Women and Journalism* (1993), point out the relationship of feminist periodicals to the male-dominated mainstream press. The oppositional status they register – where feminist magazines foster critical response to images of women created by men – is vital to an understanding of feminist publications and press politics, yet it remains only part of the story. The American feminist magazines also organized effective 'internal' protests, engaging in critical dialogue with the gender discourse and dominant formulas popularized by the 'lady editors' and writers of the era. And they utilized many mainstream conventions to accomplish their goals. Capitalizing on the formal qualities of the American magazine, the discursive strategies popularized by commercial 'ladies magazines,' and various industry conventions, the first feminist magazines helped women create and circulate progressive gender images.

The American magazine was a particularly conducive form for such a process. Although the first American magazines were inspired by Britain's popular *Gentleman's Journal* (launched 1692), *Tatler* (begun 1709), *Spectator* (1711) and *The Gentlemen's Magazine* (1731), they were also distinctly keyed to the democratizing culture in which they were produced and consumed. From their emergence in 1741, simultaneously from rival Philadelphia printers Andrew Bradford and Benjamin Franklin, American magazines were more miscellaneous, multi-vocal, and participatory than their British forbears. The *Tatler* had its 'Isaac Bickerstaff' and the *Spectator* its 'Mr. Spectator' personae to dramatize comment and organize contents; London's *Gentleman's Journal* was comprised of sequential entries linked by carefully composed editorial transitions (Foster 1917, 32).

By contrast, colonial, Early Republican, and antebellum magazines presented contributed content, typically written by amateur writers, in little apparent order on the page. As a result, American magazines were marked by polyvocality and by discursive competition between adjacent items.

Although American magazines as a group would not begin to become profitable until about the mid nineteenth century, from the first they did succeed at promoting ideological and verbal commerce among the often conflicting ideas, norms, and beliefs of its audiences.

This dynamic environment and ongoing exchange offered special advantages to women. The relatively free play of ideas and discourses promoted the innovation and revision of language. As communities of women readers together wrote and revised magazine content, they collectively negotiated new language or new meanings, and the magazine made them available for public use.

The first feminist magazines seized the opportunities of the medium – and the moment. They emerged amid the so-called 'feminine fifties,' when the popularity of the sentimental-domestic or 'woman's' novel was peaking in the US, and commercial 'ladies' magazines' were likewise attaining substantial public notice. From the rise of American women's magazines with the *Ladies' Magazine and Repository of Entertaining and Instructive Knowledge* (1792–3) to 1850, at least 110 periodicals for women were launched in cities and towns across the United States (Stearns 1930, 248; Garnsey 1954, 82). The three publishing centres at the time produced the majority of these periodicals, with New England accounting for 38, New York for 20, and Philadelphia for 24; however, the South produced at least 14 and the emerging West accounted for at least 16 (Garnsey 1954, 82).

The best-selling women's magazines of the nineteenth century – including *Godey's Lady's Book* (1830–98) published in Philadelphia by Louis Godey; *The Peterson Magazine* (1842–98), Charles J. Peterson's monthly published out of Philadelphia and later New York; and *Arthur's (Home) Magazine* (1852–98), popular novelist T. S. Arthur's Philadelphia monthly – became commercial powerhouses in this era. They boasted circulations of 40,000 by the time *The Lily* appeared, and grew to over 150,000 through the antebellum period (Okker 1995, 56). And that was just the beginning. By 1869, *Godey's* circulation had more than tripled, reaching some 500,000 readers. *Peterson's* circulation came closest to *Godey's* in the 1850s, probably exceeding it by the Civil War (Okker 1995, 56–7). Although these popular magazines were male-dominated on the business side, women controlled editorial matters. In fact, Patricia Okker has recently identified more than 600 women editors on nineteenth-century mastheads, many of whom first became active in the 1850s (Okker 1995, 167–220). The American feminist magazine was born in what might be called the 'Decade of Literary Ladydom On the Move'.

At its height in 1855, *The Lily* achieved a circulation of about 6000 subscribers – half the circulation of 12,000 averaged by all women's monthlies at mid-century, and nearly equal to the average circulation of all American monthly magazines at the time (Mott 1930, 41; Okker 1995, 56). Although the magazine struggled to survive mostly by subscription instead of advertising, the significance of *The Lily* did not lie in its numbers. The magazine was important in gender politics and media history for the ways it appropriated the popular gender discourse of the day and transformed it, producing new images and stories that were both visible to and viable in the public eye.

Discursive deployments: sentimental-domesticity and its uses in *The Lily*

Although scholars of women's media tend to highlight distinctions between feminist 'women' and sentimental-domestic 'ladies,' their competing ideas and images of women often came together in *The Lily* and her sister periodicals – and in the service of feminist goals. Both written and read mostly by the predominantly white, middle-class women who were reading domestic novels, *The Lily*'s pages were suffused with sentimental-domestic stories and rhetorical styles, but they were used in decidedly tactical ways. A visible and market-tested virtue of the sentimental voice was its capacity to arouse emotional engagement – which cried out for reader response. For *Lily* contributors, that response could be rallied to the cause of social change.

One early example of this strategic use of the sentimental is 'Maternal Influence,' contributed by Mary C. Vaughan, a president of the New York State Temperance Society and later editor of the *Woman's Temperance Paper* (1854–6). Vaughan herself experienced a change of heart and mind over the course of her interaction with the magazine. Over a two-year period of regular contribution, she was transformed from a firm believer in the gendered doctrine of separate spheres (Vaughan 1850, 94) to a commitment to dress reform (Vaughan June 1851, 43), suffrage (Nov. 1851, 91), divorce (Oct. 1851, 79), women's education (Oct. 1852, 85–6), and property rights (Oct. 1852, 85–6; see also Ginzberg 1990, 114–15).

Vaughan opens with a rhetorical question: 'a mother's influence – who can estimate its strength or its power?' (Vaughan 1851, 11) Then, she answers, and in sentimental style so heightened and intricate as to be syntactically dizzying: 'none may know in this life of the souls encouraged by maternal solicitude to heavenward aspirations; fitted by maternal prayers and examples, when death has dissolved the ties that bound them to earthly influences, to plume their flight beyond the stars, and join evermore in the grand employments of the better world' (ibid.).

Vaughan's highly emotionalized rhetoric, her spiralling structure, and her commingling of maternal success with heavenly rewards (and maternal failure with hellish punition) captures the cosmic dimensions and intensified emotionalism of the sentimental voice. But Vaughan does not deploy sentimentality with a straight face. She uses it expertly, exploiting its emotional value to rouse readers to the more earthly argument she goes on to make.

Following her evocation of 'grand employments,' she suddenly and abruptly shifts to more pragmatic prose. 'This train of thought was induced,' she announces, 'by hearing of the illness in the prison of our city of one whom I knew in other days and very different circumstances' (ibid.). 'Sometime since,' she continues, 'while under the influence of ardent spirits, he committed theft and was arrested and imprisoned' (ibid.). Vaughan then reports on the poor conditions and limited options that punish the wayward son.

But the important element of the article is not the son's story, but the writer's rhetorical strategy in telling it in print. The kind of stirring rhetoric Vaughan employs, readers are both told and shown, is a state of mind that one achieves, or is moved to, under certain kinds of compelling circumstances – circumstances that other *Lily* contributors will variously define. The sentimental is deployed to move female readers to the agitated state that an argument for action requires to hold sway.

Vaughan's strategic use of 'feminine' sentimentality becomes even clearer by her conclusion. Having written in an earth-bound, reportorial style about the lives of the mother's children, she then spirals up again, and intones: 'Oh! that mothers would take warning by such fatal examples' (ibid.). 'That they would ever remember the vast responsibility which rests upon them as they write upon the unsullied soul the characters that shall stand there at the Judgment Day, to determine the weal or woe of the beloved ones entrusted to their care' (ibid.).

Vaughan's rhetoric changes her tune as well as her trajectory. The concluding paragraph speaks more broadly than did the body of the piece, and it speaks to a wider audience as well. In the end, Vaughan speaks out, through *The Lily*, in a popular rhetoric designed for a mass audience somewhat beyond *The Lily*'s own. Her discrete final paragraph, like her lead, seems to be tacked into another kind of writing, and for the purpose of carrying woman-centered social commentary into the literary marketplace where women readers were attaining a presence and clout.

This kind of distinct concluding device – what might be called an 'authorial addendum' – appears several times in the first feminist magazines. It helped women writers accommodate the gender limitations placed upon their public speech, while enabling their politicized expression. Sometimes, as in Vaughan's case, the addendum helped to package 'un-lady-like' social and political ideas in an accredited, 'feminine' sentimental story. By 'bookending' her discussion of poverty, women's work, and the prison system with sentimental rhetoric at the beginning and the end, Vaughan can pass off unconventional copy as approved 'feminine' fare. In the process, of course, she helps change what kind of fare is legitimated for women to write.

In some other examples, 'authorial addenda' were used by women writers to convey direct address to their reading audience. Tacked onto sentimentalized, non-fiction articles, they enabled writers to politicize the implications of stories about the trials and tribulations of women's lives. Some of the 'direct authorial address'-type of addenda explicitly expanded beyond speaking only to feminist magazine readers, directing a one-paragraph plea or plan to 'men', to 'editors', or to reading leaders. Later, some authorial addenda also shifted from third-person narration into the first person, deploying the urgency and actuality of personal experience to refresh a formulation that might otherwise have become a deadened, useless cliché.

The 'authorial addendum' signals as it simultaneously creates a formal space to convey links between popular literature and life, between media and reality, all within the legitimatized and increasingly useful parameters of popular gender discourse. But early feminist magazine contributors weren't the first ones to discover its strategic value. The authorial addendum actually had a history in women's advocacy journalism going back nearly two decades, to the moral reform publications of the 1830s (see Ryan 1985, 77; and Ginzberg 1990, 113–15).

The device probably remained in print because it served two consistent needs of women writers interested in pursuing social reform. First, it connected popular stories and growing antebellum audiences of women to the public and political scene, helping to turn women readers into gendered constituencies. And, it enabled women writers to speak critically and politically without compromising their status and authority as 'ladies.' The authorial addendum helped early activist women to become political, speaking out and plugging into the immanent power of growing middle-class audiences, while also maintaining substantial, accepted sources of eloquence and power.

Another variation of the 'authorial addendum' appears in an 1 August 1856, article by another professional woman writer, Elizabeth Oakes Smith, a contributor to both feminist and popular magazines. This particular article, 'Out of Her Sphere,' was pirated from a sister feminist periodical, Anne McDowell's *Woman's Advocate*. In calling for woman suffrage – quite a radical plaint at this time – Oakes Smith brings the time-tested authorial addendum into contact with contemporaneous capacities of the antebellum literary marketplace. Her authorial addendum exploits a signal capability of mass media: she portrays an imagined reality as 'obvious,' making it so for readers. 'Women are to be found everywhere out of their sphere,' (Oakes Smith 1856, 111), she claims at the outset. Since consignment to the private realm is actually an illusion, Oakes Smith argues at length, votes for women only makes realistic sense.

Oakes Smith's sarcasm and overall polemicism throughout the body of the piece are starkly contrasted by the voice of her authorial addendum. She closes on a high sentimental note, promising that transcendent rewards will follow from women's enfranchisement. Indeed, Oakes Smith suggests that heaven-on-earth is but a hyphen away: 'Do away with this glaring inconsistency, and women have little more to ask – the eternal harmonies will roll on, human governments represent better human justice and divine love, and this senseless cry of "out of her sphere" be as little applicable to woman as to the planet which, morning or evening, lends its clear beautiful beams to gild the early or later day' (Oakes Smith 1856, 111).

Serialism, repetition and revisionism: periodical pressures and the 'woman's sphere'

This tactical deployment of the sentimental is but one strategy supported by *The Lily*'s polyvocal environment. Other, broader conditions of the magazine also worked to effect change in popular gender discourse.

The clipping of copy, a frequent industry practice through about mid-century, for example, held practical as well as political utility for *The Lily*. In America, pilfering copy had long helped to evolve and circulate emergent political ideals. In the 1740s, the concept of 'press liberty' itself was in part developed through pirating (Leder 1966, 1–16; Botein 1975, 126–225); arguments for the American Revolution were spread by pirating in the 1770s (Schlesinger 1958; Lund 1993, 16). In *The Lily*, copy was reprinted from a range of periodical types: literary magazines, newspapers, women's magazines, trade publications, books (particularly collections of essays of short stories), and other feminist or reform periodicals. Accordingly, reprinted articles were written by a range of writers, from very conservative sentimental-domestic authors, such as Mrs. Lydia Sigourney (Feb. 1850, [10–11]; Jan. 1851, 1; 15 Feb. 1855, 27); to more politically centrist woman novelists, such as Virginia Townsend (15 Sept. 1855, 133–5); to more radical or renowned women of the popular press, including Harriet Beecher Stowe (October 1850, 26), Jane Grey Swisshelm (Aug. 1851, 59; 15 Dec. 1854, 183), 'Fanny Fern,' and 'Grace

Greenwood' (whose *Lily* articles, both pirated and original, are too numerous to list). Since *The Lily* pirated by theme or sentiment rather than star power, amateur and anonymous contributions were clipped too. Although issues vary in their use of recycled copy – some contain exclusively original material, while others present 25 per cent reprinted content – overall, the dominant industry practice was used to serve some subversive ends throughout *The Lily*'s life.

Whether used as a pretext for supportive affirmation, reasoned rebuttal, or an occasion for outraged rebuke, the presence of reprinted copy helped put *The Lily* in the thick of things. It allowed a small-circulating, politically-marginal, out-of-town magazine to participate in the influential world of the nationalizing popular press. It allowed the magazine to bring in big-league news and feature copy and deal with it on editorially equal terms. Pirating also enhanced the polyvocality and breadth of the magazine's coverage. Plus, it lent the magazine an insider image, a sense of being 'in-the-know.' And, of course, pirating enabled *The Lily* to champion chosen ideas, perpetuating their public lives in print.

Clipping copy also helped sustain *The Lily*'s community of women readers. Patricia Okker has written that nineteenth-century women's magazines developed and deployed a 'sisterly editorial voice' that, 'with its gendered basis of authority and informal exchange between readers and editors, often affected not only the tone of the editorial columns but the very form of the periodical as well' (Okker 1995, 31). Linda Steiner has also noted the self-conscious use of 'sisterhood' in women's periodicals around this time, arguing, if too narrowly, that this idea characterized the women's suffrage press (Steiner 1983, 1–15; Steiner 1993, 38–65). Okker explains that sisterly connections were legitimated by popular gender ideas about woman's instincts for nurture and group support, and that this notion structured many mainstream women's magazines throughout the nineteenth century (Okker 1995, 22). In the earliest feminist magazines, 'sisterhood,' and the solidarity it expressed, affirmed women's participation even in a controversial enterprise – it strengthened PR – and therefore widened the magazine's potential for political impact.

The idea of 'sisterhood' is telegraphed by one article reprinted from Jane Grey Swisshelm's *Pittsburgh Saturday Visiter* (1848–57) and published in *The Lily*'s April 1850, issue (26–7). (Swisshelm insisted on Dr Johnson's spelling.) Entitled 'Woman's Sphere,' it was written by a professional journalist and feminist, Frances D. Gage – or as her readers knew her, 'Aunt Fanny.' To gainsay the predominant separate spheres model of the world, Gage calls upon a matriarchal line. She offers first the image of her grandmother, 'a small, delicate woman... [who] toil[ed] slowly but wearily by the precipitous and dangerous paths to the heights of the Alleghenies'; then of her mother, 'trudging by her side on foot, stick in hand, urging along the old cow'. Evoking women's ties and inheritances, Gage is able to mount a substantial challenge to the frailty that the 'woman's sphere' supposedly responded to – and reinscribed. 'Were they in there [*sic*] "sphere?"' Were they taking upon themselves 'duties assigned to them'? Or were they only trying to be "manish?"' (Gage 1850). Forcing prevailing media images – 'feminine' frailty and 'feminist' manliness – into direct conflict with one another, Gage suggests that both are foolish fictions, inconsistent with the realities of women's experiences and responses in American history. And links between women form the foundation from which this project ensues.

This reprinted piece is but one of many *Lily* articles to take up the topic of the 'woman's sphere'. Indeed, this vision of a divided world, promulgated by popular women's magazines beginning with Sarah Hale's Boston-based monthly, the *(American) Ladies' Magazine* (1828–36), was a thematic obsession of *The Lily*'s; treatments of this topic, both pirated and original, in prose and poetry (e.g., 15 Feb. 1854, 15), by amateurs and professionals, appeared.

The persistent, almost obsessive treatment of the concept of the 'woman's sphere' organized varied contributions into a progressive, serial assault. In time, contributors had rewritten, evacuated or fully exploded the concept, opening up imaginative possibilities for a freer, fairer world for women. Plus, such repeated treatment also organized various contributors into an effective, collective fighting force. It unified disparate and distanced women writers into a politicized community of, at minimum, rhetorical activists.

Formatting for feminism: re-fashioning readers and popular formulas

The various dynamics at play on the level of popular discourse in *The Lily* also operated on popular literary genres, on the level of formula. The magazine helped create critical readers, and it encouraged them to rethink the popular women's stories underwriting imaginative possibilities for women's lives.

By the time *The Lily* emerged, the sentimental-domestic or 'woman's' novel was in its heyday. Readers and critics alike recognized the style and structure of this formula, as well as its phenomenal marketplace success (see Baym 1984, 202–07). Contemporary critics of women's popular culture still debate about whether these woman-authored, woman-centered stories offered empowering or entrapping messages to audiences (see Douglas 1977; Kelley 1984; Tompkins 1985; Ryan 1985; Davidson 1986; Brown 1991). But answers to these questions depend now, as they depended in 1850, on individual insight and cultural spin, on how readers read them and how the popular media construed them.

In a letter sequence exchanged with T. S. Arthur, the popular novelist and editor of the weekly *Arthur's Home Gazette* (1850–54), soon to be absorbed into *Arthur's (Home) Magazine* (1852–98), the *Lily*'s editor, Amelia Bloomer, dons both hats: she is woman reader and magazine editor. Bloomer had previously given Arthur's new novel, *Ruling A Wife* (1850), a positive review. But as the novel continued, she changed her mind, and so felt 'obliged' to lodge her objections in a letter to Arthur published in *The Lily* (Nov. 1850, 86–7).

Sentimental-domestic novels like Arthur's have been dubbed 'woman's fictions' (Baym 1978). They are tales of social relations with a heroine at the heart. Typically, they detail the trials and triumph of this female protagonist who has been orphaned, and who is egregiously mistreated by the very authorities in her life who ought to have nurtured and protected her. The heroine's virtuous character is often demonstrated and measured against a double, another but lesser woman, who is seen to be seduced by luxury or status, for example, or who is conniving, duplicitous, or even vicious – who is somehow fallen in distinctly gendered ways. Plotlines recount the heroine's achievement of identity, of autonomy and self-respect, despite the inequitable treatment and cultural obstacles repeatedly thrown up in her way. Neither her tragic personal circumstances, nor the violent treatment to which she is sometimes subjected, nor the debilitating limitations placed on her sex prevents the female protagonist of the sentimental-domestic novel from becoming a literary heroine in both gendered and American terms: by the repeated exercise of her virtuous character, she achieves self-reliance.

Like the sentimental-domestic novel, the letter itself was also a strongly gendered genre of writing at the time Bloomer's letters to Arthur appeared. Epistolarity had long provided special opportunities for women's self-expression, interpersonal connection, and ongoing exchange – precisely the dynamics most important in the nineteenth-century feminist magazines. Since literate women garnered status and unusual expressive authority through letter-writing competencies (Goldsmith 1989), letter-writers in *The Lily* often used the genre to explore possibilities for unconventional expression or to convey critical, 'unladylike' views and responses.

Bloomer uses both the form of the letter and the formula of the sentimental story to mount critique and to construct feminist readers. She opens with 'an abstract' of the novel which appropriates its sentimental-domestic plotline in order to pan Arthur's use of it: 'Mr. Lane, the hero of the story,' she dashes, 'was an overbearing, lordly husband, who looked upon his wife as in every respect his inferior, and from whom he extracted the most perfect submission' (Bloomer 1850, 86). Mrs Lane's own anger is aroused when her maternal authority – the ideological centre of domesticity – is contravened by her husband: a 'child was given to them...and the father made it his business to direct in the nursery, and to order this and that treatment for the child, contrary to the better judgment of the mother' (ibid.).

This damage to the foundation of the domestic plot causes the narrative to collapse. Mrs Lane flees with her baby to the city, where, swindled of her last dollar, she finds herself alone. When a passer-by offers to escort her to a hotel, he instead conveys her to a house of prostitution. There, 'while they were dragging [Mrs Lane] from the room – she the meanwhile wringing her hands and shrieking for mercy – the door was open', Bloomer relates, and who but 'Mr. Lane stood before them!' she rails.

Bloomer's outrage burns all the way down to her grammar; one paragraph – long on commas and furious dashes and short on the more calming and placed periods – conveys her critique. 'You have only shown that we are weak and helpless – incapable of taking care of ourselves or keeping out of harm's way', she charges (ibid.) She summons the sentimental structure through recitation of the heroine's trials by Arthur's hand: 'No matter to how bad a man a woman is tied – no matter to how much insult she may be subjected – no matter...her feelings, and...the indifference and scorn [of] her opinions, no matter how he won her young heart with promises of...love [only to] be transformed into a demon – and her [sense of] disgust and loathing –', Bloomer barely stops to breathe, 'you have shown us that it is useless for a [woman] to think of freeing herself...[for]...should she attempt [to be freed], she will fall into the snares and dangers from which she is powerless to extricate herself, and which will speedily cause her to repent the step and sigh and return'. 'But', she rails, to the very marriage 'from which there is no escape, and [to] this same cruel man from whom she has fled', [and who], humiliatingly, 'comes to her rescue' (ibid.).

Bloomer wants and demands a different story. 'Why did you not', she presses, 'let Mrs. Lane show that she was equal to the emergency in which you placed her?' (ibid.). Pushing the boundaries of the sentimental storyline to its limits, she asks further, 'Why not let her rise superior to so dependent and so *degrading* a position?' (ibid.) 'Why not let her seek and find some honorable employment, where, if but for a day, she might support herself and her child by her own independent exertions?' (ibid.).

Bloomer's repeated challenges mount a substantive critique of the sentimental story – from a strong-minded woman reader's point of view. Such deep defilement of a lady, many middle-class women readers in 1850 would have known, compels corrective action. A second letter from Bloomer to Arthur follows, explaining that, 'I think women should...put up with many things hard to endure before resorting to the extreme step of separation' (Bloomer 1850, 87). 'Yet,' she delimits, 'I believe there is a point beyond which endurance ceases to be a virtue.' Beyond that point, she explains, 'it is both her right and her duty to seek safety and peace' (ibid.).

A critical ideological transition is written here. By conjoining 'rights' and 'duties' together under the rubric of woman's 'virtue', Bloomer uses popular notions of women's identity to pivot toward more radical hopes for female empowerment. Bloomer's letter sequence both delimits and expands – she critically redefines – the strongly-gendered notion of 'feminine virtue'. In her reconstruction, formulated within her epistolary relationship with Arthur, there is a point beyond which the seemingly endless requirements of feminine virtue cannot go without discrediting authorized gender ideology. At that moment – when virtue may be undone – a good woman is obliged to act. It becomes her duty, a 'feminine' requirement, as well as her 'right', a new feminist endowment, to act independently, and in the interest of herself. The popular gender discourse is challenged from within itself, and demands a new (yet still accredited) response. The popular mid-nineteenth-century idea of 'feminine virtue' has been transformed so as to reach – indeed require – the autonomy, self-sufficiency and strength it had previously effaced or opposed.

Bloomer's correspondence with Arthur illustrates a type of transformative pressure politics widely applied by the first American feminist magazines. The turn is accomplished by pushing domestic discourse to its own extremes, so that one element confounds or contradicts other, definitive elements, threatening to explode the whole configuration. Here the ideological shift from woman's private 'duties' to public 'rights' is mediated by pressure by a women reader to redefine the popular domestic conception of her 'virtue'.

Even more distinctive is the editorial stance. Bloomer addresses Arthur, using her status as letter-writer, magazine editor, and woman reader simultaneously. What is more, she construes herself alternately as single self and a collective speaker: her 'our' means both herself formally and her readers collectively; her 'we' conflates the editorial 'we', in widespread use at the time, and the collective 'we' increasingly used by feminist women writers. Throughout the sequence, Bloomer's confidence in the letter form helps her claim the authority to volley between 'I' and 'we', individual and collective, personal and political. These multiple and overlapping self-representations create an authorial position that is at once oppositional and collaborative, formal and personal, professional and independent. They dynamically construct a feminist editorial voice.

A range of subsequent contributors followed Bloomer's lead, apparently using their resistive reading skills to inform their contributions. An anonymous March 1855 fragment, 'The Printer Girls', (Anon. 1855, 18) for example, has re-read and so can re-write the sentimental-domestic formula suggestively. Janet Malcolm, desolate after the (formulaic) death of her mother, makes her way to 'the city'. There, rather than being deceived and deflowered, she is befriended by a dark-haired, dark-eyed young woman, the double and complement of Janet's blond and blue-eyed beauty. They open a print shop and, presumably, live fulfilling, successful lives of their own thereafter. The contribution combines the type of women's friendships authentically experienced by many nineteenth-century women and described by Carol Smith-Rosenberg (Smith-Rosenberg 1985), with the 'honorable employment' and self-support by 'independent exertions' described in Bloomer's letter to Arthur about her reading of his novel.

Such contributions to *The Lily* use familiar 'feminine' formulas to re-write stories of women's lives. A more sophisticated revision of popular formula is 'What Can Woman Do – Or the Influence of An Example' in September 1850 (65–7). It is a temperance tale by Alice B. Haven Neal, a novelist and yet another 'woman's-rights woman' who also frequently found a place to speak in the popular women's magazines of her day, including the market-leading *Godey's*. A double-narrative, half the political parable of Isabel Gray, and half sentimental-domestic story of Emily Lewis, Neal's tale juxtaposes competing discourses to make a transition into new ideas of women's identities and lives.

Offered a glass of wine at a party, Isabel attracts attention by her unexpected refusal: 'I will drink to Lucy with all my heart, but in water if you please' (Neal 1850, 65). While one guest takes her seriously, commenting that 'I do not think it a fancy – Isabel Gray always acts from principle' (ibid.), the other women cannot read her confusing behaviour. Isabel doesn't fully belong in this story. 'Why, act in such a strange way?' they ask. 'You might, at least, have touched the glass to your lips' (ibid.).

Acknowledging her visibility as both an oddity and an example, Isabel asserts that 'we none of us know the influence we exert' (ibid.). She evokes the sentimental formula when she explains that young Lewis had recently vowed temperance and 'to-night was his first trial' (ibid.). Now the women understand. They recognize this storyline. One woman even comments that Isabel 'should have been a novelist' (ibid.). While Isabel isn't, Neal is. She uses Isabel to move beyond her story. Lewis, too, is moved by the heroine. 'God bless you, Miss Gray. I confess I wavered – you made me ashamed of my weakness; I will not mind their taunting now, was all the grateful, warm-hearted man could say' (Neal 1850, 67). Marked by sentiment and the clasp of Isabel's hand, the two seem destined for domestic bliss.

But Neal interrupts the apparent romance plot, pursuing instead a feminist angle. To make the turn, she employs the structural device of the 'authorial addendum'. However, instead of positioning her political paragraph at the end of the narrative, she interjects it in the middle, signalling further development in the remainder of the story. 'Ah, my sisters', she proclaims, 'if you could but realize that all the beauty and grace are but talents entrusted to your keeping, and that the happiness of many may rest upon the most trivial act, you would not use that loveliness for an ignoble triumph, or so thoughtlessly tread the path of life!' (ibid.) Such sentiments have been articulated before, but here they are employed as a transition within a story that goes further ideologically, rather than as strategic conclusion to a sentimental tale of womanly weal or woe. By using a formula for ending in a transitional position, Neal pushes beyond the sentimental-domestic tradition in which she is working, just as her heroine does. As the narrative recommences, Isabel is centred, and is portrayed with a new authority. The hostess asks whether her niece, Emily, betrothed to Lewis, should trust his new-found temperance. 'Yes, if she chooses it', Isabel tellingly replies (ibid.).

'And so it is proved', Neal suddenly concludes. Although abrupt, her closing allows for several resolutions. As to Isabel's story, 'Robert predicts she will never marry' (ibid.). 'She is not of those who would sacrifice herself for fortune, or give her hand to any man she did not thoroughly respect and sympathize with, to escape that really very tolerable fate – becoming an old maid' (ibid.). Apparently, Isabel has been designed for public rather than private 'influence'. Hers is a different destiny. But

'Isabel Gray is always a favorite guest' (ibid.) in the conventional setting of the harmonious Lewis home.

Both Emily and Isabel provide viable conclusions to women's stories. Each is a heroine whose goals are achieved by her inner strength and 'good judgment'. And Isabel's powerful example – 'influence' in action – is here suggested to enable both the domestic harmony of the Lewis household and the feminist independence of Isabel's own story. In its dual conclusions, Neal's contribution suggests the feminist spin that *The Lily* put on 'female influence': women themselves must write its meanings, its who, what, when, where, and why. The story is reaching for a new version of popular 'female influence,' one that approaches feminist empowerment. But it isn't there yet. For Isabel's uncommon agency is curtailed at the story's close. Her 'power' is limited and partial, since it emanates only from her 'choice' of what *not* to do. Moreover, while Isabel represents choice, she also remains above communal choices herself. As the story ends, she remains solitary, and lives under the burden as much as the empowerment of exemplary status.

Transforming the popular to the political

Despite its limited idea of woman's agency, a story like Neal's uses popular formula to push through a transition and promote the evolution of new notions of women's power and place in the world (see Cawelti 1976). The moral of this story has personal, political, and also social-movement implications. The story accepts differences in women's goals and choices and enables solidarity among them, despite divergent points of view. The formulaic doubling of the two heroines, each a role model, juxtaposes gender possibilities, allowing new conceptions to cooperate with the old. Not only can new visions emerge from established constructions, but here they also coexist with these antecedents without discrediting them. The story formulates a peaceful transition from 'feminine' to feminist possibilities.

In another authorial addendum, this one attached to her final letter to T. S. Arthur, Amelia Bloomer anticipates the future trajectory of feminist-inspired writing and media in America:

> We care not what the name, or how popular the writer, who holds up the weakness of woman to public view, so long as we have a pen to write, or a voice to speak, we shall defend our sex from such libelous imputations. Woman has too long been kept in awe, and her powers of mind and body cramped and fettered by the false ideas in regard to

> her sphere, and her duty which man has heretofore so successfully impressed upon the public mind. It is time she, herself, arouse [*sic*], and teach him another lesson.
>
> (Bloomer 1850, 87)

Speaking directly to Arthur, as well as to *Lily* readers, and to an implied audience of women and men in culture at large, Bloomer warns that women will use their presence and voice in the expanding popular presses to revise gendered language and media images, re-writing women's roles in American life. Claiming credibility and collectivity, asserting visibility and critical voice, she calls for the shift from representation to self-representation that has become a dominant strategy of feminists in American media production and politics today.

Works cited

Anon. 'The Printer Girls,' *The Lily: Devoted to the Interests of Women* 9.3 (1 March 1855), 18.

Baym, Nina. *Novels, Readers, and Reviewers: Responses to Fiction in Antebellum America*. Ithaca, NY, 1984.

Baym, Nina. *Woman's Fiction: A Guide to Novels By and About Women in America, 1820–1870*. Ithaca, NY, 1978.

Beasley, Maurine H. and Sheila J. Gibbons. *Taking Their Place: A Documentary History of Women and Journalism*. Washington, DC, 1993.

Bloomer, Amelia. '*Ruling A Wife*', *The Lily: Devoted to Temperance and Literature* 2.9 (Nov. 1850), 86–7.

Botein, Stephen. '"Meer Mechanics" and an Open Press: The Business and Political Strategies of Colonial American Printers', *Perspectives in American History* 9 (1975), 126–225.

Brown, Gillian. *Domestic Individualism: Imagining Self in Nineteenth-Century America*. Berkeley, CA, 1990.

Cawelti, John. *Adventure, Mystery, Romance: Formula Stories as Art and Popular Culture*. Chicago, IL, 1976.

Davidson, Cathy. *Revolution and the Word: The Rise of the Novel in America*. New York, 1986.

Douglas, Ann. *The Feminization of American Culture*. New York, 1977.

'E. B.' 'Woman's Sphere – What Is It?', *The Lily: Devoted to the Interests of Women* (July 1854), 91.

Foster, Dorothy. 'The Earliest Precursors of Our Present-Day Monthly Miscellanies', *PMLA* 32.1 (June 1991), 225–41.

Frohock, Jane. 'Correspondence', *The Lily: Devoted to Temperance and Literature* [8].23 (1 Dec. 1856), 150.

Gage, Frances D. ('Aunt Fanny'). 'Woman's Sphere', *The Lily: Devoted to Temperance and Literature* 2.4 (April 1850), 26–7.

Garnsey, Caroline John. 'Ladies Magazines to 1850: The Beginnings of an Industry', *Bulletin of the New York Public Library* 58 (1954), 74–88.
Ginzberg, Lori. *Women and the Work of Benevolence: Morality, Politics and Class in the Nineteenth-Century United States*. New Haven, CT, 1990.
Goldsmith, Elizabeth, ed. *Writing the Female Voice: Essays on Epistolary Literature*. New Haven, CT, 1989.
Kelley, Mary. *Private Woman, Public Stage: Literary Domesticity in Nineteenth-Century America*. New York, 1984.
Leder, Lawrence. 'The Role of Newspapers in Early America "In Defense of Their Own Liberty"', *Huntington Library Quarterly* 30.1 (Nov. 1966), 1–16.
The Lily 1–9 (1849–58).
Lund, Michael. *America's Continuing Story: An Introduction to Serial Fiction, 1850–1900*. Detroit, MI, 1993.
Marzolf, Marion. *Up From the Footnotes: A History of Women Journalists*. New York, 1977.
Mather, Anne. 'A History of Feminist Periodicals, Part I', *Journalism History* 53 (Summer 1973), 82–5.
Mott, Frank Luther. *A History of American Magazines. I: 1741–1850*. New York, 1930.
Neal, Alice B. Haven. 'What Can Woman Do – Or the Influence of An Example', *The Lily: Devoted to Temperance and Literature* 2.9 (Sept. 1850), 65–7.
Oakes Smith, Elizabeth. 'Woman's Sphere', *The Lily: Devoted to the Interests of Women* 8.14 (1 Aug. 1856), 111.
Okker, Patricia. *Our Sister Editors: Sarah J. Hale and the Tradition of Nineteenth-Century American Women Editors*. Athens, GA, 1995.
Russo, Ann and Cheris Kramarae. *The Radical Women's Press of the 1850s*. New York, 1991.
Ryan, Mary. *The Empire of the Mother: American Writing About Domesticity, 1830–1860*. New York, 1985.
Schlesinger, Arthur M. *Prelude to Independence: The Newspaper War in Britain, 1746–1776*. New York, 1958.
Smith-Rosenberg, Carol, 'The Female World of Love and Ritual: Relations Between Women in Nineteenth-Century America', in *Disorderly Conduct: Visions of Gender in Victorian America*. New York, 1985, 53–76.
Solomon, Martha, ed. *A Voice of Their Own: The Woman Suffrage Press, 1840–1910*. Tuscaloosa, AL, 1991.
Stearns, Bertha-Monica. 'Before *Godey's*', *American Literature* 2 (1930), 248–55.
Steiner, Linda. 'Nineteenth-Century Suffrage Periodicals: Conceptions of Womanhood and the Press', in William Solomon, and Robert W. Chesney, eds. *Ruthless Criticism: New Perspectives in U.S. Communication History*. Minneapolis, MN, 1993.
Steiner, Linda. 'Finding Community in Nineteenth-Century Suffrage Periodicals', *American Journalism* 1.1 (Summer 1983), 1–15.
Tompkins, Jane. *Sensational Designs: The Cultural Work of American Fiction, 1790–1860*. New York, 1985.
Vaughan, Mary C. 'Maternal Influence', *The Lily: Devoted to Temperance and Literature* 3.2 (Feb. 1851), 11.
Vaughan, Mary C. 'Correspondence From Mary Vaughan', *The Lily: Devoted to Temperance and Literature* 2.12 (Dec. 1850), 94.

Ville, Belle. 'Correspondence from Bell[e] Ville', *The Lily: Devoted to the Interests of Women* [8].6 (15 March 1855), 46.
Ville, Belle. 'Woman's Sphere', *The Lily: Devoted to the Interests of Women* 7.17 (15 Sept. 1854a), 135–6.
Ville, Belle. 'Woman's Sphere', *The Lily: Devoted to the Interests of Women* [7].21 (15 Nov. 1854b), 168.
Welter, Barbara. 'The Cult of True Womanhood', *American Quarterly* 18 (Summer 1966), 151–74.

Appendix: editorial backgrounds of known editors and contributors

Elizabeth Aldrich, publisher and editor of the *Genius of Liberty* (1851–3), was also a participant in several women's rights conventions, including the National Convention, held in Syracuse, New York, in 1852. After suspension of the magazine, Aldrich edited the 10-page 'Genius of Liberty' department of *Moore's Western Lady's Book* (1849–56), a Cincinnati magazine for women.

Paulina Wright Davis was an abolitionist and women's rights activist who was a key organizer and leader in the first two national women's rights conventions, in Worcester, Massachusetts, in 1850 and 1851. She put her own money into publishing the *Una*, which advocated for radical causes including marriage reform and birth control.

Amelia Bloomer helped organize the Ladies Temperance Society in Seneca Falls that launched *The Lily*. Bloomer's radicalism soon discouraged the Temperance Society's support. Her 1851 editorial advocating the wearing of the Turkish-style pants outfit, and daguerrotype illustrations showing Bloomer and Elizabeth Cady Stanton wearing the new style, brought notoriety to the costume and her name – and a surge in subscriptions to the magazine. By 1853, the magazine's subscriptions had jumped to over 4000 and were growing steadily (Marzoff 1977, 223). Moving to Ohio with her husband in 1854, Bloomer took *The Lily* with her. Unable to manage such a large circulation in a town 300 miles from a railway and in an area devoid of sufficient printing facilities, she turned over the magazine to **Mary B. Birdsall**, former women's department editor of the *Indiana Farmer* (1851–61). Birdsall continued to publish *The Lily*, now devoted to 'Temperance and the Elevation of Women,' from Richmond, Indiana, semi-monthly until 1858. During this time, Bloomer and her husband owned and edited the *Western Home Visitor* (1853–5).

Jane Frohock's frequent, bylined contributions to *The Lily* after it was moved to Richmond, Indiana, contain much of what is commonly known about her life and work. Her contributions suggest a strong-minded commitment to temperance and other 'women's issues' pursued through the logic and language of religious revivalism. Frohock consistently wrote in terms of good and evil, and typically pivoted from a belief in woman's instinctive reformist ethics to outspoken calls for social activism and change.

Frances Dana Gage or '**Aunt Fanny**,' a native of Ohio, wrote widely in the reform and the popular press on issues ranging from abolition, to temperance, to women's rights. She lectured and organized on behalf of freed Blacks after the American Civil War, and became a paid lecturer for the American Equal Rights Association in the 1870s. After moving to Columbus with her husband, she eventually became associate editor of *The Ohio Cultivator* (1845–66), a farm and family magazine, and *Field Notes* (1861–2), a farm weekly. Gage's pen-name often encourages the modern reader to confuse her with *Ruth Hall* author, **Fanny Fern**, a frequent contributor to *The Lily* and *Una*, and the first woman syndicated columnist in the US.

Sara Jane Clarke Lippincott or '**Grace Greenwood**' was briefly an editorial assistant for *Godey's Lady's Book*, but lost the job when her contributions to an abolitionist journal, the *National Era* (1845–60), offended southern subscribers. Thereafter, she served as only the second woman Washington correspondent for the *Era*; later, she became editor of Philadelphia's short-lived *Lady's Dollar Newspaper* in 1848. After entering what became a stormy marriage with Philadelphia publisher Leander Lippincott in 1853, 'Greenwood' edited *Little Pilgrim* (1853–68), one of the nation's earliest children's magazines.

Sarah Josepha Hale is one of the most renowned American women's magazine editors of the nineteenth century. She was also a published poet and successful novelist. She began her career as editor of two Boston magazines, the *Juvenile Miscellany* (1826–36) and the *(American) Ladies' Magazine* (1828–36). When publisher Louis Godey tapped her to edit his *Lady's Book*, she folded the magazine into the larger venture and commenced a wildly successful career that spanned forty years. She stepped down from editor's chair at *Godey's* in 1877. Hale's first book of verse was *The Genius of Oblivion* (1823), and was followed by another, *Poems for Our Children* (1830), which contained the time-honored 'Mary's Lamb.' Hale wrote or edited some 50 books during her career, the most memorable of which include *Northwood* (1827), a novel, *Traits of American Life* (1835), a collection of local-colour vignettes, and *Woman's Record* (1853), a biographical encyclopedia of American women containing more than 2500 entries.

Dr Lydia Sayer Hasbrouck edited *Sybil* (1856–64) and was also editor of the long-lived *Whig Press* (1851–1928), a Middletown, New York newspaper. A doctor, lecturer, journalist, and editor, Hasbrouck advocated dress reform, temperance, 'hygeopathy' (a healthy-lifestyle programme consisting of fresh air, exercise, frequent bathing, and the avoidance of tobacco), and women's rights. She was president of the National Dress Reform Association in 1863–4 and, in the 1880s, was the first woman in the state of New York to be elected to the school board. Hasbrouck also co-founded and co-edited the *Liberal Sentinel*, a Middletown weekly reform periodical.

Anne McDowell's feminist magazine, *The Woman's Advocate* (1855–60), was the only one of the first generation to be run by and for women only. The periodical announced that it was 'produced exclusively by the joint-stock capital, energies, and industry of females.' Only women typesetters and printers were used from the very beginning, and they received the same wages as men. After the magazine expired, McDowell served as editor of the woman's department of the Philadelphia *Sunday Dispatch* and later, in 1871, of the Philadelphia *Sunday Republic.*

Sarah Towne Smith Martyn was a temperance and moral reform activist and editor, and edited New York's *Advocate of Moral Reform* (1835–41) and the *Olive Plant and Ladies' Temperance Advocate* (1841–2). During the 1840s, she also edited the *True Advocate* and the *White Banner.* Martyn retired from moral reform work to edit her own women's magazine, *The Ladies Wreath* (1846–55), a 36-page New York monthly that, at $1.00/year, claimed a circulation of 25,000 at the height of its popularity in 1853. The *Wreath* was later edited by Helen Irving, wife of the nephew and biographer of Washington Irving, between 1856–62.

Alice B. Haven Neal or **'Cousin Alice'** met her husband through her frequent publication of stories in his magazine, *Neal's Saturday Gazette and Lady's Literary Museum* (1836–53). She was a frequent contributor to commercial women's magazines in the 1850s, including *Godey's* and *Peterson's.* A popular novelist as well as a journalist, her novels include *Helen Morton's Trial* (1849); *The Gossips of Rivertown* (1850), a biting commentary she famously regretted having published; *No Such Word As Fail* (1852); and *Patient Waiting No Loss* (1853). Her last novel, *The Good Report,* was published posthumously in 1867.

Elizabeth Oakes Smith wrote historical sketches, essays, short stories, and poems for general-interest newspapers and magazines and a host of women's magazines, including *Godey's* and *Lady's Companion* (1834–44). Her work also appeared in leading literary magazines in the 1840s and 1850s, including *Graham's* (1826–58) and the *Southern Literary Messenger* (1835–59). After the Seneca Falls Women's Rights Convention in 1848, Oakes Smith became increasingly active with the women's rights cause, and published a series of women's rights articles in Horace Greeley's *New York Tribune,* later collected as *Woman and Her Needs* (1851). Her *Hints on Dress and Beauty* (1852) was shortly followed by an odd, feminist-leaning novel, *Bertha and Lily* (1854). Oakes Smith later co-edited a short-lived literary magazine, *Great Republican Monthly* (1859).

Anna W. Spencer was so unknown in periodical publishing that she published a statement, signed by friends, attesting to her character and ability to edit a magazine in the 8 January 1853, issue of the *Pioneer and Woman's Advocate.* Apparently, the voucher didn't boost recognition or circulation much, as the magazine expired the next year. Little is known about Spencer's journalistic activity thereafter; she did publish a 32-page pamphlet in 1880, *The Pen and the Sword,* a philosophical essay advocating writing over violence in seeking social change.

Elizabeth Cady Stanton was a leading activist, organizer, and the philosopher of the American women's movement. Her first published work appeared in *The Lily,* and Stanton went on to become an accomplished essayist and journalist. She contributed to other first-generation women's rights journals, including the *Una,* and was co-editor of the second-generation journal, *Revolution* (1868–72), the most radical American feminist magazine published in the nineteenth century. During her long and active intellectual and political life, Stanton contributed to monthlies including *Westminster Review, Arena, Forum, North American Review, Woman's Tribune, New York Journal,* and *American.*

Harriet Beecher Stowe became a literary celebrity of almost unprecedented proportions after writing the mega-best-selling novel, *Uncle Tom's Cabin* (1852). She also maintained an active journalistic career as a contributor to numerous feminist and women's magazines through the 1850s and 1860s. With **Mary Elizabeth Mapes Dodge**, former Assistant Editor of *Working Farmer* (1849–75) and future Editor of the children's magazine, *St. Nicholas* (1873–1943), Stowe became Co-Editor of *Hearth and Home* (1868–75), a successful New York women's magazine. Stowe also served as Contributing Editor to Stanton and Anthony's *Revolution*.

Jane Grey Swisshelm's *Pittsburgh Saturday Visiter* spoke for abolition, temperance, and women's suffrage. Although Swisshelm was often critical of many of the women and the activities of the women's movement, in her writings she pushed hard for an end to the 'woman's sphere'. After an unhappy marriage, Swisshelm owned and edited the anti-slavery paper, the *St. Cloud Visiter* (1857–8). She launched *The St. Cloud Democrat* (1858–66) after a libel suit forced her to agree never to mention the Democratic party power structure in her *Saturday Visiter* again. Later, she edited her own Republican paper, *Reconstructionist* (1865–6).

Virginia Townsend was an active woman journalist who published extensively in American magazines. She was Associate Editor of *Arthur's* from 1856–72, and remained a frequent contributor until 1882. A collection of her magazine writing, *Living and Loving*, was published in 1857, and her sentimental-domestic novel, *While It Was Morning*, appeared in 1858. She later published two successful series of novels for young girls in the 1860s and 1870s.

Mary C. Vaughan's background is surprisingly sketchy. It is known that it was Vaughan's opposition to Elizabeth Cady Stanton and Susan B. Anthony at the first annual Woman's New York State Temperance Society convention that caused the two to withdraw from both the society and the temperance movement overall after 1853. In addition, Vaughan is known to have co-edited, with Linus Brockett, *Women's Work in the Civil War*. 'Fruits of Sorrow, or an Old Maid's Story,' is in Sandra Koppelman's *Old Maids: Short Stories by Nineteenth Century U.S. Women Writers* (London: Pandora Press, 1984).

14

Coming Apart: The British Newspaper Press and the Divorce Court[1]

Anne Humpherys

Before 1857, divorce in Britain was expensive and rare, practically impossible for women.[2] Three separate legal actions, including a bill in the House of Lords, were necessary, though there was a somewhat more easily achieved judicial separation 'from bed and board'. Men could sue for divorce based on adultery alone, but for women the adultery had to be aggravated by incest, bigamy, or life-threatening cruelty.

Periodic agitation for reform during the first half of the century led to a Royal Commission of Divorce in 1850 and, finally, in 1857 a new law. Though the Matrimonial Causes Bill of 1857 was quite conservative, not touching the double standard or traditional patriarchal structures,[3] its main provision, the establishing of a Divorce Court in London, was to have a long-term impact not only on the construction of marriage and gender, but also on the press. In fact, the Divorce Court was to lead to the first governmental restriction of newspaper reporting of judicial proceedings.

The press and the Divorce Court were imbricated from the beginning. Court reporting (Quarter Sessions, Assizes, and the courts of justice at Westminster – Criminal, Bankruptcy and Insolvent Debtors Courts, and the Admiralty and Probate Courts [to which the Divorce Court was attached]) – had been a staple of newspapers since the end of the eighteenth century. The reports from the courts were like transcripts, made possible by shorthand which, as a special journalistic skill, had emerged in the early part of the nineteenth century (Keeble 1994, 82). By 1875 there was an 'official shorthand reporter' in the courts as well as other Divorce Court reporters who took more general notes (Fenn 1910, 36). The reporters sat on the main floor of the Divorce Court, and when there was no jury were sometimes allowed to sit in the jury box as the public gallery seated only 33. They sometimes served more than one newspaper; Henry Fenn, who began as Divorce Court reporter in 1875, commented in his memoirs that a messenger 'used to collect my "copy" and take it around to the various evening papers' (Fenn 1910, 264), and the papers could then edit it as they desired (Roderick 1996, 3).

By the time the Divorce Court was established in 1857, the format for reports from the courts was well established. Though the most sensational court cases, involving either particularly horrible crimes or well-known people, might be covered in the news columns, in which case they would be summarized and sometimes editorialized, for the most part the court reporting was found in transcripts from shorthand, reduced in type size, in the second half of the newspaper under a title like 'Law Report' (*The Times*) or 'Law Intelligence' (*Reynolds's Weekly*).

These law reports came after the foreign and domestic news, the transcripts from parliament, the theatre reviews and just before sports news and announcements of entertainments, thus emblemizing their liminal cultural status – part record, part entertainment.

The Times reported the largest number of divorce cases and gave the fullest transcripts with the least comment; as Ed Cohen says, its coverage was 'staid, microscopic, monologic' (Cohen 1993, 130). The weeklies and later the tabloid press selected, summarized, and editorialized.[4] In *The Times*, the whole trial was reported over a series of weeks or months from the first hearing to the judge's summing up and judgment, whereas the weeklies and later the tabloids felt under no obligation to report the trial in its entirety. In fact, this selective element in the popular press's reports from the Divorce Court worried many critics of the press's coverage, even those who believed in full disclosure. Selective reporting of trials which could go on for weeks or months inevitably highlighted sensational charges and sometimes meant that corrections of false charges might not be reported.

The problem was not limited to selective editing either; the same possibilities of uncorrected allegations resulted from the quotidian columns of *The Times*. Reporting opening statements, which inevitably contained unproven charges, released into the public domain through the press allegations that might not be corrected until weeks later. There was some effort to address this problem in 1882, but the effort had the opposite result – the Law of Libel Amendment of 1888 assured that 'any newspaper could safely publish fair and accurate reports of court proceedings as soon as it could get print to paper, irrespective of the accuracy or otherwise of the statements made in court' (Jones 1974, 64). In the back-of-the-newspaper columns in *The Times*, bias is mainly introduced in the selection of trials to cover; Gail Savage says that her sampling of *The Times*'s Divorce Court reports compared to the court records indicates that 'working class cases were greatly underreported, and cases of theatrical figures were overrepresented' (Savage 1992, 36). *The Times* claimed that really objectionable cases 'are never reported in respectable newspapers' (19 Jan. 1860, 8).

The Divorce Court officials themselves did believe in some limitations, however. In 1891, the President of the Court, Lord St Helier, is reported as saying on his retirement: 'the representatives of the Press throughout my judicial career have invariably lent me their fullest assistance by the cancelling of parts of cases, or by retiring from Court when things arose of which, in my judgment, it was desirable in the interests of decency and morality that the details should not be disclosed to the world' (Fenn 1910, 30). Henry Fenn remarked at the close of his memoir *Thirty-Five Years in the Divorce Court*, that

> the public should know that what is published in the leading newspapers regarding unsavory matter is the merest bagatelle compared with the details which are eliminated. When they are printed it is, as a rule, because they are essential, because their omission, or even their material alteration, would entail, or imply, misrepresentation of facts. Every case is and has to be sub-edited at the fountain source before leaving the hands of skilled reporters.
>
> (Fenn 1910, 291)

Sunday papers like *Reynolds's Weekly*, on the other hand, had less interest in presenting a full 'record' like *The Times*, either of the Divorce Court in general or of individual cases. It used the Divorce Court reports for two purposes: as a source of entertaining sensational stories and as a basis for political comment. The first category of cases appeared in the second half of *Reynolds's* with all other court reports under the general headline of either 'Law Intelligence' or 'Yesterday's Law, Police, etc.' in short pithy summaries without editorial comment. Such was the report of Peacock *v.* Peacock (a case involving a silk manufacturer charged with adultery) which was reported in *The Times* twice in over three months, the second report being a half column of 14 paragraphs (8 June 1858, 11). In *Reynolds's Weekly* the story was reduced to a one paragraph summary of the charges, the counter charges (condonation), and the disposition (Mrs Peacock received her judicial separation) (13 June 1858, 10), but one empty of editorial comment. In fact, in all of *Reynolds's Weekly* back-of-the-paper court reports, the only difference between them and those in *The Times* is length, those in *Reynolds's* being summaries of the facts and those in *The Times* fuller transcripts.

Though the longer transcripts in *The Times* have an immediacy and drama achieved through the courtroom procedures of testimony, counter-testimony, and judicial intervention, they are fragmentary and lack closure – the whole story emerges only over time and in a few cases, a long time, making it unlikely that the reader will remember the earlier stages. Reading these accounts is thus a frustrating experience in terms of 'story'. On the other hand the short summaries of *Reynolds's Weekly* can have real power, for the events, while not editorialized, have a beginning, middle, and end and sufficient detail to turn the extended courtroom procedures into a satisfying narrative. Consider, for example, the *Reynolds's* report of the case of Peacock *v.* Peacock:

> This was a petition presented by Mrs. Peacock against her husband, for judicial separation by reason of adultery. On behalf of Mr. Peacock, the adultery was denied, and it was further pleaded that if it had taken place it was connived at by the wife, who, it was asserted, had condoned it, and had even committed adultery herself. The petition set forth the marriage of the parties, in October 1853, the husband at the time being a silk-manufacturer, at Spitalfields. There were two children of the marriage, but one had since died. In the month of June, 1856, Mrs. Peacock went with her child to pay a visit in the country, and remained from home a month and, during her absence, it was asserted, the husband had been guilty of adultery with a servant who was living under his roof. The female in question afterwards left and gave birth to a child, and about the 2nd of May the respondent was summoned for the maintenance of the infant. The summons, however, after an adjournment, was dismissed, but upon a second application at another police court Mr. Peacock was ordered to pay half-a-crown a week towards its maintenance, and he had since continued to pay that amount. The testimony of witnesses was adduced to prove that there existed admissions by Mr. Peacock of his having been guilty of the act of adultery with the female in question. On the part of the wife it was denied that she had ever committed adultery or had condoned the act of her husband, or at any time connived at it. Drs. Swabey and Phillimore, and Mr. Hoggins, Mr. H. D. Cole, and Mr. Frere appeared for the parties. Sir H. Cresswell having summed up, the jury retired, and after an absence of more than two hours, returned into court at half past six, and gave a verdict for Mrs. Peacock on all the issues of the case.
>
> (13 June 1858, 8)

Generally speaking and not surprisingly, *Reynolds's Weekly* chose only the most sensational cases to report more fully, and given its republican political position, these were for the most part cases that revealed the aristocrats, clergy, and wealthy businessmen in the worst possible light. Some of the cases reported at length, such as Marchmont *v.* Marchmont where the miscreant husband was a clergyman, escaped the 'Law Intelligence' columns and instead were reported in the front sections, sometimes in an extended news story (5 Dec. 1858, 6) or in a column called 'Gossip of the Week' where the details were ignored in favour of heavily ironic editorial comment: 'The three consecutive cases of Marchmont v. Marchmont, Evans v. Evans and Robinson [also involving clergymen], and Keats v. Keats [the husband was the owner of Fortnum and Mason], all within one week, seem to indicate that amongst the high, the moral, the respectable, and the christian classes – the classes who alone are worthy of political powers, – adultery is in a highly flourishing if not exceedingly rampant, condition' (19 Dec. 1858, 7).

Or, when there was a sensational case with a twist of some sort, such as that of Robinson *v.* Robinson and Lane (the husband was a civil engineer and the co-respondent a hydropathic doctor), where the husband was suing for divorce based on his wife's highly evocative diary of an affair with the family friend, *Reynolds's Weekly* reported the trial in great detail (20 June 1858), though still not as fully as *The Times* (27 Nov. 1858, 10–11) – one day in two-and-a-half columns compared to five days and a total of over seven columns. The sensational was marked in *Reynolds's* not so much by changes in the language of the reports but by the ironic headlines, such as 'A Nice Couple and a Happy Marriage' (Marsh *v.* Marsh, 15 Nov. 1858, 3–4). Later in the century however, summary and transcript were completely abandoned for more florid 'news' by the new tabloids, whose coverage Ed Cohen characterizes as 'bold, sensational, [and] illustrated' (Cohen 1993, 130).

Press reporting of divorce cases in the Ecclesiastical Courts, routine by 1835, had provoked little comment from the public. The reports of the proceedings of the new Divorce Court in 1857 were to turn out to be another matter. For one thing divorce cases turned up more regularly than prior to the new law because of the new Court and because the new law made it easier for middle-class men and women to sue for divorce. Further, because the only ground for divorce was adultery, and for women adultery aggravated by cruelty or worse, the details in the press reports were by necessity sensational and scandalous. Indeed, while novels that dealt with sexual infidelity and marital brutality were denounced as shocking and indecent and circulating libraries refused to handle them, the reports of court proceedings were uncensored and available to every person who could read.

It wasn't long before this almost daily onslaught of details about the private, specifically sexual lives of many middle-class men and women in the daily press provoked a demand for press censorship. As early as 1859, just two years after the passage of the first Matrimonial Causes Bill, Queen Victoria for one protested in the strongest terms to Lord Chancellor Campbell about the press reporting of the Divorce Court's proceedings and asked if there were no way to curb the publication of them. Unfortunately, Lord Campbell replied, there was no way, at least not without a bill passed in parliament, and though the idea of such a bill was floated it came to nothing. Indeed, in 1857 Lord Campbell himself had actually ruled against any curbing of the press because 'the balance of public benefit from publicity is great' (Jones 1974, 69).

The Times itself responded to the initial dismay about the press coverage of the Divorce Court by pointing out that no one complained about the sensational details emerging from Criminal Court news reporting (8 Feb. 1860, 9), though what *The Times* didn't say was that most of the criminal cases involved the already marginalized lower classes not the middle classes. While the sexual scandals of the aristocracy had been long familiar, the press reports of the Divorce Court proceedings revealed unimagined degrees of betrayal, infidelity, and cruelty in the solid middle class, as when, in the very early days of the court, the owner of Fortnum and Mason's, and Sheriff of London, ended up in the Divorce Court as the plaintiff against his wife based on her hotel affairs with a fake Spanish nobleman (Keats *v.* Keats and Montezuma, *The Times*, 15 Dec. 1858, 12).

Even *The Times*, vigorously opposing any government control of the press reporting from the Divorce Court, admitted, 'the chief evil is the spreading abroad of the scandalous stories which are necessarily told before Sir C. Cresswell [the first Judge Ordinary of the Divorce Court]', 'the Confessor-General of England' as *The Times* called him. But reporting on 'the Divorce Court is useful and necessary; the evil lies in looking to it for a succession of exciting narratives' (19 Jan. 1860, 8), a complaint repeated intermittently throughout the century and beyond. Nonetheless *The Times* argued in 1860 'in justice to the judges and the jury, a newspaper has no right to omit important matter on the ground of its indelicacy when the omission makes the verdict or the judgment seem grossly unjust' (6 Feb. 1860, 8).

While *The Times*'s complaint about the use its readers made of this material and its implied insistence on its own high-minded innocence are disingenuous, it is not uncharacteristic of the press in general. From the beginning the commercial basis of the newspaper press was in uneasy tension with the disinterested reporting of the news (it still is). *The Times* was not Hansard after all; the reports from the Divorce Court sold papers, probably more than any of the other court reports.

The history of the press and the courts begins in 1796 with the first important legal ruling in the case of Magistrate Curry against John Walter, editor of *The Times*, for libel. This 'key decision in the relationship between journalism and justice stated that an English court of justice was open to all the world, but did not differentiate between the passing on of information by persons who might have been present in court and the publications of reports in a newspaper' (Jones 1974, 12). Three years later, Mr Justice Lawrence specifically said in a ruling that it was beneficial for the information in courts to be made public by newspapers: 'There was no system under the control of the courts by which their proceedings could be made "universally known", but newspapers were freely providing this service' (Jones 1974, 13).

In the early nineteenth century the courts did try to make a distinction between reporting preliminary hearings and court proceedings where both sides were heard, but in 1848, the Jervis Act said all courts must be open. Then came the 1857 Matrimonial Causes Bill followed by a half century of debates about the Divorce Court and the press. But still every effort at restriction soundly failed. Lord Gorell, who became President of the Divorce Court in 1907, did manage to ban sketching in court by reporters, but as stated above the 1888 Law of Libel Amendment Act gave newspaper reports of judicial proceedings statutory privilege. The turning point came at the end of the century when the reports of the Divorce Court proceedings became central to the intense competition among the cheap mass media papers. The rivalry among newspaper owners was so ruthless that, as a member of parliament said, 'the better-class of papers think themselves compelled by the competition of others less scrupulous to give more particulars than they otherwise would' (Jones 1974, 71).

Finally in 1926, in the wake of a new wave of divorces after World War I, the more than seventy years of push-and-pull over press reporting of the Divorce Court led to the first significant restraint on the modern British press. The Judicial Proceedings (Regulation of Reports) Act of 1926 stated that while public access to the Divorce Courts was not restricted, the publication of reports was. The law 'forbade the publication, in relation to divorce and similar proceedings, of any particulars other than the outline of facts – names, addresses, charges, the summing up of the judge and the result' (Jones 1974, 73).

What brought the debate to a head? First of all, though the rate of divorce remained low, over half a century of daily reports from the Court made it seem that divorce was increasing radically. Secondly, the New Journalism which helped create a different kind of press, one less visually boring, one more sensational in reporting, one more personal and emotional, opened the news columns to expanded treatments of divorce coverage, and as Janice Harris argues, more novelistic ones (Harris 1996, 30–63). And of course, the whole subject of marriage was under intense debate in the 1880s and 1890s in the press, the novel, and at dinner tables. The reports from the Divorce Court, where women were on average filing 50 per cent of the petitions for divorce and judicial separation, not only provided evidence for the 'anti-marriage league', as Mrs Oliphant called the New Woman writers, but also reinforced a general sense that marriage as people knew it was a floundering institution.

Why did it take so long to regulate what so many complained about? Obviously Divorce Court proceedings made good reading. The newspapers naturally resisted any curb on this potentially lucrative source of readership. But the strength of the resistance did not come just from the newspapers. It was shared by a wide majority of people because it was based on two deeply held convictions shared by lawmakers, the judiciary, the press, and the public alike.

First, there was the strong belief that the absolute openness of the justice system was essential for the preservation of justice itself, since observation from the outside was the main defence against miscarriages of justice. Jeremy Bentham had said that 'publicity is the very soul of justice' and Henry Hallam ranked the publicity of judicial proceedings even higher than the rights of parliament as a guarantee of public security (Jones 1974, 69). As *The Times* said in 1860, 'that publicity is the best safeguard against unfairness has passed into a Truism' (6 Feb. 1860, 8). And Lord St Helier summed it up by remarking that 'The Press is the voice of the country. Justice is a public thing, and the administration of justice should be given all publicity. If this were not done, how would the public ever know that litigants were getting their rights?' (Fenn 1910, 287).

Second, there was an equally impassioned and widespread conviction that reporting of court proceedings was one of the most powerful weapons society had in the battle against immorality. By disseminating the details of all crimes – and divorce was treated as if it were a criminal action – the verdicts, and where relevant the punishments, the courts and the press constructed the national sense of what was acceptable and unacceptable behaviour.

Specifically, in terms of the Divorce Court, the general belief was that full disclosure of the proceedings in the Court would be a deterrent to marital misbehaviour and discontent, and hence divorce. As Allen Horstman argues in his book on Victorian divorce 'national standards of Respectability would emerge' (Horstman 1985, 109) through the press dissemination of the Divorce Court proceedings. Lord Mersey, President of the Divorce Court for a year at the turn of the century, remarked 'I know from the anxiety of the parties that cases should be kept out of the papers that publicity helps to keep people straight' (Fenn 1910, 290). And Henry Fenn concluded his memoirs by stating that 'If it is desired to encourage divorce and condone the courses which lead to it, let us have no publicity. But if infidelity is an evil, and divorce a shame, let us have them out in the light of day' (Fenn 1910, 295).

Nonetheless, the news reports from the Divorce Court did not limit divorce. In fact, it is my contention that the press reports from the Divorce Court proceedings probably had the opposite effect: they helped naturalize the idea of divorce. Though according to Gail Savage's statistics, 'the annual number of divorces in England did not exceed five hundred before the turn of the century' (Savage 1992, 11), the perception was that divorce was common. By following in detail court activity with all its repetitions and hiatuses, the press was a strong factor in the gradual acceptance of divorce as a means of dealing with an unsatisfactory marriage.

However, though divorce was naturalized through the press, marriage was not undermined. The newspaper press reports from the Divorce Court ironically played a role in the idealization of marriage. By patrolling the boundaries of acceptable marital behaviour, they contributed to marriage's evolution from 'solely... an institution for sexual or reproductive bonding' to one 'as a locus for companionship and mutual support' (Shanley 1982, 369). Divorce Court and press together both lessened the tolerance for violence in the family and helped construct the notion of marriage as an equal companionate relationship between husband and wife.

Contrary to the fears of anti-divorce advocates from the Victorian period to today, divorce does *not* undermine marriage whatever its impact on families and society. If it were true that divorce undermined marriage, then most people who get divorced would never get married again. And of course this is not true now and was not true in the nineteenth century. The Registrar General, who recorded the number of divorced people who remarried in annual reports starting in 1861, reported in 1878 that the number was around 20 per cent. However, in a letter to *The Times*, the Registrar General was the first to point out that the figures were undoubtedly higher because people did not need to describe themselves as divorced when they remarried. In any case, the number of recorded divorced people remarrying steadily increased from 1861 (Horstman 1985, 102).

What press reports of the Divorce Courts undermine is not marriage but one presumption about marriage – namely that it is, and should be, forever. As long as marriage was considered undissolvable,[5] the inevitable disappointments and problems that come with the long-term effort to live 'as one' loomed large in the culture's sense of the institution. As one of the judges in the Ecclesiastical Court ruminated when denying a petition for a judicial separation, the couple had 'as much happiness as married life usually yielded' (*The Times* 26 April 1857, 11).

Though the details of the broken marriages that emerged from the press reports of the Divorce Court proceedings certainly reinforced the sense of the difficulties of marriage, the ultimate social impact on the idea of marriage could be, surprisingly, the opposite, precisely because so many of these cases *ended* the misery by divorce. Thus, the unhappily married newspaper reader reading these daily and weekly proceedings might have thought, not that marriage as an institution was a failure but rather, if I could get out of this marriage, the next time I could get it right. The news about divorce made the idealization of marriage possible. Combined with evolving values of companionate marriage and the linking of romantic love with marriage in the novel, the newspaper conveyed the double message that marriage is both ideal and expendable – ideal because expendable. As George Bernard Shaw once said 'divorce, in fact, is not the destruction of marriage, but the first condition of its maintenance' (Rubenstein 1990, 295). Perhaps then the press's creation of the sense of an increasing availability and acceptance of divorce also played its role in the dissipating away of the marriage debates of the 1880s and 1890s. People didn't want to end marriage, as Mrs Oliphant thought, but to have the opportunity to get it right the second time.

In the present day, we take divorce for granted. And we take our selected divorce news reports in the sensationalized and summarized versions created by the popular press from *Reynolds's Weekly* to the twenty-first century, and we work them into our perennial fascination with the lives of the rich and famous. Ordinary citizens, whose stories of marital breakdown filled the back pages of the nineteenth-century newspapers, are no longer newsworthy; they have become our friends and our relatives. But nonetheless, like the Victorians, we are still trying to work through the idealization of marriage that the nineteenth-century press's dissemination of the news about the availability of divorce helped solidify. Perhaps then the most important effect of the imbrication of the press and the Divorce Court in nineteenth-century Britain was not the naturalization of divorce nor the limits on the press, but rather the role these press reports played in reconstructing marriage as the tenuous locus of personal fulfilment and individual happiness.

Notes

1. The research for this chapter was supported in part by a PSC-CUNY grant.
2. Only four women had successfully petitioned parliament for a full divorce before 1857, only three between 1827 and 1857.
3. The double standard in divorce was not abolished until 1923, though Gail Savage has found that after the 1857 Bill 'both wives and ordinary working families managed to bring their cases to the court despite the obstacles' (1992, 19). In fact, 'wives initiated 40 per cent of divorce petitions and 92 per cent of judicial separation proceedings filed before the Divorce Court from 1858–1868' (1992, 26). Poor women were given some relief by the 1878 and 1886 'Maintenance of Wives' acts, which allowed a battered or deserted wife to obtain a temporary maintenance order from a local magistrate. Grounds other than adultery were admitted only in 1937.
4. In the 1890s, the Sunday weeklies included *Reynolds's Weekly, The People, News of the World, Weekly Dispatch, Lloyd's Weekly Newspaper, Illustrated Police Gazette*; morning dailies included *The Times, Morning Leader, Morning Chronicle, Daily Telegraph*; evening dailies included *Evening Standard, Star, Echo, Evening News, Pall Mall Gazette, Westminister Gazette, St. James's Gazette, Globe*. Harris discusses the Edwardian tabloid coverage of the Divorce Court in Chapter 1 (Harris 1996, 30–63).

   Cohen remarks about the newspaper reports of the Oscar Wilde trials that 'the texts are always pastiches, comprised of direct and indirect quotation, paraphrase, description, commentary, and quite crucially, omission, in varying proportions', and he categorizes these accounts 'into three distinct groups according to narrative mode. One group [is] signalled by their manifestly unanalyzed "transcription" of opening speeches, examination, and cross examination, by their lack of narrative interruptions, and by their almost fetishistic attention to detail.... A second group, while also purporting to provide complete accounts of the proceedings, had a much more highly dramatized and intrusive narrative voice which both set the scene – graphically and interpretively – and foregrounded the significance of particular elements in the case with well-developed editorial asides and commentary, interpretive subheadings, and illustrations.... The third group, while not professing to provide detailed accounts, instead offered digests of the proceedings either with ... or without ... editorial commentary' (Cohen 1993, 252 n. 6).

5. Because life expectancy was lower than it is today, a great many Victorian men and women lost their spouses through death (sometimes more than once), and hence were married to more than one partner during their lives. As Stone says, 'Statistically speaking, marriage has today in many ways merely reverted to a pattern that existed before the sharp decline in adult mortality in the late nineteenth and early twentieth centuries' (Stone 1990, 410).

Works cited

Cohen, Ed. *Talk on the Wilde Side: Toward a Genealogy of a Discourse on Male Sexualities*. London and New York, 1993.

Fenn, Henry Edward. *Thirty-Five Years in the Divorce Court*. London, [1910].

Hammerton, A. James. *Cruelty and Companionship: Conflict in nineteenth-century married life*. London and New York, 1992.

Harris, Janice Hubbard. *Edwardian Stories of Divorce*. New Brunswick, NJ, 1996.

Horstman, Allen. *Victorian Divorce*. New York, 1985.

Jones, Marjorie. *Justice and journalism: a study of the influence of newspaper reporting upon the administration of justice by magistrates*. Chichester, 1974.

Keeble, Richard. *The Newspapers Handbook*. London, 1994.

Oliphant, Margaret. 'The Anti-Marriage League', *Blackwood's* 159 (1896), 135–49.

Roderick, Anne Baltz. '"Only a Newspaper Metaphor": Crime Reports, Class Conflict, and Social Criticism in Two Victorian Newspapers', *Victorian Periodicals Review* 29 (Spring 1996), 1–18.

Rubenstein, Helge, ed. *Oxford Book of Marriage*. London and New York, 1990.

Savage, Gail. '"Intended Only for the Husband": Gender, Class, and the Provision for Divorce in England, 1858–1868' in Kristine Ottesen Garrigan, ed. *Victorian Scandals: Representations of Gender and Class*. Athens, OH, 1992. pp. 11–42.

Shanley, Mary Lyndon. '"One Must Ride Behind": Married Women's Rights and the Divorce Act of 1857', *Victorian Studies* 25 (1982), 355–76.

Stone, Lawrence. *The Road to Divorce*. London and New York, 1990.

15

Saint Pauls Magazine and the Project of Masculinity

Mark W. Turner

In 1867, Anthony Trollope was the founding editor of *Saint Pauls Magazine*, which he continued to edit until 1870. Although aiming at the general shilling monthly reader, the periodical under Trollope was largely a male space – as indicated by the prospectus, the political discourse of the non-fiction, and the constructions of masculinity in Trollope's own serializations. I wish to consider how issues raised in Trollope's *Phineas Finn* (serialized from 1867–9) intersect with broader, public debates about masculinity and 'manliness' at a time of political and social transition after the second Reform Bill. In my reading, the serial is a male *bildungsroman* in which the hero is, importantly, an Irish outsider at the centre of the British establishment. The fictional subject merges intertextually with the magazine's non-fiction in the many political articles on the Irish question published alongside the serial. Ultimately, the failure of *Saint Pauls* to engage women readers led to its demise, and a move to popularize the periodical after Trollope's departure was too little, too late.

Politics and male space

As Trollope recalled, 1867–8 were watershed years for the author-editor. Looking back over this period, writing in his *Autobiography*, he believed it to be the 'busiest in my life' (Trollope 1989, 322): he left his beloved job at the Post Office, travelled to America, wrote five novels, hunted three times a week each winter, lost a parliamentary election, and began editing *Saint Pauls Magazine*. The publisher, James Virtue, had for several months discussed a new periodical for Trollope to edit, and the first issue of *Saint Pauls Magazine* came out in October 1867. Virtue had initially wanted *Saint Pauls* to be called *Anthony Trollope's Magazine*, but the humble author balked at that notion. Trollope preferred to call it 'The Whitehall Magazine,' a name which would have signalled the political nature of the project. Other titles suggested by Trollope include 'The Monthly Westminster' and 'The Monthly Liberal,' both of which point to the specificity of Trollope's view of the magazine. At a late date, the convention of using a London place-name was agreed, and in doing so, *Saint Pauls* followed magazines such as *Cornhill* and *Temple Bar* in using a London location. The fashionable use of place-names, however, had a real significance: just as Mary Braddon's *Belgravia* was a middle- and lower-middle-class periodical inviting readers to buy into the notion of 'society,' *Saint Pauls* indicated a distinctly male part of London poised between the city and the courts, in the heart of the publishing world. Conventional practice though it was, the name denotes a particularly male readership.

As the advertisements indicate, the new shilling monthly was dominated by Trollope. This is interesting because in the same month that he resigned from the Post Office and launched *Saint Pauls*, he was experimenting with anonymity by publishing *Linda Tressel* anonymously in *Blackwood's*. While he was testing the power of his 'brand name,' trying to create a second literary identity, adverts for his own periodical were anything but modest about the attractions of 'Anthony Trollope'. In a full-page advert which appeared in the *Athenaeum* and *Saturday Review*, for example, his proper name and his role as editor are each used three times (figure 18). Clearly, this periodical was to be sold on the strength of the popular author's name, no doubt deemed necessary in a competitive shilling monthly market. Trollope was different from other well-known editors in that he demanded full editorial control in deciding on contributors and their payments. But such control was not enough to make the venture lucrative – 'publishers themselves have been the best editors of magazines,' he writes in the *Autobiography* (Trollope 1883, 288) – and certainly Trollope never made the magazine pay. *Saint Pauls* never reached the 25,000 circulation target which Virtue had expected, and he sold the magazine to Alexander Strahan in May 1869. In January 1870, Trollope was informed that the publishers could no longer afford his services and by July, he was out and *Saint Pauls* was left to 'edit itself' (Strahan's phrase) in the manner of *Blackwood's Magazine*. Four years later, *Saint Pauls* came quietly to an end.

Trollope's dedication to editing a political magazine is one primary reason that this shilling monthly never reached the wide audience who read his fiction. It is significant that while editing *Saint Pauls*, Trollope was considering yet another career move. In the autumn of 1868,

A NEW MONTHLY MAGAZINE, EDITED BY ANTHONY TROLLOPE

On the FIRST of OCTOBER, Number I., price ONE SHILLING,

SAINT PAULS.

A Monthly Magazine

OF

FICTION, ART, AND LITERATURE.

EDITED BY

ANTHONY TROLLOPE,

AND

ILLUSTRATED BY J. E. MILLAIS, R.A.

The FIRST NUMBER, to be published on TUESDAY, October 1st, will contain:—

AN INTRODUCTORY PAPER. By the Editor.

A LEAP IN THE DARK; or, a Glance at What was Done Last Session.

ALL FOR GREED. Chapters I., II., III., IV. A Novel. By the Baroness de Bury.

THE ETHICS OF TRADE-UNIONS.

THE TURF, its PRESENT CONDITION and PROSPECTS.

ON SOVEREIGNTY.

ON TASTE. By Henry O'Neill, R.A.

PHINEAS FINN, the IRISH MEMBER. Chapters I., II., III., IV. By Anthony Trollope.

London: VIRTUE & CO. 294, City-road.
Publishing Office: 26, Ivy-lane, Paternoster-row.

Figure 18 *Saint Pauls,* advertisement. *Athenaeum,* 14 September 1867.

Trollope bid unsuccessfully to become an MP for the borough of Beverley in Yorkshire. *Phineas Finn, the Irish Member,* one of the serials which inaugurated *Saint Pauls* and the first of his novels to deal primarily with the machinations of men in parliament, voiced the thwarted writer's political views: 'As I was debarred from expressing my opinions in the House of Commons, I took this method of declaring myself' (Trollope 1883, 317). *Saint Pauls* was an almost wholly political undertaking for Trollope. Fiction, poetry and miscellaneous articles were an important part of the periodical, but as he states frankly in an 'Introduction' in the first issue, 'Saint Pauls, if it be anything, will be political,' and 'the good old Liberal cause still needs support' (Trollope 1867, 4 and 5). Just after the passing of the second Reform Bill in 1867, such statements were avowedly party political, not least because the Bill was passed by a Tory government seen to have hijacked Liberal policies. At the core of *Phineas Finn* is a struggle for the soul of Liberalism and many political articles on post-Reform Bill questions, and some by Trollope, are resonant in the serialization of this parliamentary novel.

As he had done when writing for magazines in the past, Trollope was again writing for a distinct market in *Saint Pauls*: *Phineas Finn* is a political novel in a magazine Trollope declares to be political. '*St. Paul's*,' John Sutherland asserts, 'has a strong aroma of the club smoking-room about it' and 'if nothing else, *St. Paul's* gives a reflection of the mid-Victorian, Liberal gentleman's cultural and physical recreations' (Sutherland 1980, 119). The serial fiction by Trollope, with its emphasis on politics, is bound up in the larger project of the magazine devoted to male subjects and a male reader. It is interesting that in his 'Introduction' in the first issue, Trollope constructs the magazine as rather ordinary: 'SAINT PAULS MAGAZINE is not established, on and from this present 1st of October, 1867, on any rooted and matured conviction that such a periodical is the great and pressing want of the age' (Trollope 1867, 1). His hopes for the shilling monthly were not radical or innovatory (as they had been when founding the *Fortnightly Review* in 1865) and a lesser assertion to introduce a new periodical in an intensely competitive market could hardly be imagined. By contrast, as we have seen, the adverts had already loudly announced that Trollope and *Saint Pauls* were arriving at booksellers and newsagents. There is a divergence or conflict between what Trollope says in his gentle introduction and what the editor and publisher had actually to do commercially. In the full page *Athenaeum* ad (figure 18), *Saint Pauls* is called 'A Monthly Magazine of Fiction, Art, and Literature' and politics is not mentioned overtly. The political bias announced in the 'Introduction' and contents, Trollope's primary interest in the magazine, remain unadvertised, unspoken outside the text, and what is to the fore is the fiction which actually sold magazines to women readers. The fiction is a lone bid for women readers, and the anonymous sensation novel *All For Greed*, written by Marie Pauline Rose, a.k.a. Baroness Blaze de Bury, who signed 'A.A.A.', seems particularly included to attract readers of popular fiction. Alas, it did not. Furthermore, an article in the first issue, 'Taste' by the artist Henry Nelson O'Neil, surely would have discouraged (and perhaps even offended) many potential women readers. After arguing that women are prone to following fashion by noting the number of red-haired women to be found after the appearance of the Pre-Raphaelites, O'Neil asserts that women do not have 'a true appreciation of the beautiful':

> They have all the feeling necessary for the possession of taste, but they want judgment, and while having the sensibility to admire what is pretty or pleasing, they lack the discrimination to select what is really beautiful. And what removes their feeling from taste is its absence of critical power. Moreover, though women are keen in perception, they have less reflection and are more precipitate than men.
>
> (O'Neil 1867, 99)

Would an editor seriously interested in bidding for women readers include such an article in the very first issue? The way to attract women readers is *not* to say that they are ill-judged, undiscriminating, timid, and uncritical – not if you want them to buy the second issue. The lack of women readers was perceived as one of the magazine's weaknesses by Strahan, after he had taken over as publisher and Trollope had left as editor in the summer of 1870. Under the new publisher, *Saint Pauls* tried to shift its target market in a number of ways: signature, less politics, and women's advice columns indicate a move toward a more conventional, female reader of shilling monthlies (Malcolm 1984, ch. 4). Adverts for the post-Trollope February 1871 issue tend to list *only* the serial fiction, and not non-fiction articles on 'Convivial Pauperism,' 'Hints to Army Reformers,' or 'The Gamut of Light' (figure 19) Despite the presence of sensation fiction early on, *Saint Pauls* was a male space, as the non-fiction and Trollope's fiction suggests. The author-editor's goal was to produce a significant forum for political discussion within the popular magazine market. And politics, of course, was a male domain; consider the ways Thackeray had gendered his family-reader shilling monthly, *Cornhill Magazine*, by excluding overt discussions of politics and religion from its pages when it was founded in 1860. Furthermore, the gendering of politics is a fact particularly well illustrated in *Phineas Finn* by the case of Lady Laura, whose political savvy and ambition are impeded because she is a woman.

The year 1867 was as much a time of transition for the country as it was for Trollope. The second Reform Bill, despite having been passed by a

WALKS IN ROME.

Now Ready, in 2 handsome crown 8vo vols. price 21*s*. 'WALKS in ROME.'
By AUGUSTUS J. C. HARE.

From THE PALL MALL GAZETTE.—"The best handbook of the city and environs of Rome ever published.....Cannot too much commended."

From THE DAILY NEWS.—"This book is likely to be very useful, and seems to be correct on all points. It is thoroughly ctical, and is the best guide that has yet been offered."

From THE SPECTATOR.—"Mr. Hare's book fills a real void, and gives to the tourist all the latest discoveries and the est information bearing on that most inexhaustible of subjects, the City of Rome..... It is much fuller than 'Murray,' and one who chooses may now know how Rome really looks in sun or shade."

THE BISHOP OF WINCHESTER.

In the Press, the FOURTH EDITION of HEROES of HEBREW HISTORY.
By SAMUEL WILBERFORCE, D.D., Bishop of Winchester.

"Dr. Wilberforce presents Old Testament story with such force and vigour, in a style so clear and adorned with such rich chaste ornament, that while his high and unyielding orthodoxy will delight all High Churchmen, his rendering of Scripture ory will be accepted by the veriest sceptic as a rare intellectual treat—an instance of eloquence, earnestness, and dramatic er which must attract the reader, quite irrespective of the opinions bound up with it."—*The Times.*

Brilliant and picturesque sketches of most of the more famous personages of the Old Testament, so arranged and connected h each other as to give almost a continuous history of the Hebrew nation in the most attractive form in which history can exhibited.....The Bishop's graphic power of description sets before us a clear image of the stage upon which the hero acts, le his keen analysis of the human heart reveals often unsuspected depths of meaning in characters with which we had fancied selves familiar.....All will welcome them as a new treasure."—*The Guardian.*

GINX'S BABY.

The SEVENTH EDITION is now ready of 'GINX'S BABY: His Birth and other Misfortunes.' Crown 8vo. 5*s*.

"The unknown author of 'Ginx's Baby' has made good his claim to be ranked as one of the truest and profoundest nourists."—*Westminster Review.*

"A remarkable work.....The most powerful political satire since the time of Swift.....We should all read and think about h a book as this when we can get it, which is seldom."—*Henry Kingsley, in the Edinburgh Daily Review.*

"The strangely-clever and tragical history of 'Ginx's Baby.'.....The anonymous author of this remarkable book speaks derfully to the purpose."—*Blackwood's Magazine.*

"It is long since we met with so honest or seasonable a piece of work.....It is one which it will do us all good to read—icals and Tories, Liberals and Conservatives."—*Standard.*

R. H. HUTTON.

w Ready, in 2 vols. post 8vo. price 24*s*. 'ESSAYS, THEOLOGICAL and LITERARY.'
By R. H. HUTTON.

VOLUME I.

The Moral Significance of Atheism.
The Atheistic Explanation of Religion.
Science and Theism.
Popular Pantheism.
What is Revelation?
Historical Problems of the Fourth Gospel.
The Incarnation and Principles of Evidence.
Renan's Christ.
Renan's St Paul.
The Hard Church.
Romanism, Protestantism, and Anglicanism.

VOLUME II.

Goethe and his Influence.
Wordsworth's Genius.
Shelley's Poetical Mysticism.
Mr. Browning.
The Poetry of the Old Testament.
George Eliot.
Arthur Hugh Clough.
Nathaniel Hawthorne.

THE LATE DEAN ALFORD'S WORKS.

NEW TESTAMENT. Authorized Version Revised. Long Primer Edition, crown 8vo. 6*s*.; Brevier Edition, fcap. 8vo. 3*s*. 6*d*.; Nonpareil Edition, small 8vo. 1*s*. 6*d*.

SAYS and ADDRESSES, chiefly on Church Subjects. Demy 8vo. 7*s*. 6*d*.

YEAR of PRAYER; being Family Prayers for the Christian Year. Crown 8vo. 3*s*. 6*d*.; small 8vo. 1*s*. 6*d*.

YEAR of PRAISE; being Hymns, with Tunes, for the Sundays and Holidays of the Year. Large type, with music, 3*s*. 6*d*.; without music, 1*s*. Small type, with music, 1*s*. 6*d*.; without music, 6*d*. Tonic Sol-Fa Edition, crown 8vo. 1*s*. 6*d*.

STERTIDE SERMONS. Small 8vo. 3*s*. 6*d*.

HOW to STUDY the NEW TESTAMENT. Part I. The Gospels and the Acts.—II. The Epistles (first section).—III. The Epistles (second section) and the Revelation. Small 8vo. 3*s*. 6*d*. each.

The QUEEN'S ENGLISH: a Manual of Idiom and Usage. New and Enlarged Edition. Small 8vo. 5*s*.

MEDITATIONS: Advent, Creation, and Providence. Small 8vo. 3*s*. 6*d*.

LETTERS from ABROAD. Crown 8vo. 7*s*. 6*d*.

POETICAL WORKS. New and Enlarged Edition. Crown 8vo. 6*s*.

NEW WORK BY C. FRASER-TYTLER.

ow Ready, in crown 8vo. price 5*s*. 'JASMINE LEIGH.' By C. FRASER-TYTLER.

HENRY HOLBEACH.

In a Few Days will be published, in 2 vols. crown 8vo. 'SHOEMAKERS' VILLAGE.'
By HENRY HOLBEACH.

"UNA AND HER PAUPERS."

eparing for immediate Publication, in One Volume, crown 8vo. MEMORIALS of AGNES ELIZABETH JONES. By Her SISTER. With an Introductory Sketch by FLORENCE NIGHTINGALE.

"I call her Una, for when her whole Life and Image rise before me, so far from thinking the story of Una and her lion a h, I say here is Una in real flesh and blood—Una and her Paupers, far more untameable than lions."
FLORENCE NIGHTINGALE.

STRAHAN & CO. LUDGATE-HILL.

THE
CONTEMPORARY REVIEW,
Theological, Literary, and Social.
2*s*. 6*d*. MONTHLY.

CONTENTS for FEBRUARY.

POPULAR *versus* PROFESSIONAL ARMIES.
By W. R. Greg.
The LONDON SCHOOL BOARD.
By W. M. Torrens, M.P.
KNOWING and FEELING. Part III. Speculative Thought.
By William Smith, Author of 'Thorndale; or, the Conflict of Opinion.'
The WAR and GENERAL CULTURE. Conversations.
By the Author of 'Friends in Council.'
Nos. VIII.—IX.
SPECULATIVE THEOLOGY and the CHRISTIANITY of CHRIST.
By the Rev. John Hunt.
The FUTURE of CAPITAL and LABOUR.
By George Potter.
SOME WORDS of COUNT BISMARCK.
By C. E. Maurice.
The RELATION of the WILL to THOUGHT.
By Archbishop Manning.
The SOLDIER'S DUTY.
By Sir E. Strachey.
DEAN ALFORD
By the Dean of Westminster and the Rev. E. T. Vaughan.

CONTENTS for JANUARY.

IMPERIAL FEDERALISM.
By the Author of 'Ginx's Baby.'
The DEBTS of THEOLOGY to SECULAR MOVEMENTS.
By the Rev. J. Ll. Davies.
The MISTAKE of HONEST DEMOCRATS.
By W. R. Greg.
NATIONAL DEFENCE.
By Major Noel.
The PRESENT RELATIONS of PHYSICAL SCIENCE to MENTAL PHILOSOPHY.
By Professor Calderwood.
The WAR and GENERAL CULTURE. Conversations.
By the Author of 'Friends in Council.'
Nos. VI.—VII.
PRAYERS for the DEAD.
By the Rev. M. Maccoll.
MUSIC and MORALS.
By H. R. Haweis. Part II.
The ENGLISH CHURCH and DISSENTERS.
By the Rev. J. Baldwin Brown.
FRENCH UNITY. By William Stigand.

THE ST. PAULS MAGAZINE.
ONE SHILLING MONTHLY, ILLUSTRATED.

HANNAH,
The New Story by the Author of 'John Halifax,'
Is begun in the FEBRUARY Number.

WILFRED CUMBERMEDE,
By GEORGE MACDONALD,
Is now appearing from Month to Month.

RALPH THE HEIR,
By ANTHONY TROLLOPE,
Is now appearing from Month to Month.

STRAHAN & Co. Ludgate-hill.

Figure 19 Advertisement. *Athenaeum* 1871.

Tory government, renewed hopes for Liberalism. A whole new class of voters was ushered in at a swipe, altering the political topography of Britain forever. However, alongside such an optimistic reading is another more pessimistic one. Matthew Arnold's late 1860s *Cornhill* articles attacking a mid-Victorian society of Philistines were collected as the socio-political tract *Culture and Anarchy* in 1869. And as an article in *Macmillan's* on 'Social Disintegration' put it, the growing gap between rich and poor was part of 'an evil of ever-increasing magnitude in its influence on the lives and characters, the moral and physical well-being, of each member of what should be one body politic and religious' ([Rathbone] 1867, 31). *Phineas Finn* contains both the sense of progress and the sense of socio-economic anxiety.

Many of the social anxieties of the late 1860s, relieved to some extent by the second Reform Bill are present in the pages of *Saint Pauls*, and it is significant to note how a number of contributions are connected inter-textually within the periodical. Numerous articles on reform issues, the Irish Question, and the working classes intersect with issues raised in *Phineas Finn*, which is set (and was written) during the second Reform Bill debates. And there are other interesting moments of self-referentiality within the periodical. For example, an article on the Irish Church refers to 'the diocese of Killaloe' and in the same issue, a *Phineas Finn* chapter tells us that 'Phineas returns to Killaloe' (Stack 1868, 569). In the following issue, an article written by a man posing as an old maid advises that 'the study proper to womankind' is 'to make yourself the best possible wife for your best possible husband' ([Cayley] 1868, 735). The advice runs concurrently with chapter 23 of *Phineas Finn* in which Laura's unhappiness at her dismal marriage becomes increasingly difficult to take. In January 1868, Trollope's article on garrotting in London suggested that street violence was not as severe as it was often made to seem. But in an instalment of his serial in May, Phineas saves Mr Kennedy from garrotters (see Trollope 1868). These examples demonstrate that within the periodical (in a single issue and across issues) there is a network of references and understandings, allowing the reader in on the intricacy of the magazine and often challenging him or her with contradictions.

Such a network also may indicate how an author-editor can control and shape the periodical text. One of the most interesting intersections expresses the tensions in 1867–8 about gender, particularly about masculinity and man's proper role in society. *Phineas Finn* serializes the development of a young man in at least two ways: the movement of an Irish outsider within the British establishment and the education of a youth learning to become a man. Definitions of manliness were much discussed and disputed in the 1860s, and one of the discourses of the serial expresses a young man's anxiety as he seeks to fit into the male culture of British parliament. Phineas undergoes a number of male rites of passage as he learns what it means to be a middle-class man; but, ultimately, he returns to Ireland, away from the centre of power and the dominant English culture. What makes reading *Phineas Finn* in its serial context so poignant is recognizing the ways the fiction and non-fiction overlap by focusing on middle-class masculinity and Westminster politics.

Transitional masculinities in the 1860s

As Phineas was growing and changing throughout the year-and-a-half of the serialization, so too was British culture. It is significant that Phineas arrives in London from Ireland, a marginalized space within Britain, during the Reform Bill debates. He participates in the political progress of the day while learning how to be a man. The 1860s was a decade of ideological debate and contradiction with regard to gender issues, and the various aspects of the Woman Question – about education, employment, marriage – were continually under public scrutiny in the periodicals and newspapers. But, as one critic discussing Victorian culture puts it, 'the question of the young male who belonged to the newly recognized middle class and his maturation and development in society' needs to be examined in connection with work on the Woman Question (Tuss 1992, 2). For Phineas Finn, development and maturation means fitting in to a man's world and learning what it means to be manly.

Most accounts of Victorian manliness describe a general shift from a religious to a physical manliness, from morals to muscles.[1] In the 1850s, especially after the publication of Thomas Hughes's *Tom Brown's Schooldays* (1857), definitions of manliness signalled the beginning of a burgeoning boy's own culture. A number of critics focus on boys' literature as a site for examining the development of manliness as a term and ideal. Claudia Nelson maintains convincingly that an androgynous spirituality defined early nineteenth-century notions of manliness and that manliness did not become associated with masculinity and physical strength *exclusively* until later in the century, perhaps after the 1870s. She asserts that in the 1860s, the cult of the physical had not yet taken over as a model for boys (Nelson 1989, 526). However, I think both discourses of manliness – the physical and the spiritual – were available to readers at the time of *Phineas Finn*. The number of children's periodicals founded in the 1860s that distinguish a gendered reader is illuminating. A far greater number of boys' magazines were started in the 1860s than in any previous decade, and the numbers would increase even more in the 1880–90s. One reason for the growth in boys' periodicals was the belief that boys and girls ought to read different material.[2] *A Book About Boys* (1868) crystallized feelings about the growing cult of boys in the 1860s; for the author A. R. Hope, boys' love and friendship surpassed even that of men.[3] Manliness became a characteristic of ideal male-ness and masculinity in the pages of boys' periodicals, partly because it was in youth that one learned how to be a man.[4]

We might also consider the discourses surrounding the mid-century aesthete, challenging norms of masculinity, and the often vociferous reactions to the aestheticized man. Matthew Arnold's 'Culture and Its Enemies' in the *Cornhill* in July 1867 (what would become the basis of 'Sweetness and Light' in *Culture and Anarchy*) was ridiculed for its dandyism by Frederic Harrison in a humorous *Fortnightly* article in November. Walter Pater had recently published a homoerotic essay on the eighteenth-century art historian, 'Winkelmann,' in the *Westminster Review* in January, 1867. Richard Dellamora reads 'Winkelmann' as a response to Arnold, and maintains that Pater, unlike Arnold, 'eroticizes aesthetic discourse' (Dellamora 1990, 15). Dellamora also argues that 'homoeroticism was central to Winkelmann's sensibility and to his ideal of culture,' and this is partly what appeals to Pater (Dellamora 1990, 110). Pater's discourse, and that surrounding aestheticism in general, give us a glimpse into male–male desire in the 1860s. However, there are other, less high-culture examples of tangible and diverse sexualities. For example, in 1870, penny pamphlets sensationalized the cases of Boulton and Park, two men arrested at the Haymarket Theatre for transvesticism. So, at the time *Phineas Finn* was serialized, definitions of masculinity were contested, not least by the nascent gay culture of which Pater and others were a part.

Masculinity is not a fixed category. The use of the term should be historically specific, for, as I have indicated, masculinity meant something different in the early nineteenth century from what it meant by the end of the century.[5] In 1867, the notion of manliness had not yet taken on the imperative of an imperialist show of strength defining both man and nation, although as early as 1859 we see guidebooks connecting manliness with the notion of national progress.[6] The late 1860s was, however, a period of transition, in which the term manliness appropriates both spiritual and physical meanings. At any given moment, there is not a single, unified definition of manliness or masculinity since there are always competing forms of masculinity, however much there may be a hegemonic, normative definition of it. Phineas Finn's version of masculinity is middle-class, urban (perhaps specifically metropolitan)[7] and upwardly mobile; this, I would suggest, is the type of masculinity also cultivated and disseminated by *Saint Pauls*.

According to David Newsome, ideals of manliness stemmed from two traditions of thought: the spiritual from Coleridge and Thomas Arnold; the physical from Charles Kingsley and Thomas Hughes (Newsome 1961, 197).[8] Both discourses were available to readers in 1867. Samuel Smiles's *Self-Help*, first published in 1859 but reprinted in 1866, preaches that 'truthfulness, integrity, and goodness – qualities that hang not on any man's breath – form the essence of manly character' (Smiles 1866, 385). The religious line, linked to a Christian manliness, was taken up in lectures, sermons and guidebooks often aimed at young men of Phineas Finn's age and younger, at a transitional moment between youth and manhood. John Caird, preaching to Glasgow University students in 1871, declares that 'religion is the manliest of all things,' but he recognizes the sexual awareness of the young men he addresses:

> How shall one with youth's warm passions tingling in his veins, its eager impulses, its sensuous susceptibilities, its impatience of restraint, its new zest of life's pleasures, its joyous use of unwonted freedom, undergo unharmed the transition from the guarded security of early life to the freedom and independence of manhood?
>
> (Caird 1871, 29 and 27)

Equally, manliness is associated with masculine strength in a number of lectures and sermons. Believing that to be manly is to be Christlike, Rev. Hugh Stowell Brown discusses how to put manliness into practice. Theatre is unmanly, and most other amusements are 'silly and effeminate' or 'demoralising'. Already in 1858, manliness implies not-feminine, and Brown lists the following, mostly middle-class sports as manly: fencing, wrestling, skating, curling, cricket, shooting, coursing, hunting, fishing, yachting, and rowing. Excessive drinking, betting, singing obscene songs, laughing at virtue and chastity are unmanly in the extreme. And already one can determine manliness by observing physical appearance. Tailored men in monocles with jewels and walking sticks and other such fops 'that go by the name of men' were hardly men, let alone manly (Brown 1858, [14–18]). Brown had precedents in this, particularly in reactions to the Tractarian Puseyites which often focused on the physical attributes of manner and dress.[9] Categories of dress, voice, and gait continue to be used as determinants of manliness, and Lynne Segal's work on masculinity argues persuasively that notions of physicality and manliness were used to stigmatize homosexuals in the late-nineteenth and early-twentieth centuries (Segal 1990, 139). Brown's views on manliness which include both spiritual and physical elements were reiterated often verbatim by John Brookes's *Manliness: Hints to Young Men* in 1859, 1875 and 1877.

Lectures and sermons in the 1850–60s, based on homophobia, send conflicting signals which place importance on both religion and physicality, as in *Tom Brown's Schooldays*. Sport increasingly played an important part in preparing boys and young men to be manly and masculine. In 'Culture and Its Enemies,' Arnold recognizes the 'essential value' of physical strength but advocates the 'subordination to higher and spiritual ends of the cultivation of bodily vigour and activity' (Arnold 1867, 43). Between 1860–80, 'games became compulsory, organised and eulogised at all the leading schools,' and men like Leslie Stephen at Cambridge took up athletics with a vengeance (Newsome 1961, 222, 216; Weeks 1981, 40). For Stephen, who became an atheist, sport and physical ruggedness were integral to manliness which had nothing at all to do with religion.

During the serialization of *Phineas Finn*, several articles about sport appeared in *Saint Pauls*. Stephen contributed two anonymous articles, about rowing and alpine climbing. He was an avid oarsman at Cambridge, and his nostalgic memories of his all-male experience in rowing certainly question just what it meant to be manly:

> To be in the same boat with a man is a proverbial expression implying the closest conceivable bond of union.... But the bond established for the time being between the members of a racing crew, is perhaps the closest known, with the single and doubtful exception of marriage.
>
> ([Stephen] 1867, 329)

Stephen, like Caird in his sermon, is aware of the sexual eagerness of young undergraduates; but for Stephen, the cure for such passion is to get close to several other young men in a narrow boat and stroke vigorously in unison – a curious exercise in manly behaviour. It is worth reading his fine appreciation of young and virile male bonding alongside his anti-aesthete response to Pater's *Renaissance* in the *Cornhill*.[10] And it is also interesting to note that Stephen found Trollope a particularly manly individual, as he notes in a letter when he says that Trollope was 'as sturdy, wholesome, and kindly a being as could be desired' (Annan 1984, 303–4). Another *Saint Pauls* article, in an Iron John-like attempt to define masculinity through man's relationship with his untamed nature, asserts of the 'manly and genuine sportsmen' who enjoy shooting that:

> So natural is the taste for wild and adventurous shooting, which is innate in every man who is worthy of the name of man, and which it takes a long course of luxury, and of battue-shooting and of hot luncheons among the brown fern, to finally eradicate.
>
> ([Lawley] 1868, 548)

The writer, Francis Lawley, clearly prefers wild shooting to other domesticated forms, and wild shooting, in fact, renews one's manliness from the dangers of luxury. Lawley adds that the best wild shooting is found in places such as Ireland, Scotland, Africa, India and North America; clearly, there is an imperial overtone in his assertions that in other lands, one can revivify one's masculinity ([Lawley] 1868, 559). Other sports articles which appeared include pieces by Trollope on hunting, and by others on fishing, yachting, cricket and horse-racing; only horse-racing does not appear on Rev. Mr. Brown's list of acceptable manly activities noted above. The *Saint Pauls* articles, collected and edited by Trollope as *British Sports and Pastimes* in 1868, show just how important manliness was to the discourse of the magazine. As the title signals, all of the articles emphasized the essentially national character of the sports so that if the sports were manly and vigorous, so too were the English men who practised them. Hunting and shooting are important to the male social network of politics in *Phineas Finn*, and as in many of Trollope's novels, a whole chapter is devoted to foxes and hounds.

Trollope's *Saint Pauls* men

The great optimism in liberal progress, some argued, was to be found in the triumph of individualism (as Samuel Smiles suggests in *Self-Help*, for example), a creed given particular attention in the pages of *Saint Pauls* in *Phineas Finn*. The struggle to be an individual is part of Phineas's dilemma as he tries to break free of his dependence on his father and assert his position as an MP and man of the world. He must try to negotiate the border between party loyalty and self-integrity. As he embarks on his career, Phineas tells one of his political allies that, '"I wouldn't change my views in politics either for you or for the Earl, though each of you carried seats in your breeches pockets"' (Trollope 1985, 57). About the same time, Phineas's legal mentor, Mr Low, warns the youth that no good can come of entering parliament at such a young age:

> I see nothing in it that can satisfy any *manly* heart. Even if you are successful, what are you to become? You will be the creature of some

> minister; not his colleague. You are to make your way up the ladder by pretending to agree whenever agreement is demanded from you, and by voting whether you agree or do not.
>
> (Trollope 1985, 87, my emphasis)

Low associates manliness with self-assertion and independence. One of the more annoying things about Phineas is indecisiveness which leads him to renounce his principles: for example, he lives off his father and accepts Whig patronage, which go against his instincts. Part of his education and maturation involve a growing independence apart from the influence of others.

Phineas Finn is a divided subject. He is both Irish and British, both boy and man. He is at a transitional crossroads in history when, among other things, his home country fights for basic democratic rights documented in numerous articles on the Irish questions published in *Saint Pauls*. Edward Dicey and John Herbert Stack, for example, wrote articles in the magazine asserting that a point of agreement for all Liberals was the disestablishment of the State Church in Ireland. Like the country from which he comes, Roman Catholic Phineas tries to establish himself as an independent subject at the very centre of power in a dominant culture. But Phineas's maturation from boy to masculine/manly self is complex and full of anxiety and uncertainty. John Rutherford suggests that 'masculinity is overly dependent for its coherence upon external public discourses. In consequence, men will experience periods of social and cultural transition as a disturbance to their identities' (Rutherford 1992, 15). In the most public and most male of discourses (politics) at a time of cultural disturbance (after the Reform Bill) Phineas tries to become a man and an individual. *Saint Pauls*, a new magazine making its own way in the world, dedicated to the politics of Liberalism, was just the place to explore the discourse of male growth.

Phineas represents a new generation of political men which emerges at a political crossroads in 1867. The second Reform Bill, certainly as it is constructed in the novel, acts as a pivotal moment for political generations. In Trollope's political world, the ballot becomes an issue which appears to define how far an older generation of politicians is willing to go in relinquishing established principles. John Sutherland notes that 'all parties shared a common excitement that society was about to take "a leap in the dark"' (Sutherland 1985, 10); indeed, this was a catchphrase for reform, as it had been around 1832. The first article in the first issue of *Saint Pauls* is 'A Leap in the Dark' by Edward Dicey, and *Punch* also used the slogan in cartoons. Some hoped that new generations of

MPs, represented at their youngest by Phineas, would continue the momentum of change. Michael Roper and John Tosh, in *Manful Assertions*, argue that,

> one of the most precarious moments in the reproduction of masculinity is the transfer of power to the succeeding generation. ... The key question is whether the 'sons' take on the older generation's gender identity without question, or whether they mount a challenge, and if so how.
>
> (Roper and Tosh 1991, 17)

Part of the challenge for Phineas in becoming an individual is in trying not to replicate the sins of the fathers. In this respect, Phineas's ultimate return to Ireland, among other things, demonstrates an unwillingness merely to mimic the patterns of political manoeuvring of a previous generation of party politics.

That Phineas is a young boy at the beginning of the novel is made clear a number of times. Trollope insists that we see his development from youth to adulthood, and Phineas is described as both 'a plunging colt' and 'a gosling' (Trollope 1869, 59 and 60). His most important relationship is with his political adviser, Lady Laura Standish, who becomes both a surrogate mother and unrequited lover to Phineas. 'I will take upon myself the task of mentor,' Laura declares to Phineas, who in her eyes has the fault of 'being impetuous, – being young, – being a boy' (Trollope 1985, 113). This early exchange between Laura and Phineas establishes the mother–son strand of their relationship. Later we are told that 'Phineas had been, as it were, adopted by her as her own political offspring, – or at any rate as her political godchild' (Trollope 1985, 279).

Laura as both mother and lover to Phineas occupies an ambiguous position in the young man's psychological development. Phineas becomes intimate not just with Laura, but also with her father and brother. Laura's father acts as a political patron and surrogate father, and her brother becomes both brother and rival for Phineas. The break from Laura, of which she feels the sting deeply, is part of Phineas's progress away from the nurturing mother/unconsummated lover to a more adult, but all-male world. Lynne Segal argues of masculine development that male rites of passage signify the move from dependency in childhood to 'a new sense of belonging to a distinct world of adult males': 'it is insufficient for the "men" to be distinguished from the "boys"; the "men" must be distinguished from the "women"' (Segal 1990, 131 and 132). In order to become an adult male, with a developed sense of masculinity, Phineas must abandon Laura and enter fully into the male domain of politics and parliament. Segal's argument about separation is important to keep in mind when we remember the point made earlier about children's magazines in the 1860s and the increasing distinction according to gender. *Saint Pauls*, although importantly including serial fiction, wished to distinguish itself as a political (male) magazine, and the failure to secure a substantial female readership was undoubtedly one reason for the periodical's lack of success. The shilling monthly market was a hybrid one and *Saint Pauls* did not offer a full enough range of interest to attract the all-important woman reader. On the whole, it was a magazine written by men, to men, about male discourse.

In addition to establishing his place with other men, Phineas undergoes other male rites of passage. Perhaps the supreme hurdle in Phineas's process of maturation is the need to speak in Parliament, to become a fully-realized speaking subject. For an MP, a maiden speech in the House is a rite of passage; but, more broadly, the ability to assert oneself in front of other men forms part of masculine, manly development. When approached and given the opportunity to speak during the ballot debate, Phineas has a panic attack:

> The chamber seemed to swim round before our hero's eyes.... The task was exactly that which, of all tasks, he would best like to have accomplished, and to have accomplished well. But if he should fail! And he felt that he would fail.... At this moment, so confused was he, that he did not know where sat Mr Mildmay, and where Mr Daubeny. All was confused, and there arose as it were a sound of waters in his ears, and a feeling as of a great hell around him.
>
> (Trollope 1985, 218)

Phineas drowns in a sea of impotence, unable literally to rise to the occasion and deliver his speech. Because the worth of MPs in the novel is determined largely by speeches in the House, Phineas, in not having a voice, is politically impotent and experiences acute anxiety.

After his initial failure, Phineas does make other more successful attempts at speech. It is crucial that Phineas find his voice because it is through a speech in parliament that the young boy finally becomes a young man and asserts his individuality. This occurs during Irish land reform debates in which Phineas cannot agree to support his party and government. Having resigned his position at the Colonial Office, Phineas waits for his opportunity to speak and recalls his first unrealized attempt to speak in the House. He waits impatiently but 'his audience was assured to him now, and he did not fear it' (Trollope 1869, 701). In delivering such a speech, Phineas finds his voice and becomes a fully-realized speaking subject. The divided subject of before, uncertain of his identity, is now an assured and confident member of parliament able to vocalize his principles to his male peers. He has gained a sense of nationality, and as Julian Wolfreys believes, 'his resignation is an affirmative refusal of English identity' (Wolfreys 1994, 307). Phineas is an outsider in English society whose position on the margins is reinforced: the Irishman in England, the boy in a man's world, the humble man in high society. Early on, observing English politics at work, he realizes that 'he, at any rate, could not be one of them' (Trollope 1869, 125),[11] but in the end he emerges ahead of the pack. In his farewell speech we see a synthesis of the strands of his maturation: he speaks out publicly and decisively for personal principle and embraces an understanding of his own identity. That this development relates to his masculinity is made clear in the conclusion; Phineas resolves that in the move back to Ireland, 'he would make the change, or attempt to make it, with *manly* strength' (Trollope 1985, 712, my emphasis). Part of Trollope's boldness in the novel is to put an Irishman at the centre, so that at a time when the Irish Question was passionately debated, the serial acts, in accordance with the Liberal politics of the periodical, as a political statement about Irish identity within British culture.

'In writing *Phineas Finn*, and also some other novels which follow it,' Trollope records in the *Autobiography*, 'I was conscious that I could not make a tale pleasing chiefly, or perhaps in any part, by politics. If I wrote politics for my own sake, I must put in love and intrigue, social incidents, with perhaps a dash of sport, for the sake of my readers' (Trollope 1989, 317). Trollope was aware, then, that he had at least to try to make his novels (and his periodical) appeal to a wider audience than a strictly political novel could do. He faced this problem again in *Ralph the Heir*, the other serial he contributed to *Saint Pauls*, published as a supplement to the periodical beginning in January 1870. Trollope left as editor in July, 1870, but the novel was written to be serialized in the periodical and the discourses of *Ralph the Heir* are attuned to the political, male discourses of the magazine generally. For Trollope's money, 'it was in part a political novel; and that part which appertains to politics, and which recounts the electioneering experiences of the candidates at Percycross, is well enough' (Trollope 1989, 343). As in *Phineas Finn*, then, we see Trollope wanting to write a political novel, to engage generally with the other discourses in *Saint Pauls*, but filling it out with love interests, presumably aimed at readers of (especially his own) popular domestic fiction.

In the introduction to his edition of *Ralph the Heir*, John Sutherland makes the point that the novel was always intended to be a man's story. The working title for the novel was 'Underwood,' indicating that the focus was to be on Sir Thomas Underwood, the old lawyer who spends most of his time in chambers, supposedly working on a biography of Francis Bacon. Sutherland argues that Sir Thomas was imagined as a Lear figure, and that Trollope only introduced the romantic love plots around the two Ralphs when it became possible that the novel might not be serialized in the failing *Saint Pauls* (Sutherland 1990, xvi). Sir Thomas is the most interesting character in the novel, and in the love plots, we have conventional romance characters and no exceptionally well-wrought heroines on the par of, for example, Lily Dale in *The Small House at Allington* or Lady Glencora in *Can You Forgive Her?* However, *Ralph the Heir* complements *Phineas Finn* in a number of ways, which further demonstrates the male project of *Saint Pauls*.

Most obviously, the two Ralphs remind us of Phineas and his rival Lord Chiltern, on the verge of passing from youth into manhood. Both Ralphs share the same lover, and like Phineas and Chiltern, Ralph Newton (the heir) 'was nearly driven wild with the need of deciding' who to marry and how best to take responsibility for his life (Trollope 1991, bk 1, 208). Ralph is particularly close to Chiltern in his spendthrift attitude and passion for hunting and its attendant men's pleasures. Much of the novel is taken up with his need to resolve whether or not to marry Polly Neefit, a breeches maker's daughter with a large fortune, in order to pay off the substantial debts from his lavish bachelor's life of pleasure. The other Ralph Newton (a bastard cousin of the heir's) is described as having 'a double memory, and a second identity' (Trollope 1991, bk 2, 68), just as Phineas is said to have two separate identities. This Ralph is isolated by birth in a similar way that Phineas's Irish identity sets him apart from the clique of Liberals who form his circle of friends. Both Ralph the heir and Ralph the bastard, then, would have been familiar in some ways to readers of *Phineas Finn*.

Sir Thomas brings up other discourses about men, particularly about male responsibility and work and about the separate spheres that men and women inhabit in the novel. Something which Sir Thomas must comfort is that his life, especially in later years, has been unproductive, that, literally, he does nothing of use or importance. Most ridiculously, he has been preparing a Life of Bacon for thirty-five years without having written a single word, and, finally, he must accept his own failures:

> Why had he not contented himself with having his children around him; walking with them to church on Sunday morning, taking them to the theatre on Monday evening, and allowing them to read him to sleep after tea on the Tuesday? He had not done these things, was not doing them now, because he had ventured to think himself capable of something that would justify him in leaving the common circle. He had left it, but was not justified. He had been in Parliament, had been in office, and had tried to write a book. But he was not a legislator, was not a statesman, and was not an author. He was simply a weak, vain, wretched man, who, through false conceit, had been induced to neglect almost every duty of life!
>
> (Trollope 1991, bk 2, 285)

As this passage attests, the novel is, in Sutherland's words, 'suffused with a premature but overwhelming sense of age's growing impotence and exile from all that previously gave life savour' (Sutherland 1990, xii). It particularly registers male fear – fear of domesticity, of aging, of failure, of neglected duty. If *Phineas Finn* depicts the growth of a young man on the threshold of a brilliant career, *Ralph the Heir* depicts an old man whose achievements were only ever illusory. Phineas's promising political career is cut short, while Sir Thomas is left to wonder how he has so profoundly misjudged his life. The novels share similar concerns but use men at different times of life to discuss them. In this sense, *Saint Pauls* serialized the spectrum of middle-class development from youth into maturity, from maturity into old age.

There are other connections between Trollope's two novels serialized in *Saint Pauls*, such as the definition of what it means to be a gentleman in the late 1860s and the debates about worker's rights and trade unionism. A cluster of issues dealing specifically with men's lives and with men's concerns dominate these two serials. *Phineas Finn*, as the inaugural novel by the editor, helped define to a large extent the magazine's tone and position, and the male focus was reiterated and rearticulated in the second novel published as a supplement. But neither of these fictions brought in the popular readers *Saint Pauls* badly needed.

Collectively, then, *Phineas Finn* and *Ralph the Heir* represent a range of stories about men: particularly about male development. Trollope was often praised by reviewers for his sympathetic portrayals of women, but in *Saint Pauls* – a magazine he avowed to be political from the outset, a magazine sold on the strength of his name – his fiction focuses on the lives of men rather than women. *Saint Pauls* did carry other fiction, as indicated earlier, even sensation fiction, but it was Trollope, the

'Anthony Trollope' emblazoned across the adverts, who defined the mood and tone of this politically-focused shilling monthly.

Notes

1. The seminal work on manliness in Victorian culture is Newsome. More recent works on manliness which amend some of Newsome's conclusions include: Roper and Tosh, Weeks, Segal, and Gay. For recent work on masculinity in general, see Rutherford.
2. See Weeks 50.
3. See Anon, 1868, 501–2, which agrees with the author on the importance of boys.
4. See Boyd 1991, 147: 'Manliness is one of the central goals of character-formation, and the boys' story paper offered readers a place to see this in the making'.
5. See Gay 115–16.
6. See Brookes.
7. My thanks to Lynda Nead for suggesting that Phineas's masculinity is metropolitan.
8. Note that several critics do not accept Newsome's lumping of Hughes with Kingsley. This partly depends on how one reads *Tom Brown's Schooldays*. See articles by Nelson and Hilton. Although I do not generally accept that Hughes was a muscular Christian in the same vein as Kingsley, my point here is to distinguish two strands of thinking about manliness.
9. See the well-known letter from Samuel Wilberforce, bishop of Oxford, to a friend in 1858 in Newsome 1961, 208. See also Adams 1992, 456.
10. See [Stephen] 1875, 91–101. Stephen's argument is that morality is an essential element of all art, and that Pater's *Renaissance* (although not mentioned directly) encourages, indeed glorifies, immorality. Note that Stephen read the first edition which contained the notorious and homoerotic 'Conclusion,' suppressed in the 1877 edition.
11. Later in the novel, Phineas 'found himself among them as one of themselves,' (Trollope 1869, 379). However, this clearly is not the case, as the end of the novel insists.

Works cited

Adams, James Eli. 'Gentleman, Dandy, Priest: Manliness and Social Authority in Pater's Aestheticism,' *ELH* 59 (1992), 441–66.

Annan, Noel. *Leslie Stephen: The Godless Victorian*. London, 1984.

Anon. Review of A. R. Hope, *A Book About Boys*, *Saturday Review* 26 (10 Oct. 1868), 501–2.

Arnold, Matthew. 'Culture and Its Enemies,' *Cornhill Magazine* (July 1867), 36–53.

Boyd, Kelly. 'Knowing Your Place: The tensions of manliness in boys' Story Papers, 1918–39,' in Michael Roper and John Tosh, eds, *Manful Assertions: Masculinities in Britain since 1800*. London, 1991, 145–67.

Brookes, John. *Manliness: Hints to Young Men*. London, 1859.

Brown, Rev. Hugh Stowell. *Manliness: A Lecture*. London, 1858.
Caird, John, D. D. C. 'Christian Manliness. A Sermon preached before the University of Glasgow.' Glasgow, 1871.
[Cayley, George John]. 'On Matrimony,' *Saint Pauls Magazine* 1 (March, 1868), 732–7.
Dellamora, Richard. *Masculine Desire: The Sexual Politics of Victorian Aestheticism*. Chapel Hill, 1990.
[Dicey, Edward]. 'Our Programme for the Liberals', *Saint Pauls Magazine* 1 (March, 1868), 659–74.
Gay, Peter. *The Cultivation of Hatred*. London, 1993.
Harrison, Frederic. 'Culture: A Dialogue', *Fortnightly Review* 2 (Nov., 1867), 603–14.
Hilton, Boyd. 'Manliness, Masculinity and the Mid-Victorian Temperament', in Lawrence Goldman, ed., *The Blind Victorian: Henry Fawcett and British Liberalism*. Cambridge, 1989, 60–70.
Hope, A. R. *A Book About Boys*. Edinburgh, 1868.
[Lawley, Francis]. 'On Shooting', *Saint Pauls Magazine* 1 (Feb., 1868), 546–59.
Malcolm, Judith Wittosch. 'Trollope's *Saint Pauls* Magazine'. Unpublished Ph.D. dissertation, University of Michigan, 1984.
Nelson, Claudia. 'Sex and the Single Boy: Ideals of Manliness and Sexuality in Victorian Literature for Boys', *Victorian Studies* 32(4) (Summer, 1989), 525–50.
Newsome, David. *Godliness and Good Learning: Four Studies on a Victorian Ideal*. London, 1961.
[O'Neil, Henry Nelson]. 'Taste', *Saint Pauls Magazine* 1 (Oct., 1867), 92–102.
Pater, Walter. 'Winkelmann', *Westminster Review* 87 o.s., 31 n.s. (Jan. 1867), 80–110.
'Prospectus' [for *Tinsley's Magazine*]. *Athenaeum* advert (13 July, 1867), 62.
[Rathbone, WM] 'Social Disintegration', *Macmillan's Magazine* 16 (May 1867), 28–35.
Roper, Michael and Tosh, John, eds. *Manful Assertions: Masculinities in Britain since 1800*. London, 1991.
Rutherford, Jonathan. *Men's Silence: Predicaments in Masculinity*. London, 1992.
Segal, Lynne. *Slow Motion: Changing Masculinities, Changing Men*. London, 1990.
Smiles, Samuel. *Self-Help*. London, 1866.
Stack, John Herbert. 'The Irish Church', *Saint Pauls Magazine* 1 (Feb., 1868), 564–77.
[Stephen, Leslie]. 'About Rowing', *Saint Pauls Magazine* 1 (Dec., 1867), 319–33.
——. 'Anonymous Journalism', *Saint Pauls Magazine* 2 (May, 1868), 217–30.
——. 'Art and Morality', *Cornhill Magazine* 32 (July, 1875), 91–101.
Sutherland, John. 'Trollope and *St Paul's* 1866–70', in Tony Bareham, ed., *Anthony Trollpe*, London, 1980, 116–37.
——. 'Introduction' to Anthony Trollpe, *Phineas Finn, The Irish Member* (1869). London, 1985.
——. 'Introduction' to Anthony Trollope, *Ralph the Heir* (1870). London, 1990.
Trollope, Anothony (1883). *An Autobiography*, eds. Michael Sadlier and Frederick Page. Oxford, 1989.
[Trollope, Anthony]. 'Introduction', *Saint Pauls* 1 (Oct., 1867), 1–7.
Trollope. Anthony (1869). *Phineas Finn, the Irish Member*, ed. John Sutherland. London, 1985.

[Trollope, Anthony]. 'The Uncontrolled Ruffianism of London, as Measured by the Rule of Thumb', *Saint Pauls Magazine* 1 (Jan., 1868), 419–24.
Trollope, Anthony (1871). *Ralph the Heir*, ed. John Sutherland. London, 1990.
Tuss Alex J. *The Inward Revolution: Troubled Young Men in Victorian Fiction. 1850–80.* New York, 1992.
Weeks Jeffrey. *Sex, Politics and Societgy: The Regulation of Sexuality since 1800.* London, 1981.
Wolfreys, Julian. 'Reading Trollope: Whoose Englishness Is It Anyway?', *Dickens Studies Annual* 22, 1994. 304–29.

16

The Agony Aunt, the Romancing Uncle and the Family of Empire: Defining the Sixpenny Reading Public in the 1890s

Margaret Beetham

Introductory

In the 1990s feminist history and the history of imperialism have been ruitfully brought together, a relationship that Clare Midgely describes as uncomfortable but productive (Midgely 1998, 1). This chapter seeks to extend the complexity of that dialogue by bringing into it a third element, the cultural history of the metropolita popular press, represented here by two particular periodicales, *Women at Home* and *Longman's Magazine*. In 1882 the publisher Longman's announced the launch of a monthly magazine costing sixpence and bearing the house name. Like many of his contemporaries, Charles Longman argued that the 1870 Education Act had produced an 'immense class' eager for good reading matter at 'a reasonable price'. The success of his own firm's sixpenny reprints was taken as evidence both of this new reading public and of what it was prepared to pay (Longman's *Notes on Books*, Sept. 1882). *Longman's Magazine* was to last twenty-three years, a period marked by increasingly 'severe' competition between publishers to define the immense new reading public and to capture its loyalty (Anon. 1905, 518). These were the decades of the 'New Journalism'. The arrival from America of magazines like *Harper's* with plentiful quality illustrations and above all the launch of Newnes's *Strand Magazine* in 1891 shifted the ground on which the competition for the sixpenny reader was fought.

As we reach our own *fin-de-siècle*, there are dangers in searching the 1890s for our late twentieth-century concerns. The late nineteenth-century debate around the 'new' reading public must be understood in relation to the specifics of 1890s metropolitan periodical publishing and through close analysis of its texts. However, the competition for the sixpenny magazine market was part of a wider struggle over the nature of the English reading public which still resonates today. 'The New Journalism' and the attempts by Longman's and others to define a 'mass' periodical press went with a series of revolutionary changes in the discourses and institutions of publishing, including a bitter debate about the nature of fiction, the collapse of established patterns of fiction publishing and changes in the profession of writing marked by the establishment of writers' organizations.[1] One hundred years later our print and other media are still shaped by these transformations and by the cultural politics of that struggle over the meaning of popular reading.

In defining the sixpenny reader, the position of monthly magazines like *Longman's* at the metropolitan centre of Empire was crucial. Here 'reading' and 'Englishness' were linked, and inflected in complex ways with class and gender. Yet these relationships tended to be taken for granted. The sixpenny reader was simultaneously defined as 'the common man' or 'Everyman' *and* located very specifically in terms of nation, class and gender. This dynamic between universal claims and specifics in the creation of new identities is my topic.

My method is close textual analysis of material from two sixpenny monthly magazines, *Longman's* (1882–1905) and *Woman at Home* (1892–1920), the second of which the publishers, Hodder and Stoughton, launched specifically as a *Strand Magazine* for women (Darlow 1925, 111). These two offer useful case studies because they span the spectrum of the sixpenny monthly. *Longman's*, unillustrated, aimed at a general readership and stressing 'high quality literature', claimed to carry forward the traditions of mid-century publishing. By 1890 it was already looking old-fashioned but it remained a powerful cultural presence, particularly through its association with the critic, Andrew Lang, whose gossipy column or causerie was a regular feature from 1885. *Woman at Home*, by contrast, was explicitly new journalistic, copiously illustrated, and targeted specifically at women, with advice on love, marriage, cookery and fashion. It, too, was associated with a well-known writer, the romantic novelist, Annie S. Swan. Yet both publications adopted the miscellany form, appeared monthly and concentrated on 'quality' fiction. Both were in competition for 'the sixpenny public' which they claimed represented the new 'mass' readership.

The word 'mass' here is relative. Sixpence was exactly half the price of the *Cornhill* which had led the field of middle-class monthly fiction magazines since its launch in 1860. However, sixpence was still a good deal more than Newnes's weekly *Tit-Bits* (at tuppence) or the cheapest fiction papers which sold for a penny or a halfpenny. Though he claimed to address the Board school leavers, Longman was not targeting the average working-class man. Similarly, when *Woman at Home*'s publishers said their new magazine aimed to catch the masses, they meant the middle class (Darlow 1925, 111, see also *Woman at Home* 1 (1893), 62 and *passim*). The 'sixpenny' public was always implicitly located in the lower-middle or middle class and this assumption pervaded every aspect of these journals.

Another element which these two apparently very different magazines had in common was their deployment of that new journalistic device, the regular causerie or correspondence column to which readers were invited to send contributions and where the same writer dealt month after month with a range of topics.[2] Through the personae of their two star contributors, Andrew Lang in *Longman's* and Annie S. Swan in *Woman at Home*, the magazines offered their readers not only information, advice and entertainment but models of the self.

Of all ephemeral forms, the causerie and correspondence column seem the most ephemeral. Unlike some of their rivals, neither Lang nor Swan ever published a selection from their columns in volume form (Green 1946, 57, 251). Tucked away towards the end of their respective magazines, these two columns seem utterly marginal to the culture of the period. This is a mis-reading. Because of the particular nature of the periodical as simultaneously of its moment and part of a series, the regular column in which readers and writers return month after month to the same topics is a significant cultural presence. Moreover, it is a space in which readers are invited to become writers. To varying degrees both Lang and Swan dramatized in their columns the process of negotiation between writer and readers through which collective meanings are forged in print. Both these columns occupied an important place not only in the specific magazines in which they appeared but in the wider print culture. Forgotten though both are today, they were at the centre of the process of defining the new reading public of the 1890s. It is on these columns, therefore, that I focus in my discussion of the way new identities were created from the metropolitan centre at the *fin de siècle*.

The column and the causerie

In the first of his columns, Lang in a characteristic verse introduction promised his readers 'a stall of bric-a-brac' (*LM* 7 [1885–6], 317). The miscellaneous nature of the causerie made it an epitome of the whole magazine with its variety of different subjects and voices. For most readers Lang's column was representative of *Longman's* in a more specific sense since they assumed that Lang was the editor. This belief gained credence because the title of his column, 'At the Sign of the Ship', referred specifically to the logo of the firm and because the real editor, Charles Longman, was invisible. So persistent was this claim that Lang was forced more than once to deny it (*LM* 9 [1887–8], 351). However, the denial was somewhat disingenuous since Lang not only read manuscripts and advised Charles Longman generally on what he should publish, but he also took an active role in shaping the other contents of the magazine and specifically the fiction (Blagden 1963, 19).

Annie S. Swan's column 'Over the Teacups' in *Woman at Home* was not a causerie but an 'Advice to Correspondents'.[3] It was perceived by readers to be 'the titbit' of the magazine (Black 1906, 319). This phrase suggested not only the column's importance in the magazine in which it appeared but the way it represented the new kind of journalism of which Newnes's magazine *Tit-Bits* was one of the first and most important examples. Certainly *Woman at Home* was openly modelled on the new journalism of which Newnes was a pioneer, though taking his sixpenny *Strand*, rather than the tuppenny *Tit-Bits* as its model.

As with Lang, so with Swan, readers so thoroughly identified the magazine with their favourite columnist that they regularly assumed that she was the Editor of the magazine, an assumption encouraged by its sub-title 'Annie S. Swan's Magazine'. Despite this, *Woman at Home* was never edited by Mrs Burnett Smith, the writer whose pen-name was Swan. However, she did work closely with the first editor and magazine's originator, her friend Robertson Nicoll, and she played a crucial role in shaping the rest of the magazine, especially its fiction, much of which she wrote herself (Beetham 1996, 165, 171–3). Both Lang and Swan, then, not only represented their respective magazines to their readers, but they actively shaped the contents and tone of the whole. This was particularly true of the fiction, which continued to form the staple of the monthly miscellany.

Both Lang and Swan were journalistic stars with substantial reputations before they embarked on these particular columns and with substantial publications outside the magazines during their time as columnists. Lang's facility as a writer and his great productivity led to the rumour, which he jokingly endorsed, that he was in fact a syndicate (Green 1946, 463). In the causerie, therefore, he did not so much create a persona as provide a space in which the already mythical 'Andrew Lang' could appear. In that space he simultaneously laid out his 'stall of bric-a-brac' and regulated his readers' relationship with him, constructing a mixture of intimacy and distance which is now so familiar in relation to media stars. On one hand Lang represented himself as just like his readers and indeed dependent on them for examples of folk-stories, psychic experiences, and poetry – especially ballades, which he printed and commented on. At the same time, he constantly complained of being besieged by unwanted letters (especially from women) on precisely those topics: 'May I ask ladies not to write to me about worthless old books in their possession? Ballade mongers are piteously implored not to send me any ballades' (*LM* 9 (1887–8), 351–2).

Another of Lang's recurrent complaints was of correspondents who wanted him to help them get published and whom he systematically discouraged (e.g. *LM* 9 (1887–8), 569 and 35 (1899–1900), 377). A typical Lang verse 'To Correspondents', prefaced with the Latin tag, 'Eheu fugaces, postmane, postmane' ran:

> My postman, though I fear thy tread
> And tremble as the foot draws nearer,
> 'Tis not the Christmas Dun I dread,
> *MY* mental foe is much severer,
> The Unknown correspondent – who
> With indefatigable pen
> Perplexes literary men.
>
> (*LM* 8 [1886–7], 446–7)

The literary man and his correspondent, the sixpenny reader, are separated by the immense gap between the professional writer and his admirer, yet both are characterized by the 'indefatigable pen'. The tone – so typical of Lang – defuses the complaint and establishes a sense of shared values around the pleasure of the literary joke. That derives not only from the skill with which Lang works his rhyme and rhythm but also from the shared knowledge of Latin grammar and literature, which the reference to Horace and the dog Latin address to 'the postman' implies.

Annie S. Swan's persona as a writer was serious when Lang's was light and her reputation rested on her work as a romantic novelist rather than across the range of journalism and scholarship where Lang was famous. However, she too brought a body of devoted readers and correspondents into the magazine. Indeed one version of the story of the magazine's launch was that it was intended to provide a space in which Swan could regulate an unmanageable body of correspondence from her readers (Black 1906, 319). She explicitly positioned herself as sharing her readers' sorrows and joys in the intimacy created 'Over the Teacups'. She never berated correspondents for writing to her but did complain that they constantly assumed she could get them into publishing (*WAH* 5 [1895–6], 67). Like Lang she resisted the idea that writing was an open profession in which anyone who could hold a pen could make a living. For Swan, even more than for Lang, her regular column brought into sharp focus the paradoxical relationship of the professional writer to the new sixpenny readership. Distanced from her readers by her position as a media personality, she positioned herself as essentially *like* her readers. Her access to public print, she argued, enabled her to create the space in which their shared values could be articulated.

As professional writers in several genres, Swan and Lang were both engaged in trying to find new forms of writing which would be appropriate for the end-of-the-century readership. For both of them the magazine with which they were so closely associated provided an important place for that engagement, particularly in relation to fiction. Debates over the meaning of 'the new fiction' were entangled with debates about the new reading public.

The agony aunt and the search for new forms

Annie Swan understood very well that her readers were no longer happy with the old 'goody-goody' stories, what she called the 'pabulum', of an earlier generation (*WAH* 1 [1892–3], 62). However, she resisted the 'new' fiction just as she resisted the demands of the 'new' woman. Rejecting the experimental genres in which George Egerton or Olive Schreiner sought to redefine feminine fiction, Swan was forced back to the old models. Though some of her fiction in *Woman at Home* moved away from the traditional pattern of the love and marriage narrative, particularly in the series about a lady doctor, Elizabeth Glen, it never moved very far (Beetham 1996, 171–2). The dominant model of the stories in the magazine, whether serialized novels or short pieces, remained the traditional female romance.

This is hardly surprising. Swan, like the magazine, addressed her reader as the 'Woman at Home'! Whatever else characterized the new reading public, gender difference was crucial to it. Men wrote for 'her' magazine and apparently read it; they certainly sent in letters quite regularly (ibid., 214). Nevertheless, the magazine positioned its readers 'as women'. Closely linked to this was the assumption that femininity was always domestic, which meant woman was always placed in relationship to men in the family (Darlow 1925, 111). For Swan the new woman's rejection of traditional femininity was a rejection of the family, the basis of a social identity which encompassed both mutuality and hierarchies of difference (between husband and wife, parents and children, and master/mistress and servants).

This ideology shaped Swan's advice to her correspondents and also her search for appropriate modern female fiction. Not surprisingly, then, she returned to the romance narrative with its final insertion of the heroine into her role as wife and mother. For Swan, the family provided a model of the social which could be carried into the new era, and the romance provided a narrative pattern which could be re-made for the new reading public. The two were utterly entangled.

The context in which this tradition had to be reworked was not confined to Britain but encompassed the British imperial possessions and an increasingly international market in print. In 'Over the Teacups' Swan regularly printed letters from readers and correspondents 'in the colonies' (e.g. *WAH* 5 [1895–6], 870). The constant tenor of Swan's responses was that, across the world and against all evidence of difference, she and her readers were alike. 'Family characteristics', she wrote, 'are the same everywhere' (*WAH* 6 [1896–7], 711). However, the vocabulary of family and the pattern of romance resonated differently in the colonial context. For her colonial correspondents, Annie Swan represented 'The Woman at Home' not only as the domestic woman but also as the metropolitan arbiter of values. This was most clearly evident in the new connotations which 'Home' carried in the colonial context where it meant not only the domestic but also Britain, the 'motherland' of the Empire.

For Swan and her readers, the metaphor of family provided a model for understanding and living out relationships between women in England and their 'sisters' overseas. It functioned also as a way of managing the relationships between English women abroad and the 'natives', who were frequently described as metaphorical children, in need of discipline and control by the benevolent parent (e.g. *WAH* 5 [1895–6], 468). In this dynamic Annie S. Swan occupied the family role of 'auntie', the older, wiser woman who tells stories, answers questions and encourages her young female relatives to take on appropriate feminine identities. Despite the long tradition of advice to correspondents in women's magazines, Annie S. Swan was arguably the first modern magazine 'agony aunt'. In terms of the search for a form of print appropriate to the new reading public, this was perhaps her most significant achievement. Though the popularity of her romantic fiction helped re-make that genre for a mass market, her novel writing was formally conservative. However, in her correspondence column she instituted a thoroughly 'modern' form of print and mode of address through her persona as older woman, at once intimate and distant from her readers. 'Over the Teacups' offered female readers a model of identity in 'Annie S. Swan'. It also provided practical advice and fantasies embodied in familiar narrative patterns to aid them in achieving such an identity themselves.

In arguing that 'family characteristics are the same everywhere', Swan assumed that femininity was – at least potentially – also 'the same everywhere'. Her readers, whether 'at home' or overseas, could become like her, embodiments of true womanhood. In making such universalized claims Swan sought to make invisible the particular class and metropolitan location which were essential to the identity she represented. The difficulties of this are apparent in her column. As a professional writer she was not simply the domestic woman constructed by her column. As a Scot who had moved to London but maintained she was still at heart a Scot, she denied the importance of that move to the metropolis.

For her correspondents, too, the difficulties of living the identity she offered them were enormous. Her column was marked on one hand by the 'bitter cry' of the woman who had no man to support her financially 'at home' and on the other by silences and absences (*WAH* 5 [1895–6], 67). The mill girl, the servant, the colonized woman did not fit easily into the model of 'The Woman at Home'. Yet the identity of the middle-class woman in 'Over the Teacups' depended directly on the work of the domestic servant, and the metaphorical tea at Swan's table was evidence of the centrality of 'the colonies' to the material as well as the discursive economy of the ordinary English household.

In constructing a universalized female identity for her readers as 'women at home' Swan continued to make invisible those women who did not fit her model but on whom the English middle-class family depended. At the same time she claimed her column as the space where readers could become writers and enter as equals into the making of print identities. This did allow for the occasional letter from one of these 'other' women. A lively exchange about and from 'business girls', that is shop assistants, a letter from an Indian Christian or from a mill girl presented Swan with difficulties because they disrupted the identity she constructed (*WAH* 5 [1895–6], 468, 469). However, these disruptions were contained – at least in the 1890s.

Andrew Lang, the romancing uncle

Compared to Lang, Swan was a minor figure, but then there were few of his contemporaries who could compete with Lang in terms of literary output and influence across the range of scholarly and popular reading. He was not only one of the founders of British folklore studies, a highly regarded translator of classical works, and editor of collections of fairy tales that shaped nursery reading for more than one generation, he was also the most prolific and perhaps most influential literary critic of his day (Maurer, 1955; Green 1946, 157–74). A favourable mention by Lang could make an author's name (Elwin 1939, 241, 271). Yet despite the apparent gap between them in literary esteem and position, Lang wrestled with the same problems of defining the sixpenny public as did Swan and resolved them in analogous ways. Like Swan, he turned to romance as an appropriate form for the new reading public; he, too, worked consciously within a colonial context; he, too, assumed a model of human nature which was universal but divided by hierarchies of gender and race. These similarities were not accidental but structural. They structured the new forms of reading and the new identities offered the sixpenny reader.

Lang's importance as a dictator of end-of-the-century literary taste was most dramatically evident in his espousal of 'romance' over 'realism'. 'Romance' here meant not the traditional female narrative but the male quest story, of which Haggard's best-selling novel *She* remained for him the best example. Lang was well aware of the need to find a form of fiction appropriate to the new reading public. He consistently argued that the choice was between novels of adventure like Haggard's and the work of 'the Russian novelists', or of writers like Henry James and Thomas Hardy which he condemned as dreary and depressing realism. Though he admitted that fiction had many mansions, his own position was clear:

> The dubitations of a Bostonian spinster may be made as interesting, by one genius, as a fight between a crocodile and a catawampus, by another genius. One may be as much excited by trying to discover whom a married American lady is really in love with, as by the search for the Fire of Immortality in the heart of Africa. But if there is to be no *modus vivendi*, if the battle between the crocodile of Realism and the catawampus of Romance is to be fought out to the bitter end – why, in that Ragnarok, I am on the side of the catawampus.
>
> (*CR*, 52 [1887], 693)

This view informed Lang's reviewing and was a constant theme of his journalism. His support was crucial in establishing the reputations of Robert Louis Stevenson, the poet Newbolt, and above all of Rider Haggard, whose work he constantly puffed and with whom he collaborated in writing a novel. Rudyard Kipling, Stanley Weyman, Conan Doyle and Anthony Hope were others whose works he promoted. In *Longman's Magazine* Lang's theories on fiction informed not only the writing in his causerie but the policy on what kind of fiction the magazine carried (Blagden 1963, 19). Throughout its publishing history, *Longman's* was dominated by what Lang called 'romance': works like Haggard's *Allan Quatermain* (the sequel to *King Solomon's Mines*) and Stanley Weyman's historical *A Gentleman of France* (Maurer 1955). One of the few substantial exchanges on literary matters in the magazine's publishing history was precipitated by Robert Louis Stevenson's 'Gossip on Romance' (*LM* 1 [1882], 69–79).

In his causerie, Lang celebrated 'The Restoration of Romance' in a verse, dedicated jointly to Stevenson and to Haggard, whose serial was then running in the magazine:

> King Romance was wounded deep.
> All his knights were dead and gone,
> All his Court was fallen asleep,
> In a vale of Avalon...
>
> Then you came from South and North
> From Tugela, from the Tweed
> Blazoned his achievements forth.
> King Romance is come indeed.
>
> (*LM* 9 [1887], 554)

In advocating Romance as the ruler of the new realms of popular fiction, Lang represented himself as the sixpenny reader. Romance, he argued, appealed at a universal level and intellectuals who pretended to despise it were simply hypocrites. In his *Longman's* causerie, Lang assumed the mantle of the ordinary man:

> But does one read Dante on a train? Alas I fear one reads a silly sensational story... One might be reading the Florentine: one is perusing *Captain Kettle* and a very entertaining sixpence worth the Captain provides. Is it not so,
>
> Hypocrite lecteur, mon semblable, mon frere?
>
> We are all miserable sinners, intellectually ruined by the fruits of popular education.
>
> (*LM* 38 [1901], 95)

Lang typically both invoked and denied his own privileged position. His anti-intellectualism was articulated with all the resources of the scholar whose education had been anything but popular. Henry James was not speaking only out of personal pique when he wrote on Lang's death;

> Where I can't but feel he *should* be brought to justice is in the matter of his whole 'give-away' of the value of the wonderful chances he so continually enjoyed... give-away, I mean, by his *cultivation*, absolutely of the puerile imagination and the fourth-rate opinion.
> (Letter to Gosse, quoted Maurer 1955, 175; italics in original)

In defining the sixpenny reader, Lang was open to James's charge, but this 'give-away' should not be read only in terms of his personal failure or of pure cynicism. For Lang 'puerile imagination' was not simply a term of abuse and we can only understand the importance of the ideology he articulated if we take seriously his judgements, not least his sense of his own work as second best: 'If one cannot be a soldier, a missionary or an explorer, a man of one's hands: if one has to live a tame life and die "a cow's death", then one prefers to scribble' (Green 1946, 214).

In advocating the life as well as the fiction of adventure, Lang was articulating a crucial aspect of end-of-the-century culture. His arguments about contemporary fiction were worked through in the context of a wider set of theoretical debates about the character of culture. As a pioneer in the developing disciplines of folklore and anthropology, Lang was engaged in trying to explain the place of story in various cultures. In particular he addressed the problem of how it was that certain patterns of narrative recur in societies which are widely separated geographically and historically. Lang was the leader of those who argued that this coincidence was not the result of 'diffusion' from one ancient centre but rather was the product of 'a vast common stock of usage, opinion, and myth, everywhere developed alike by the natural operation of early human thought' (Dorson 1968, 304). From these early and primitive models, each race and nation developed its own particular forms (ibid.). Lang, therefore, believed in a universal human nature which everywhere followed the same patterns of evolution.

In this evolutionist model of culture, though different nations were at different stages of development, some traces of the primitive were evident even in the most 'advanced'. As with the biological parallels of phylogeny and ontogeny, so in cultural evolution, even the most sophisticated reader was still at heart a schoolboy. It was to these primitive elements that the story must appeal. For Lang, therefore, the taste for fiction was itself puerile, and the best fiction openly acknowledged this rather than trying to do the work which was rightly that of the treatise or the polemic: 'Not for nothing did nature leave us all savages under our white skins: she has wrought thus that we might have many delights, among others "the joy of adventurous living" and of reading about adventurous living' (*CR* 52 [1887], 689).

Like Swan, therefore, Lang committed himself energetically to the argument that human nature was much the same everywhere. However, his model of human identity was also deeply divided and hierarchical. The context for all Lang's scholarly and journalistic work was the attempt to define Britain's imperial role in the aftermath of the 'scramble for Africa'. Within that, he worked out his theory of cultural development, which divided different racial groups into the more and less advanced. This underpinned Lang's folklore studies, which were a recurrent theme of his causerie, as well as his popular verse and literary criticism.

At best this theoretical framework led Lang to assume that 'the advanced' races had available to them a greater variety of cultural forms, since they had access both to the primitive and to the more developed. At worst, this model endorsed an explicit racism, evident in his causerie in verses like one about the lazy African, who must be beaten to instil in him the discipline of work (*LM* 11 [1887–8], 11). This cliché of the white construction of the African embodied also the implicit idea of the disciplining parent, who must use whatever means he can to induce the virtues of civilization. Moreover, Lang subscribed to a hierarchy which was not only racial but specifically national. However universal the reign of King Romance, for Lang he returned in the guise of the Arthurian once and future king, that is specifically in terms of the matter of Britain and the knights of the round table.

The hierarchy of racial difference was entangled with that of gender. For Lang, the distinction between male and female lay at the most primitive level of our humanity (*LM* 40 [1901–2], 91). He sometimes claimed that the adventure stories he praised so highly appealed in the same way to girls as to boys, but elsewhere he conceded that this was unlikely (*LM* 9 [1887], 553). In Haggard's popular fictions, for example, the focus is always on a group of men, a band of brothers, who confront and defeat the fearsome power of the female principle, whether that is embodied in 'She-who-must-be-obeyed' or in the white queens of the lost tribe of Africa.

The masculine imperial identity constructed in the fiction Lang favoured was encoded also in the persona he created in his causerie.

The manly sports of cricket and fishing were among the persistent themes of Lang's writing, what he called his King Charles's heads (*LM* 7 [1885], 446: on cricket see *LM* 6 [1884], 420; 8 [1886], 221–2, etc.). For Lang cricket and criticism mimicked each other in the pleasures he offered the sixpenny reader. Robert Louis Stevenson identified this in a joking verse not intended for publication:

> My name is Andrew Lang,
> Andrew Lang
> That's my name,
> And criticism and cricket is my game.
> (quoted Green 1946, 177)

The importance of games like cricket in the construction of a masculinity appropriate to end-of-the-century Britain has been well documented.[4] Developed in the public schools, cricket had a quite different cultural significance from the working-class game of football because it was the sport of gentlemen. It thus encoded an identity which was not only gendered but also class-specific. Moreover, it was the game above all others which gave the future rulers of the Empire those qualities which were deemed appropriate to their imperial role. Cricket in this period came to be identified with 'Englishness'. In the colonies, as in the metropolitan centre, teaching team games – particularly cricket – to those other men, working-class and black or brown, became a crucial way in which the values of the 'English gentleman' could be disseminated.

Lang did not need to explain the links between the stories, sports and racialised identities which were the subjects of his writing. He simply placed side-by-side the promise of gossip about books and

> How Cambridge pulled, How Oxford bowled,
> Wild lore of races white and black
> Of these shall many a tale be told
> On this our stall of bric-a-brac.
>
> (*LM* 7 [1885], 317)

His other sporting passion was angling and this too was a recurrent theme of the column. Fishing – like team games – provided both good stories and metaphors though which other subjects could be discussed, not least writing and reading (*LM* 8 [1886], 221–2). The class and national connotations of fishing were not as explicit as for cricket.

Lang associated the delight of angling with his childhood in Scotland. Here again, as with the tale of adventure, the sport was associated with the boyish. Yet angling in Scottish rivers was the pastime of the English upper-class man and this dream of a return to careless boyhood depended on material exploitation of Scottish land and water. Lang identified himself as a Scot. Yet he was thoroughly the Oxford man. The rhyme of his fellow Scot, Robert Louis Stevenson, to which I referred above concluded:

> With my eye-glass in my eye
> Am not I,
> Am not I
> A la-dy, da-dy, Oxford kind of Scot.
>
> (Quo. Green 1946, 177–8)

The Englishness which Lang represented was not – ironically – incompatible with being a Scot, albeit an 'Oxford kind of' one. It was part of that process by which metropolitan Britain became 'England' or rather 'England' became a metonym for the United Kingdom. This metonymic slide was not simply linguistic. Though the careers of Lang and Swan were very different, both were formed by the material power which London and Englishness exerted in the institutions of print as well as in the discourses of Britishness. Location at the metropolitan centre was essential to both, yet both continued to claim their identity as Scots. Lang's angling stories helped create a persona in which Scottishness became an element in his identity as the English gentleman at play.

Lang's column was itself a kind of boyish game. Readers were invited to share pleasures which were presented as easy, intellectually undemanding and universally appealing. Lang offered himself as at heart a (well-read) boy and his ideal reader was positioned likewise, as an English public schoolboy (see Blagden 1963, 19). To read 'At the Sign of the Ship' as a woman was – and is – to experience exclusion.

Lang's biographer and bibliographer relates how Lang was utterly incapable in all domestic and practical matters. His extraordinary literary output depended increasingly on the organizing skills of his wife, who would jokingly describe him not so much as a boy as a child of three (Green 1946, 191). That dependence on the 'woman at home' was invisible in the column where Lang constructed the identity of the writer and of his likeness, his brother and 'semblable', the sixpenny reader. Indeed Lang owed something even more important to another woman than his wife. His interest in folklore and story were rooted in the stories told him in the nursery by his Scottish nurse (ibid., 1). The ways in which Freud wrote out the presence of the working-class nurse in the family romance of the middle-class boy has been brilliantly explored by Stallybrass and White (1986). In Lang's work and in the romances he favoured an analogous erasure took place.

In Lang's causerie the domestic is not just invisible, it is a positive threat to literary activity. This danger is embodied in the figure of 'the domestic', that is 'the maid', who 'tidies away' crucial papers and thus disrupts the scene of writing (*LM* 13 [1889], 659–60). The disturbing power of domestic femininity is embodied here not in the bourgeois woman but in the working-class servant. That threat appears in comic form in the column but it is recognizably the same threat as that which powers the male quest romance. The desire to escape from the domestic and thus from femininity unites such diverse end-of-the-century heroes as Haggard's Holly and Conan Doyle's Sherlock Holmes. Both of these fictional heroes owed much of their literary success to Lang's critical support. Another of Lang's literary sons, Robert Louis Stevenson, produced, in *Dr Jekyll and Mr Hyde*, a powerful end-of-the-century parable of a world without women in which masculinity is represented as in a crisis of division and self-destruction.[5]

We do not need Stevenson's disparaging verse about the 'la-dy, da-dy' Scot to alert us to the anxiety about his own masculinity which haunted Lang's work. Unable to live the life of adventure which was that of the manly man, he was afraid that the work of writing was implicitly feminized and feminizing, a 'cow's' life. Bombarded by letters and contributions from lady readers who wanted to be like him, he had to ensure that he did not become like them.

The cult of the domestic and the ideal of the family were crucial – if hidden – aspects of male as well as female metropolitan identities and 'an indispensable element of the industrial market and the imperial enterprise' (McClintock 1995, 5 and *passim*). Annie S. Swan made explicit use of the metaphors of family in the context of imperialism. Lang, like Haggard and the other writers of imperial romance, both explicitly and implicitly denied the place of the domestic and of the feminine as they tried to find new forms of masculine writing and new forms of masculine identity. However, the imperial romances, like Lang's column, implicitly mobilized metaphors of 'family' in relation both to the Empire and to the shared humanity of 'man'. The family defined as a 'natural' social unit validated the hierarchies of difference between public men and domestic invisible women and between 'advanced' or adult races and the more primitive childish ones. For Haggard and for writers like Henty the Empire was not only the place where men proved their manhood, it was where they learned to be fathers to 'their' people (McClintock 1995, 235). If Lang could not play the role of patriarch in this story of the family of man, he was always the romancing uncle whose stories provided appropriate models of the self for his (masculine, English) readers.

Resistance

A chapter such as this, which focuses on defining centres, is in danger of re-inscribing precisely those inequalities of power which it seeks to destabilize through making visible how they work. The power of the metropolitan centre to define the new reading public was undeniable. Lang's influence extended beyond the immediate pages of his column through the writers whose work he brought to public attention and who themselves exerted power into later generations. Lang's protégés played an important role in the education of a new generation of twentieth-century readers.[6] However, as I have tried to suggest, that power was not absolute, nor were the identities it produced for the new reading public monolithic. The urgent need to find a new masculine role in the 1880s and 1890s was directly related both to the demands of 'New Women' for changes to the definitions of femininity and the emergence in this period of public discourses of male homosexuality. Both of these unsettled the traditional role of father in the family. The New Woman in particular was a constant presence shadowing the identities produced by Swan and Lang.

Moreover, if the Empire was a family what happened when the children grew up, as they must? In his book *Beyond a Boundary*, C. L. R. James takes Lang's favourite sport, cricket, and writes from the point of view of just such grown-up children who can now beat their parents at their own game. In the 1980s, the Conservative politician, Norman Tebbit, suggested 'the cricket test', that is which side you supported in international cricket, as the test of English identity. The energy mobilized by the idea of the 'cricket test' is indicative both of the continuing power of the late nineteenth-century formations I have been discussing and also of the ways in which that power does not remain static or flow one way.

In *Mythologies* Roland Barthes attacked the concept of the 'Family of Man', through his critique of an American photographic exhibition with that title (Barthes 1983, 100–2). The object of his attack was both American imperialism and the imperialism of a humanism which assumes that universal stories can be told. As a recent discussion by Bob Chase suggests, perhaps the moment has come in the late 1990s to reconsider Barthes' critique (Chase 1996, 72–3). I would want to resist late twentieth-century attempts to revitalize the concept of 'the family' and its identities as the central metaphor and organizing principle of our narratives and our social policies, but the struggle to define the 'immense' reading public continues. We still need to create forms of narrative and through them models of identity which will both recognize the particularities of our different positions and address our common involvement in an increasingly global culture.

Notes

1. On changes in the professional role of writers see Cross (1985, 204–40). Fiction writing in the 1880 and 1890s was marked by the debate around the role of the young girl as reader of fiction, the collapse of the three-decker novel, and arguments about realism: see Coustillas's edition of George Moore's *Literature at Nurse* (1976), Griest (1970); on relation to New Woman see Ardis (1990), Pykett (1992).
2. The causerie, a regular column centred on the personality of the writer, was a popular genre in magazines at this period. e.g. those by 'Q' and Zangwill.
3. I trace the history of the 'advice to women' as a journalistic genre and discuss other aspects of this particular magazine in Beetham (1996).
4. See e.g. Mangan (1981). For one example of how cricket was taken up into the culture of the colonies see Beckles and Stoddart (1995). I am grateful to Maria Delgado for her coaching on this.
5. Stevenson's *Dr Jekyll and Mr Hyde* was offered first for publication in *LM* (Cox and Chandler 1925, 39).
6. Henry Newbolt, for example, who was brought to public attention when Lang published his poem, 'Admirals All' in *Longman's Magazine*, went on to write the 'Report on English' in schools.

Works cited

Books and Articles

Anon. 'Editorial Note', *Longman's Magazine* 44 (1905), 518.

Ardis, A. *New Woman; New Novels; Feminism and Early Modernism*, New Brunswick, NJ, 1990.

Barthes, R. *Mythologies*, London, 1983.

Beckles, H. McD and Stoddard B. (eds). *Liberation Cricket: West Indies Cricket Culture*. Manchester, 1995.

Beetham, M. 'The Reinvention of the English Domestic Woman', *Women's Studies International Forum* 21.3 (1998), 223–33.

Beetham, M. *A Magazine of her Own? Domesticity and Desire in the Woman's Magazine 1800–1914*. London, 1996.

Black, H. *Notable Women Authors of the Day*. London, 1906.

Blagden, C. 'History and Post-structuralism: Hayden White and Frederick Jameson' in Schwarz, B. (ed.), *The Expansion of England; race, ethnicity and cultural history*. London, 1963.
Chase, B. 'History and Post-structuralism: Hayden White and Frederic Jameson' in Schwarz, B. (ed.), *The Expansion of England*. London, 1996.
Cross, N. *The Common Writer: Life in Nineteenth Century Grub Street*. Cambridge, 1985.
Coustillas, P. (ed.). George Moore's *Literature at Nurse or Circulating Morals: A Polemic on Victorian Censorship*. Sussex, 1976.
Cox, H. and Chandler, J. *The House of Longman, 1724–1924*. London, 1925.
Darlow, T. H. *William Robertson Nicoll; Life and Letters*. London, 1925.
Dorson, R. *The British Folklorists: A History*. London, 1968.
Elwin, M. *Old Gods Falling*. London, 1939.
Green, R. L. *Andrew Lang; A Critical Biography with a short-title bibliography*. Leicester, 1946.
Griest, G. *Mudie's Circulating Library and the Victorian Novel*. Indiana, 1970.
James, C. L. R. *Beyond a Boundary*. London, 1963.
McClintock, A. *Imperial Leather: race, gender and sexuality in the colonial context*. London, 1995.
Mangan, J. A. *Athleticism in the Victorian and Edwardian Public School*. Cambridge, 1981.
Maurer O. 'Lang and Longman's', *University of Texas Studies in English* 34 (1955), 152–78.
Midgely, C. (ed). *Gender and Imperialism*. Manchester, 1998.
Parker, W. M. 'Lang and Longman's', *Scots Magazine* (March 1944), 451–63.
Pykett, L. *The Improper Feminine; The Women's Sensation Novel and the New Woman Writing*. London, 1992.
Stallybrass, P. And White, A. *The Politics and Poetics of Transgression*. London, 1986.

17

'Gay Discourse' and *The Artist and Journal of Home Culture*

Laurel Brake

> No distich in the Greek anthology is quite so full of Hellenic simplicity and Hellenic sentiment as Plato's famous two lines on Agathon: 'As I was kissing Agathon I felt my soul upon my lips even as though it were about to depart.'
>
> (*The Artist*, 1 July 1889, 195)

> Guessed you but how I loved you, watched your smile
> Hungered to see the love-light in your eyes –
> That ne'er can wake for me – Would wild surprise
> Or sheer disgust at passion you deem vile
> Be your response?...
>
> In all this world this thing can never come...
> You shall not hear me even breathe your name.
>
> (*The Artist*, 1 April 1890, 113)[1]

The Artist and its readers, 1880–1895

In *The Artist*, a 6d monthly published between 1880 and 1902, may be seen a profile of a late Victorian periodical which addresses a telling succession of dominant reader groups, in which gender and diverse categories of 'artist' are primary variables. The profile has emerged from the identification of one 'centre' at a specific period (1888?–1894),[2] under a particular editor, in which the journal seeks to integrate and establish a visible gay discourse, a gay tradition, and a gay interpretative community of readers *before* 1895 and the Wilde trials. Although *The Artist* is distinctive among nineteenth-century sites of homosexual discourse, editor-led as it is, it is not unique. Early numbers of *The Studio*, for example, under the editorship of Gleeson White,[3] are similarly oriented, but author-led (and dispersed) gay periodical discourse such as that by John Addington Symonds and Walter Pater[4] is far more common in the thirty years 1865–95. Scrutiny of the publishing history of *their* work shows clearly that mainstream periodicals circulated gender-marked, homosocial and even homoerotic 'gay discourse'. Pater and Symonds thus publish both coded, and fairly explicit, same-sex material in the *Contemporary Review*, the *Fortnightly*, *Macmillan's Magazine*, the *New Review*, the *Westminster Review*, and even *Cornhill*[5] and *Harper's New Monthly Magazine*. These reviews and magazines offer a range of contents characteristically 'general,' including arts, politics, science, religion, history, and philosophy.

By contrast, *The Artist* falls into a category of the nineteenth-century press designated 'class journalism,' so named because of the specialist nature of its contents and the appeal of its titles to a specific niche of readers, here 'artists.'[6] If between 1888 and 1894 *The Artist*'s inclusion of gay discourse and journalists of the 'new culture' – such as Lord Alfred Douglas, John Gray, J. G. Nicholson, Andre Raffalovich, Fr. Rolfe, J. A. Symonds, Gleeson White, and Theodore Wratislaw – which figure amid a more general exposition of 'Decadence' found in its pages, distinguishes it among class journals, it is only because this particular group was not publicly acknowledged as a 'class' readership, unlike women for example.[7] But under Kains-Jackson, a gay audience was an important backup to its dominant address to its 'artist' readers. More generally, the preponderance of other types of ('straight') discourse in *The Artist* made it a typical example of art journalism: it takes its place beside other art periodicals of the day such as *The Magazine of Art*, the *Art Journal*, and *Studio*, with one other exception: it is not illustrated until late in 1894. That unexpected characteristic of a magazine of the visual arts, and the nature of its layout and contents attest to its close links with categories of the press other than that of the art or 'class' magazine, to wit the general weekly, in particular the *Athenaeum* from which, generically, it appears an outgrowth.

From the outset and in the early years it was clear that the gender of potential readers had been configured to include women so as to maximize readers. No. I carried a report of the 'Female School of Art' in Queen Square (15 Jan. 1880, 3), and lady artists figure in several departments including the letters (15 Jan. 1880, 9). In the leader of No. 2 the topical issue of art education for women is lodged in one of *The Artist*'s recurring attacks on the Royal Academy, and a new section is introduced, 'Art in the House,' with the explanation that 'No feature of the late

REEVES & SONS,

MANUFACTURERS OF

ARTISTS' COLOURS

AND MATERIALS.

Established 1777.

SPECIALITY FOR ART STUDENTS AND AMATEURS.

"THE ART STUDENT'S" OIL COLOUR BOX,

PRICE 5s.

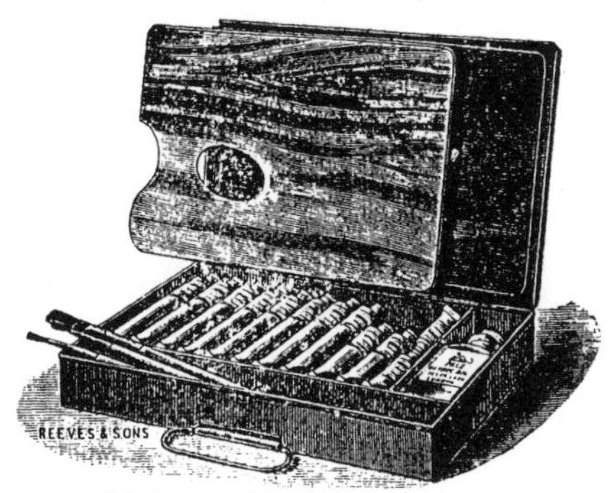

8 in. long × 5¼ in. wide × 1¼ in. deep.

Containing the following Twelve Oil Colours in Collapsible Tubes:– Burnt Sienna, Flake White, Pale Chrome Yellow, Ivory Black, Light Red, Prussian Blue, Vandyke Brown, Yellow Ochre, Brown Pink, Naple Yellow, Cobalt Blue, Crimson Lake.

Bottle of Pale Drying Oil Three Hog Hair Brushes, and Indiana Wood Palette.

The Box is japanned Tin, finished in the best style, with corrugated platform for the tubes to rest upon, and a well lid to prevent any colour on the palette from sticking to the lid.

The Colours being of good quality and low in price are specially adapted for Art Students. They can be replenished at a cost of 2d. per tube.

Figure 20 Advertisement. *The Artist,* 1 January 1889.

reawakening to the beautiful in England, we think, is more encouraging than the fact that our art is getting more and more domestic' (14 Feb. 1880, 34). By 15 March, the gendered element of 'Art in the House' – which appears on page one – is explicitly stated: 'It is intended to be, in a measure, the Ladies' Column of the paper' (15 Mar. 1880, 65). Its remit will include 'the minor developments of Art and Taste, such as Dress, Furnishing, Art Needlework, Tapestry, Artistic Stationery, Jewellery, Articles of Domestic Ornament, Fancy Goods, Design in Wall Papers, Painting on Textile Fabrics, &c' (ibid.). This element is further promoted in January 1881, when the sub-title *Journal of Home Culture* is added to the masthead. It is accompanied on the front page by 'A Picturesque Dining Room', the third article in a series of 'Outline Sketches for Furnishing' (1 Jan. 1881, 1). The formal incorporation of this element of domestic art legitimates features of the journal distinct from visual art, such as a monthly music department which in this case directly echoes its original publisher's trade in sheet music. William Reeves, manufacturer of fine art materials, printed and published music and art prints as well as *The Artist* and, as other nineteenth-century publishers used their magazines as a source of free advertising for their monthly book lists, so Reeves placed whole page adverts for paint boxes and fine art prints in *The Artist.*[8] Other advertisers of products for 'home culture' might similarly be attracted. The sub-title also extends the breadth of *The Artist*'s appeal to women consumers, beyond art practitioners to that broader group of middle-class women interested in aesthetic domestic culture more generally.

These decisive signals to women, which address them as part of the art industry and as potential consumers of *The Artist*, abate in the course of the 1880s. There is a change in the gender mix of the reading community which the journal defines, addresses, and creates. By 1888 when Charles Kains-Jackson comes to be associated visibly with the journal, the address to women readers and artists has nearly disappeared; areas of domestic art such as dress and interior design tend to be associated with men rather than women, with figures such as Oscar Wilde, Frederic Leighton, and Beardsley. 'On Art in Dress' for example, in 1890, stages a discussion of the male body and bathing costumes in a 'review article' of Cassell's *Book of the Household*; while the female body is invoked glancingly (in terms of the representation of the relation of shoulders to waist), male flesh is evocatively detailed:

> a man in well fitting clothes suited to the heat or cold of the day can and does easily forget the accident of clothing so far as he himself is concerned. What brings home to him the degradation of modern clothing is the hideous spectacle which his fellow men present. The light and tender colour of flesh, the graceful curves of the body, the exquisite unity between it and the limbs, these facts of which every morning's bath reminds him and which every night's

> undressing tells him are the real facts of one's nature, these graces and harmonies for which the soul has a natural appreciation and a perpetual yearning, are taken from him violently in the dress of every person he meets.
>
> ('Quill', 1 Sept. 1890, 259)

This is a piece in which dress[9] is demonized as part of a call for more representation of the nude in contemporary art. While the case is presented formally and technically, a homoerotic element is clear. In the period of Kains-Jackson's editorship (ca. 1888?–1894), the address to gay men is a distinctive and reiterated discursive strain in among the preponderance of copy addressed to the practising male artist who is now, by implication, defined as 'straight.'

The curve of the shift of address over the run of the periodical from a readership which includes women as a recognized minority in a primarily male readership to one which principally comprises straight and gay men to one which excludes gay men and reinstates women is signalled formally in October 1894 when there is a change of title to *THE ARTIST, Photographer and Decorator. An Illustrated Monthly Journal of Applied Art*, as well as that of format and publisher. The original sub-title foregrounding 'home culture' is displaced by wording which subsumes the new main nouns – Artist, Photographer, and Decorator – into the category of Applied Art,[10] an apparently scientific and professional nomenclature which apparently excludes the domestic women at home – the amateur – from the pages of the new *Artist* while welcoming professional practitioners. However, the shift to illustration undermines this apparent exclusion of the non-professional. Moreover, these shifts of format, title and publisher in 1894 signal an abrupt change of definition of the imagined community of male readers: the address to masculinity lurches from a mixture of heterosexual and homosexual to heterosexual and homophobic. Retrospectively, it appears that in April 1894 *The Artist* underwent the kind of homophobic purge suffered by the *Yellow Book* in April 1895 in the wake of the Wilde trials; if Kains-Jackson was dispensed with for writing a specific article, while Beardsley was sacked to eliminate the suspicion of the 'mob' (both in-house and external), the effect on the two periodicals was comparable. Neither of the journals recovered. The parallel suggests that before the trials – the alleged defining moment when male homosexuality entered the cultural consciousness of nineteenth-century Britain – such defining discourse was registered, circulating, consumed publicly, and eventually stopped, not in or by the courts, but in the public sphere of the press.

The lurching transformations of *The Artist* between May and October 1894 suggest that crises of both finance and identity were provoked by the sacking of its editor. Initially Kains-Jackson was replaced with no explanation other than the addition of a strap headline[11] added to an abbreviated masthead in May: 'Edited by Viscount Mountmorres'.[12] Nor is it announced at this point that there has been a change of publisher, although the address of the periodical office is altered to 14 Parliament St, the premises of Cassell. The substance of 'A PRELIMINARY TRUMPET-BLAST BLOWN BY OURSELVES' on the same page (ibid.) is a cosmetic attempt to adjust to the altered state of new (periodical) journalism, largely in terms of higher production standards – more pages, stiff cover, large type, better paper and printing. The Trumpet-Blast also signals *The Artist*'s turn towards applied art, and promises to notice, but not at this point include, illustrations. The tacit adjustment, which is nowhere explicit, is the transformation of the nature of the copy, which not only excludes all manifestations of the homosocial and homoerotic, but also includes sporadic swipes at Wilde and Beardsley which intensify during and after the Wilde trials. These signal the utter transformation of gender discourse which *The Artist* is undergoing. The important point in this context is the *continuum* between the pre-trial measures and the post-trial scourge; they are undoubtedly the result of the same phenomenon, homophobia.

The other material changes to *The Artist*, implemented over the summer months of 1894, take effect in the October issue. They are explained in September in an anguished prospectus which articulates the full measure of the purge: in addition to the new title, the list of new features is a catalogue of the characteristics of the new journalism: illustrations, photographs, enlargement, interviews, and prizes (Sept. 1894, 322); of these the most pervasive articulation is the change of culture from the 'beach'/'opium'/'valleys' to the 'rage' of the 'regions of money and trade' (Sept. 1894, 323). The new project – 'It awakes to business' – is associated with the city, the market, and a battle for survival: 'Henceforth will *THE ARTIST* push, and crowd, and hustle its way through the tumultuous herd of practicality and monetary values' (ibid.). Mountmorres too is sacked: buried in the gossip paragraphs of 'Studio and Personal' is the information that '*THE ARTIST* has once more changed hands, and, owing to the character of the paper being materially altered, the new proprietors have to seek a new Editor' (Sept. 1894, 338). The anonymous author of the emotional prospectus remains unknown, but his pessimism and distaste for the new project mark the profundity of its difference from the past. The prospectus ends with a long riff which likens the history of the journal to a piece of music which ends with a climax:

> A crashing, hideous discord; a flare of staring colours; a mad whirling round and round: after that a monotonous wail, a livid greeny white, a nauseous falling headlong downwards.
>
> Then the silent, black, still unconsciousness.
>
> Lastly comes the deadly, bitter, unwelcomed return to life and reality, and you rise and set out to haggle and barter in the crowded mart. Such is *THE ARTIST*'s awakening. It has passed to new hands and will henceforth preach of textiles, and applied art, and house decoration.
>
> (Sept. 1894, 324)

This anguished end to an ill-judged prospectus completes the narrative of unravelling in *The Artist* begun by the sacking of Kains-Jackson. However, if the boast is that 'THE ARTIST will be an ART TRADES' PAPER without commercial puffery' (Sept. 1894, 322), the misgivings of this conclusion supply a sub-text which in the event the revamped, illustrated *Artist* echoes; genteel and up-market, it seems geared to the sitting-room browser as well as to the studio practitioner. This final iteration of fear that fine art will give way to applied, the male artist to the commercial designer of whatever gender, and a specialist male audience to women at home to whom the new *Artist* will 'preach' is borne out by the glossy page and the poor reproduction of the illustrations. As in the parallel debate about the censorship of fiction in 1894, which resulted in the diminution of the three-volume/circulating library system of distribution of first editions, the trope of fear is that of feminization, not '*literature* at nurse' which was George Moore's phrase, but fine art.

The Artist under Kains-Jackson

I want to look at *The Artist* in the period 1888–1894 under Kains-Jackson who deploys a series of strategies to accommodate gay discourse which any editor in political journalism would recognize. Utilizing extant departments of the periodical such as 'Art Literature', 'Local Art Notes', the leader column 'From Month to Month', reports of exhibitions, and the gossip of paragraphs of 'Studio News' to insert gay material, he also commissioned articles or wrote them himself. Anonymity and pseudonyms typify some of this material, though, in keeping with the explicit

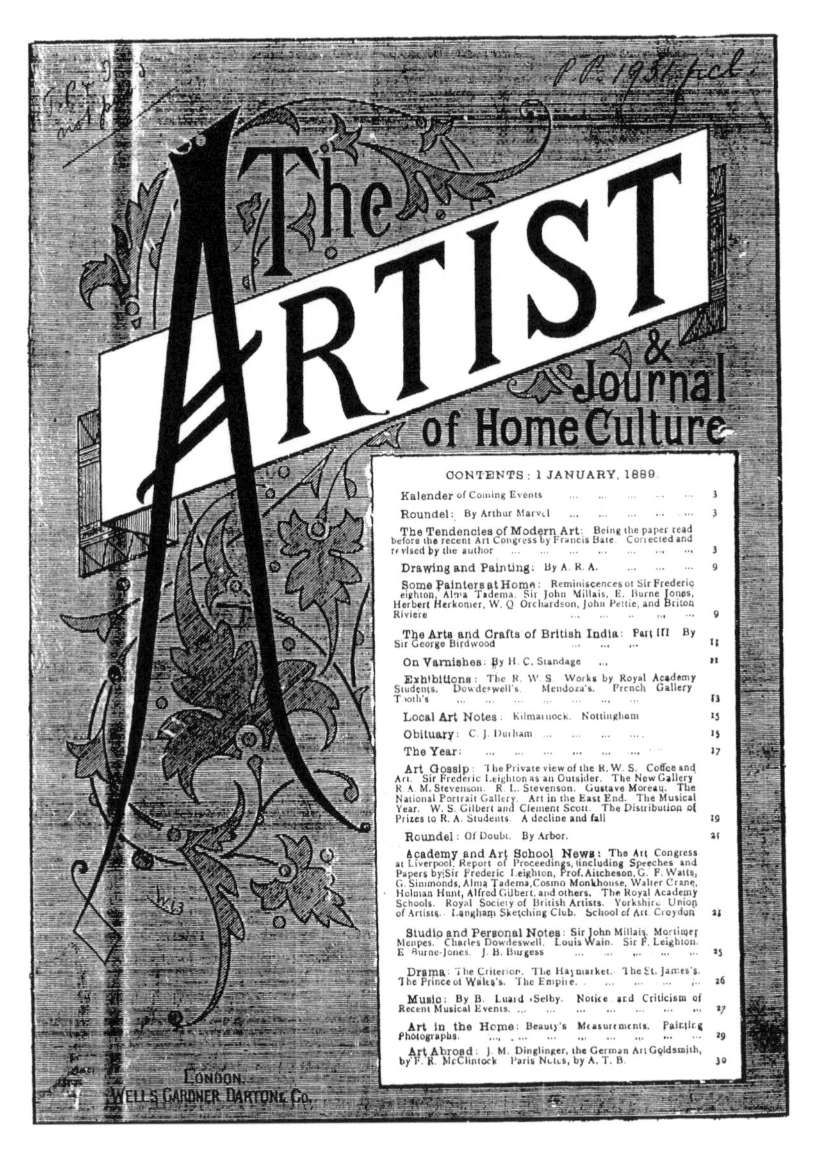

The Artist & Journal of Home Culture

CONTENTS: 1 JANUARY, 1889.

London. Wells Gardner Darton & Co.

Figure 21 Front cover, *The Artist*, 1 January 1889.

nature of much of the copy and the attempt to situate the periodical as a defining centre of a gay community of readers and discourse, signature and initials are frequently supplied in the text.[13] These are supplemented interestingly by additional attributions in the full, topically arranged annual Index, where it is available.[14] Kains-Jackson's strategies are informative more generally of the ways particular interests, of whatever kind – party political, trade, or sexual orientation – can 'construct' news and its discourses; what is notable in this period are both the impunity with which Kains-Jackson constructed and circulated a pink paper within the 'artist' community *and* the evidence, provided by the intensity and multi-vocality of the discourse, of a community of self-identifying 'gay' readers and writers in dialogue in public. Few reading *The Artist* regularly in these years could remain unaware of this strain in the periodical, due to its explicitness and level of permeation, and this must include the proprietor.[15] That it was not illustrated seems crucial to its survival for over five years in this phase of its existence. While the 'blindness' of the public to the possibility of male homosexuality which is judged to have played a central role in the Wilde trials in 1895 is improbable in the cluster of producers and consumers of *The Artist* in this period just before, the presence of (homo)erotic illustration might well have attracted unwelcome attention from a censorious public which could not be ignored.

It may be remembered that in the manifesto of 1880 William Reeves described the new journal as a 'literary organ' and the literary nature of the project, despite the visual subject, is further enhanced by the absence of illustration. Kains-Jackson's conduct of *The Artist* exploits this space for literature and the literary. Most visibly, the number of poems per issue increases markedly as he moves into the editorship. By their number, authorship, subject matter and quality, it is clear that unlike in many Victorian periodicals, poems in *The Artist* do not function primarily as filler. Many figures of the decadence are identifiable as authors of poems here, including Horatio Brown, Alfred Douglas, John Gray, Laurence Housman, J. G. Nicholson, A. Raffalovich, Fr. Rolfe, Edward Sayle, J. A. Symonds, H. S. Tuke, and Gleeson White. There is a decided preponderance of subjects central to gay discourse such as bathers, boys at play, classical figures such as Hyacinthus, Christian saints such as Sebastian, male dress, renunciation, and artists and authors such as Michelangelo, Tuke, Walter Pater and J. A. Symonds, whose life and/or work had same sex associations. One anonymous, seasonal 'Ballade' for example, dedicated to G. W. (Gleeson White), begins:

> Sing us a song of young lads' faces
>   Sing us a song of the morning row,
> Down to the sandy bathing places.

It continues in stanza three:

> Sing us a song of the feints and chases
>   Feet found nimble and limbs not slow,
> Lithe sweet forms, and budding graces
>   Of manly beauty at youth's full flow;
>   Hellas never had braver show,
> Nay, or Theocritus' Sicily,[16]
>   Sing us a song 'ere August go,
> A ballade of boys by a southern sea.
> (Sept. 1890, 255)

In October of that year an anonymous sonnet on St Sebastian permits a celebration of the representation of the wounded male body:

> This is the ultimate flower of perfect art
>   Wherein consummate beauty of body and soul
>   Is blended evermore, wherein the whole
> Of life is perfected; Love bids depart
> By martyrs' way.
> (Oct. 1890, 303)

The path by which such matter is introduced into the pages of *The Artist* is characteristic of this particular department of this magazine. The sonnet is attached to a painting which, named 'On the St. Sebastian of Francia (No. 277 Hampton Court Gallery)', legitimates its inclusion. There are countless other poems on such themes.[17] One which is notable for the combination of explictness and signature is Theodore Wratislaw's 'To a Sicilian Boy' which appears, again somewhat topically, in August 1893.[18] It begins

> Love, I adore the contours of thy shape
> Thine exquisite breasts and arms adorable;

and ends

> Ah let me in thy bosom still enjoy
> Oblivion of the past, divinist boy,
> And the dull ennui of a woman's kiss!
> (Aug. 1893, 229)

But the other literary space which Kains-Jackson exploits spectacularly is the section called, curiously, 'Art Literature.' This is the site of some of the most ingenious manoeuvring in order to introduce books he wishes to draw to the attention of gay readers. Works featured here range from the appropriate *Journal of Hellenic Studies* to a novel by Vernon Lee, *Euphorion* whose 'brutality of language' is both castigated and then quoted, and whose links with the Greek, the Platonic, and J. A. Symonds are noted (April 1886, 125).[19] In this department Kains-Jackson hails and extols Shannon and Ricketts's *Dial*, Wilde's *Woman's World*, and uses the occasion of a review of *Harper's [New] Monthly Magazine* to reprint in full a long poem, 'Giton' by C. W. Coleman, from the July *Harper's*, which begins 'Beautiful boy, the world is old' (Aug. 1889, 249). Four similarly suggestive poems by Charles Lefroy are reprinted on the occasion of a review of J.A. Symonds' *In the Key of Blue* which contains an essay on Lefroy, who is identified by Kains-Jackson as an author of interest to readers of gay discourse:

> His sympathy with youthful strength and beauty, his keen interest in boyish games and the athletic sports of young men, seem to have kept his nature always fresh and wholesome. These qualities were connected in a remarkable way with Hellenic instincts and an almost pagan delight in nature. But Lefroy's temperament assimilated from the Christian and the Greek ideals only what is really admirable in both: discarding the asceticism of the one and the sensuousness of the other.
>
> (April 1892, 122)

J. A. Symonds's two-volume life of Michelangelo, although listed first in 'Art Literature', is reviewed separately and accorded mammoth amounts of attention, being treated serially over two issues (Dec. 1892, 360–1 and Feb. 1893, 39–40); moreover, a further 'short series of articles on the actual life and work of Michael Angelo as revealed to us by Mr. Symonds' is promised in December but never materializes, it being explained in February that space does not permit this. In case the reader misses the point of the massive coverage, Symonds's chapters in which Michael Angelo's relations with Vittoria Colonna and Tommaso are respectively denied and confirmed as sexual are singled out (Feb. 1893, 38), and the painter is described favourably as 'the first man of the Renaissance who was also in many ways the first decadent' (Dec. 1892, 361). Symonds's book is trailed in November, the month before it is initially reviewed, and in another department of the Feb. 1893 number, 'Studio and Personal Notes', the volumes are said, in a single sentence reference, 'already [to be selling] at two and a half times its published price!' (Feb. 1893, 55).

One typical way in which Kains-Jackson further hones his book reviews to the remit of the journal is to comment on the design and production of the books, and Part I of this review of the Michaelangelo volumes published by Nimmo are discussed and recommended on this account; when books of poems by one of Kains-Jackson's favoured authors, with no relation to 'art,' are imported into Art Literature they are often legitimized by treatment of their design and production; *Salome*, for example, is reviewed twice, once in terms of Beardsley's illustrations (April 1894, 100–1), and once as Wilde's drama (April 1893, 120), and Charles Sayle's volume of poems, *Erotidea*, which is not favourably reviewed is nevertheless brought to notice on the strength of its production: 'Daintily bound in white buckram, and printed on paper which leaves nothing to desire, the chaste little volume now before us...' (June 1889, 169).

Other departmental spaces serve Kains-Jackson's interest in gay discourses; one is the selection of artists to feature in reviews of exhibitions and in the art gossip pages, and another is the choice of features. There are many examples of visual artists whose work, journeys, houses, projects, friends, parties, and books recur regularly in these pages, among them, predictably Leighton, Ricketts, Shannon, Beardsley, and H. S. Tuke. Tuke especially is a recurring name, and he is associated with markers of coded gay discourse – boys, the seaside/nautical, flesh, and classical Greece. The comments of 'Critias' on 'The Bathers', in an article on The New English Art Club, provide an illustration of the context for numerous shorter allusions to Tuke that typify the visual arts departments: the picture is represented by markers of coded homosocial discourse – boys, the nautical, swimming, flesh, and classical Greece.

> 'The Bathers,' by H. S. Tuke, is wonderfully full of light and air, and is excellent in tone. Three boys on the deck of an old barge form the subject; one is preparing to plunge. It is doubtful if barge decks are often painted this beautiful celadon colour which forms such a perfect contrast with the flesh tones and so perfect a harmony with the colour of the sea. It is also, perhaps, to be wished that the artist had found a plunger of somewhat less boeotian face. However, the whole tableau is one which Pindar might have celebrated and which Pericles would probably have bought.
>
> (May 1889, 128)

Comparison of *The Artist*'s attention to Tuke with other reviews of this exhibition in the *Saturday Review*, for example, show this picture to be referred to parenthetically if admiringly in one review in the *Saturday* (4 May 1889, 530) and not otherwise mentioned in the volume.

Certain feature articles unmistakably constitute gay discourse. These are pieces outside any regular department but they are nevertheless often serialized, to keep the reader coming back month after month. One such serialized piece, in July and October 1889, is almost entirely dedicated to inscribing the classical homoerotic tradition; it is included in the pages of *The Artist* by naming it 'Subjects for Pictures'! This adumbrates a whole cast of suitable characters and themes, among them Perseus and Andromeda (with the emphasis on 'the enterprise of a brave, adventurous youth in the rescue of the dedicated victim' (1 July 1889, 194)), Hyacinthus and Apollo, Narcissus, Plato and Agathon (Plato's kiss), Alecytyron (Mars's boyfriend who is changed to a cock), Tenes (a Phrygian youth), lolas (Hercules' favourite), and Diana Orthea. Of this last, it is explained that the annual whipping of Spartan youths for the festival of the goddess is

> a privilege reserved to Spartan lads of the best class. Their parents frequently, their lovers always, attended them to the altar of trial, and those who bore the lash – which was applied till the blood was drawn without uttering a cry, received a handsome award. A painting would . . .have to show us a lad grasping the top of the altar and facing us. The expression of steady and determined endurance under pain in a young face would, however, task the talent of the artist sorely, as would the complex emotions of pride, compassion and love in that of the father, elder brother or lover who would be the lad's natural aid or attendant at the sacrifice.
>
> (July 1889, 196)

A comparable serialized feature is 'Realism in Poetry and Fiction' (Sept.–Nov. 1889), the third part of which is entirely confined to salacious examples from classical literature of Greece and Rome. In October 1890 another anonymous feature article, '"Les Decadents" No. I' begins typically with a short introduction on contemporary decadence but then takes the opportunity to discuss decadence in ancient Greece. Starting with the present – 'We have not, through the entire summer, been able to escape for a single week from this new enquiry which in public and in private, friends and correspondents keep addressing to us, "Who are the Decadents?" and what is the meaning in art and letters of "the Decadence?" (Oct. 1890, 292) – it is a clear example of the use of a topical issue to introduce yet another piece on ancient Greek art and culture,

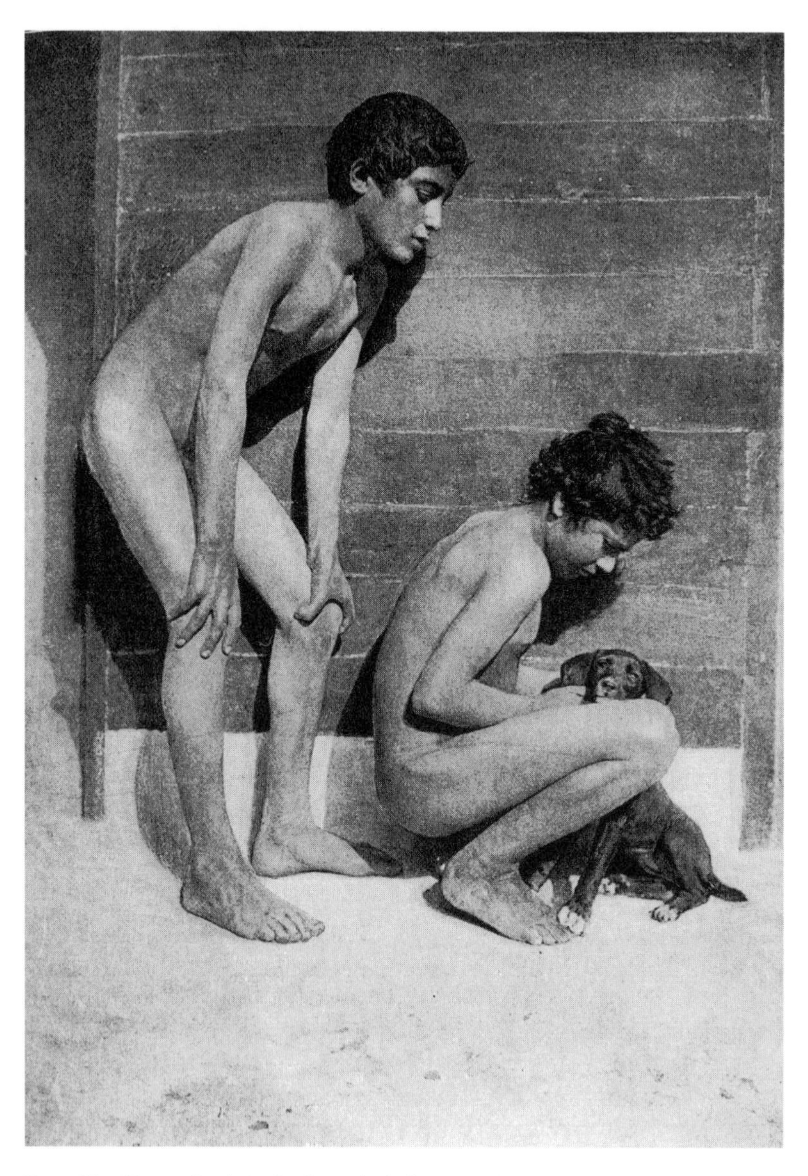

Figure 22 'From a Study in the Open Air', photograph. W. von Gloëden, *Studio*, 1893.

which is what this article turns out to be. And although *The Artist* does not allude directly to the Cleveland Street scandal of 1889/90 which involved the exposure of a male brothel, there are two leaders in Nov.

1889 and Dec. 1890 denouncing the suitability of the courts to adjudicate on 'indecency.'

By 1893 there is a detectable boldness and permeation of material which might be counted as gay discourse or contain reference to it. Under 'Local Art Notes' in June 1893, for example, two books of poems, 'apropos' of 'Oxford' and 'Rugby', are noticed in language which merely insinuates same-sex love: under Oxford appears 'the little volume just published by Mr. Blackwell for a young poet who has many friends up here. "Love in a Mist" is a quaint little square paper covered book of a hundred pages' (June 1893, 191); 'Rugby' begins 'The recherché little press of Overs which has already given us the minor masterpiece of Charles Sayle and of Norman Gale has just put forth a fresh flower of youthful verse' (ibid.); from the first book *The Artist* publishes 'To-day he loves me' and from the second a poem, 'Reminiscence,' a sonnet addressed to a lover of otherwise undisclosed gender. In December 1893, however, Kains-Jackson prints a splenetic attack by 'Cyril M. Drew' on Theodore Wratislaw and 'the new [homosexual] cult, *i.e.*, a small body of unimportant and self-opinionated young gentlemen who fancy that they can invent Nature, and the human passions' (2 Dec. 1893, 363). Citing with disgust Wratislaw's line, 'the dull ennui of a woman's kiss' from 'To a Sicilian Boy' (Aug. 1893, 229), Drew asserts 'For one man to kiss the lips of another is a positively repellent idea, but for a lover to kiss the lips of his beloved is the purest and most natural means of expressing his love' (2 Dec. 1893, 363). By publishing this explicit denunciation, Kains-Jackson appropriates it as a reverse discourse which defines and publicizes what it condemns, while advertising the identification of *The Artist* with 'the new cult.' That *The Artist* is, by this time, so identified is indicated by the nature and number of pieces on these themes published in 1892 which are undistinguished, whether by quality or authorship,[20] and appear to be selected from unsolicited submissions.

It has perhaps been difficult to glean any sense of the frequency or intensity of gay discourse in any one number from the topical way I have discussed the run. Clearly the intensity varies from number to number, but at its fullest (as for example in Feb. 1893) the articulation of the sexual orientation of *The Artist* is clear. This one number of February 1893 contains three items which explicitly introduce homosexual authors, artists, and discourse. The number reviews two books by J. A. Symonds, separately and at length, and also contains a poem by him. Beside endorsing Symonds's view of Michelangelo's sexuality (see above), the reviewer provides explicit details of the censorship of the sonnets addressed to men 'wherein the male is altered to the female'

(Feb. 93, 38). This is followed by a 'Roundel', 'In the key of blue', which is followed in turn by a review of Symonds's collection of prose and poetry of that name, from which the poem is taken. This eclectic collection contains a homoerotic autobiographical essay, 'Clifton and a Boy's Love', which the review singles out for attention, quoting a lyric of Symonds's ('He was all beautiful') and an analogous example from ancient Greece, when 'Critobulus spoke thus eloquently of Cleinias, and [at] an age when he too might have been an Oxford undergraduate from Harrow' (Feb. 1893, 40),[21] to make visible to readers the tradition of which Symonds's essay is a part.

How and why then was Kains-Jackson ousted in April 1894, a year before the trials? The termination of the editor's contract was undoubtedly provoked by his authorship and publication in April of a particularly explicit article, 'The New Chivalry,' which was part of a current debate on 'The New Hedonism' prompted by Grant Allen's piece in the *Fortnightly Review* in March.[22] Signed with initials,[23] 'P.C.' argued in *The Artist* that since overpopulation rather than underpopulation is now the problem, same-sex social bonds were to be preferred on a number of counts to heterosexual coupling; the Greek model is then explained and recommended. However, the publication of 'The New Chivalry' might best be viewed as the *occasion* for the eradication of gay discourse in *The Artist*, as its publication faithfully represents cumulative editorial policy under Kains-Jackson rather than any new departure. Moreover, there may have been other, non-editorial, commercial *reasons* which contributed to Kains-Jackson's departure.

By April 1894 the proprietors of *The Artist* would have been uncomfortably aware of the success of *The Studio*, a new, equally cheap, rival but illustrated monthly edited by Gleeson White, one of *The Artist*'s own and regular contributors. It is notable in this connection that the early numbers of *The Studio*, which coincided with White's editorship (1893/4) and immediately preceded Kains-Jackson's dismissal from *The Artist*, directly competed for the pink market niche by including generously illustrated pieces on Leighton's clay modelling of the (largely male) nude, the (male) nude in photography illustrated with sumptuous photos by Rolfe and Wilhelm von Gloeden[24] of Sicilian youths, and a pioneering piece on Beardsley's art. The time for non-illustrated art magazines had passed, and the demand of the new readership for the formats and tricks of the new journalism was irresistible. By October *The Artist* became an *illustrated* monthly of applied art, like *The Studio*, and in the new year it changed its format to folio, like *The Studio*. It is indeed possible that the personal risk Kains-Jackson took in not only publishing

'The New Chivalry' but writing it himself was in part prompted by the commercial rivalry between the two monthlies for an important sector of their common market. His piece and his dismissal are analogous in ways to the Wilde trials; they confirm, and semiotically sign and own the gay discourse and community of consenting readers which had characterized *The Artist* between 1888 and 1894, and of which Kains-Jackson's piece is a part. Gay discourse was not halted, however, even in the years immediately following the trials: one has only to look at the literary and pictorial contents of Gleeson White's and Charles Shannon's first number of the annual, *The Pageant*, of December 1896, to see that.[25]

For *The Artist*, however, the break with homoeroticism and decadence after April 1894 is signalled if erratic. In August, Beardsley is attacked in a notice of vol. 2 of the *Yellow Book* for 'diluting his bitter morbidness with sickly commonplace' (Aug. 1894, 275), but in 1895 in January and February, perhaps in fear of losing their former readers, illustrations from Michelangelo's Sketch Book appear, and in April a Beardsley frontispiece, accompanied by only mild, implied disapproval. But the same number, alert to the contemporaneous Wilde trials, also dismisses William Morris and Wilde in a triumphant article, 'The Decay of the Decadence' (April 1895, 169–71). After the verdict, however, in the numbers of May, July and August, the tone is one of homophobic denunciation without any mention of the trials. In May Beardsley's 'Black Coffee' is 'a stupid insipidity' (May 1895, 188) and in August *The Artist* takes advantage of a misprint in a daily's comment on the *Yellow Book* to emphasize their position on Decadence: 'Everyone must by this time be familiar with our attitude towards this decadent publication, but in our most irritated mood we never called it "stinking," whatever we may have thought' (Aug. 1895, 321). A July review of a re-designed, illustrated periodical, the *Evergreen*, affords an opportunity to deny the ethos of 'The New Chivalry' most strongly, and to clarify the new position of *The Artist*:

> There is nothing of the decadence about 'The Evergreen,' for truths and evergreens are perpetual.... That faith may be had still in the friendliness of fellows; that the love of country is not a lost cause; that the love of women is the way of life; and that in the eternal newness of every child is an undying promise for the race.'
>
> (July 1895, 271)

With its eye on *The Studio*, however, this is accompanied by a prominent, Beardsleyesque illustration of 'Apollo's School-Days' which depicts a prepubescent, naked Apollo splayed on the lap of a lascivious Satyr who is teaching the youth how to play the pipes.

The years of the run of *The Artist* make it a useful vantage point from which to consider the debates of our own period about the history of gay subjectivity. Echoing Foucault in historicizing the homosexual, queer theory seeks to deny that homosexual as the term is used now by critics – people with same-sex preferences and 'personality' – is an appropriate category of sexuality in the nineteenth century until after the transformations in public perception and ideology brought about by the Wilde trials in 1895. Thus, it is claimed by some that there is no publicly visible 'gay discourse' before 1895 in the absence of gay subjectivity as a signifying mark of gender identity. In an impressive range of readings of effeminacy, faultlines and fissures, and silences, the contours of representation, repression, and denial have been mapped by critics who include Joe Bristow, Ed Cohen and Alan Sinfield. It is in this context that I offer the profile of *The Artist* in the years before the Wilde trials, and particularly between 1888? and 1894 under Charles Philip Castle Kains-Jackson/Philip Castle/ P.C./P. K. J./[anon.], Kains-Jackson or any other of his multiple identities. It is a semiotics which goes with the turf, whether it be that of the editor function, gay discourse, or the new journalism of personality and signature.

Notes

1 The poem is signed 'G.G.' who is probably George Gillet. Given the gist of the poem, it is unlikely that the initials of its title 'To W.J.M' are attributable. G.G. contributes other poems to *The Artist* in this period.

2 It is uncertain when Charles Kains-Jackson began to edit *The Artist*; 1888 is the date usually given (e.g. D'Arch Smith and Fletcher), but its provenance is unclear. The archives of William Reeves, the first publisher of the journal, in BL show that as early as April 1886, 'C.K.J.' received, with 'Crowdy', 81 copies out of a print-run of 1500.

3 See for example Emmanuel Cooper, 'The Love that dare not speak its name' and Stephen Calloway, 'The Dandyism of the Senses' in Spens, pp. 48–53 and 55–63 respectively.

4 For Symonds see Grosskurth, and for Pater see, for example, Brake, Inman, and Shuter.

5 *Cornhill* was well known as a family magazine, with a recipe of contents which included several serial novels per issue, and normally excluded articles on politics, religion, and philosophy. Symonds's unlikely spate of contributions to *Cornhill* includes a remarkable two-part article on 'Antinous' in February and March 1879, of which the following is characteristic of Part I which is, as a whole, especially explicit: 'Hadrian had loved Antinous with a Greek passion in his lifetime. The Roman Emperor was half a god. He remembered how Zeus

had loved Ganymede, and raised him to Olympus.... He, Hadrian, would do the like' (Symonds 1879, p. 210).

6 *Punch*, for example, in 1887 is placed by Fox Bourne in the category of 'class journalism' as both a comic and illustrated paper. See Bourne 1887, II, 117–18, 298. It is notable that even the *Athenaeum*, which prints 'culture' news but not political news, is placed in this category in Bourne.

7 See Beetham.

8 See 1 Jan. 1889, front cover verso, and Dec. 1890, 380.

9 This piece is 'apropos' of the advice in Cassell's *Book* that dark bathing costumes are best.

10 It is important to note that *The Studio* also used the phrase 'applied art' in its sub-title. If publishers and editors of *The Artist* were conscious of competition, it was from the new *Studio* rather than from the stalwarts, the *Art Journal* (folio, illustrated, monthly, at 1s/6d in 1894) or the *Magazine of Art* (folio, illustrated, monthly, at 1s/4d in 1894).

11 Neither Kains-Jackson nor any other editor of *The Artist* had been named before in the journal, in this way or any other.

12 Mountmorres was William Geoffrey Bouchard (*b.* Sept 1872) and part of the Irish peerage. He was 21 when he stepped in to edit *The Artist* for five months between May and September of 1894, having attended Balliol in 1890–91. He married in September of 1893, became a member of the London County Council for Mile End in 1895, and served as Metropolitan and Provincial Secretary of the Primrose League in 1895–6.

13 Kains-Jackson himself writes under a number of signatures, as Philip Castle, as P.C., as Charles Kains-Jackson, and as C.K. J. His full name was Charles Philip Castle Kains-Jackson.

14 Unfortunately these are not always bound in to extant copies. The BL run includes annotations with queries to the effect that some of the missing annual Indices were 'not published'.

15 That William Reeves's list included from 1898–1960 *Flagellation and the Flagellants. A History of the Rod*' by 'William M. Cooper' [James Glass Bertram] shows that the firm was open to publication of gender-specific and erotic material. The title was first published by Hotten from 1870. Correspondence and ledger books in the Reeves archive in the British Library indicate the tenacity of the firm with respect to this title and glimpses of its publication history. One would-be blackmailer of the firm on 3 August 1935 suggested that Reeves had made a good deal of money from the publication: 'Finding an old volume which had a reasonable chance of passing the British censors, he speculated in the copyright and issued a new edition. At first the book was offered with discression [*sic*] but, as no complaint was made, sales became more and more open and the book is now displayed prominently in most book shops'.

16 Theocritus, in whose work same-sex love between men figures, was a common classical allusion in gay discourse of this period. See J.A. Symonds's translation of Theocritus' second idyl in the *Fortnightly Review* in February 1891:

> Delphis
> Loves: be it girl or boy for whom he pines in his passion,
> Surely, she could not say. (April 1891, 551)

17 Another St Sebastian poem, 'On a Picture', is published in February of the same year. Stanza two begins 'Stripped we behold thee, beautiful / As youthful pagan gods of yore' and the poem goes on 'See in thy suave and tender side/ The arrows tremble, as athirst/ They drink, as little mouths not wide, / The life blood where the heart has burst' (Feb. 1892, 50); its alliance with Anglo-Catholicism is indicated by an inscription of a date in the following form: '20th January 1892, FEST. S. SEBASTIAN'. See Kaye 1999.

18 It is topical because of the season, but also because photographs of Sicilian boys by Corvo and W. von Gloëden had appeared in the June 1893 number of *The Studio*.

19 This article antedates the alleged date on which Kains-Jackson became editor of the magazine. He contributes signed articles in 1887, so it is known that he is about before 1888. This review (from 1886) is anonymous.

20 See, for example, two poems which appeared in August and September 1892 (227, 277), 'The Last Secret' by E. Bonney Steyne which begins 'From young Greek lips a whisper fell' and 'On Edward Fitzgerald' (anon.) which includes the couplet 'Hurrah for him who dared to say/ "We want no more Aurora Leighs"' and goes on 'We would recall the Persian's lay/Shrined in Fitzgerald's cadences'.

21 See Xenophon's *The Dinner-Party*, 241–4. Clinias (*sic*) is represented as the object of Critobolus' obsessive love in a discussion of the value of good looks ('I would rather be blind to everything else than to this one person, Clinias'). According to Waterfield, Clinias is the son of Axiochus and the cousin of Alcibides; he also appears in Plato's *Euthydemus* (241, n.2). I am indebted to Professor Jim Coulter, Columbia University, for help with this reference..

22 Kains-Jackson's was one of the earliest responses. Three more articles appeared in *The Humanitarian*, in August, September, and October 1894, by respectively the Reverend Bonney (highly critical), Grant Allen (riposte to Bonney, with the addition of an important qualification repudiating same-sex relationships and closing a loop-hole in his *Fortnightly* piece), and George Ives (in defence of the homosexual, the 'new culture'). See Brake 1995, 34–8.

23 We know independently from letters in the William Andrews Clark Library that Kains-Jackson was the author of 'The New Chivalry' under the guise of P.C.

24 For von Glöeden see Aldrich 1993, 143–52 and D'Arch Smith 1970, 62–4.

25 The contents of *The Pageant* include Laurence Housman's beautiful and homoerotic drawing 'Death and the Bather' as well as a number of literary texts and art works produced by contributors such as Ricketts, Shannon, Verlaine, and John Gray, formerly part of Kains-Jackson's and Gleeson White's network of Decadents as well as contributors such as Yeats, Max Beerbohm, and Lionel Johnson from the *Yellow Book*, Frederick Wedmore from the *Savoy*, and Michael Field.

Works cited

Aldrich, Robert. *The Seduction of the Mediterranean*. London 1993.

Anon. 'The Nude in Photography: With Some Studies Taken in the Open Air', *The Studio*, I (June 1893), 104–8; photographs 103, 109.

The Artist and Journal of Home Culture, 1880ff.
Beetham, Margaret, *A Magazine of her Own?*. London, 1996.
Bourne, H. R. Fox, *English Newspapers. Chapters in the History of Journalism*. 2 vols. London, 1887.
Brake, Laurel. *Walter Pater*. Writers and their Work. Plymouth, 1994.
——'Endgames: The Politics of *The Yellow Book* or, Decadence, Gender and the New Journalism', *The Endings of Epochs*, ed. Laurel Brake. Essays and Studies. Woodbridge, Suffolk, 1995, 38–64.
Cooper, Emmanuel. *The Sexual Perspective*. London and New York, 1986.
D'Arch Smith, Timothy. *Love in Earnest*. London, 1970.
Drew, Cyril M., 'Mr Le Gallienne and his Critics', *The Artist* 14(2 Dec. 1893), 362–4.
Fletcher, Ian. 'Decadence and the Little Magazines', *Decadence and the 1890s*, ed. Ian Fletcher. Stratford-Upon-Avon Studies 17. London, 1979, 173–202.
Grant, James, *The Newspaper Press*. 3 vols. 1871–2.
Grosskurth, Phyllis, *John Addington Symonds*. London, 1964.
Inman, B.A, 'Estrangement and Connection: Walter Pater, Benjamin Jowett, and William M. Hardinge', in *Pater in the 1990s*, eds. Laurel Brake and Ian Small. Greensboro, NC, 1991, 1–20.
Kaye, Richard A., '"Determined Raptures": St. Sebastian and the Victorian Discourse of Decadence', *Victorian Literature and Culture* 27.1 (1999), 269–303.
Pittock, Murray. *Spectrum of Decadence*. London and New York, 1993.
'Quill, John', 'On Art in Dress', *The Artist* 11 (1 Sept. 1890), 258–9.
Shuter, William, 'The Outing of Walter Pater', *Nineteenth Century Literature* 48. 4(1994), 480–506.
Spens, Michael, ed. *High Art and Low Life: 'The Studio' and the 'fin de siecle', Studio International* Special Centenary Number 201. 1021/1023 (1993).
Stokes, John. *In the Nineties*. London, 1990.
The Studio, 1893ff.
Symonds, J. A., 'Antinous', *Cornhill Magazine*, 39 (Feb. March 1879), 200–12, 343–58.
——, translator, 'The Second Idyl of Theocritus. "Incantations"', *Fortnightly Review* 49 n.s. (April 1891), 545–52.
Xenophon, 'The Dinner-Party', *Conversations of Socrates*, trans., Hugh Tredennick and Robin Waterfield, ed. by R. Waterfield. Harmondsworth, 1990, 219–67.

Part V

National and Ethnic Identity

18

Bad Press: Thomas Campbell Foster and British Reportage on the Irish Famine 1845–1849

Leslie Williams

Newspapers and periodicals are privileged as primary sources for historians because of their immediacy as an archive of passing events and opinions regarding those events. But these sources provide traces of agency as well as archive. Victorian publications were also disseminators of ideas, bulwarks of values, normative gatekeepers, and social barometers. There is certainly in the editorial content of these periodicals a desire for suasion, for effectiveness of rhetoric which, in regard to the Irish famine, may have been lethal.

Victorian periodicals were partisan and polemical. Many papers had begun and still functioned as press organs for political parties. Righteous defence of a particular truth was the norm as were pitched editorial battles between rival papers. The press thrived on emphasizing exclusionary identities for their opponents to the detriment of political flexibility or compromise. In this environment, the Irish, especially, were presented as Other in a series of identity-altering stereotypes in which the colonizing rhetoric of ridicule, disparagement, and negation was employed.

This rhetoric was particularly evident in the metropolitan reportage of the potato blight of 1845–9 in Ireland. Protectionists viewed reports of the blight as politically deceptive and denounced mention of the expected scarcity of food as 'Peel's Panic'. On the other hand, those favouring free trade, such as James Wilson of the *Economist*, used reports of the blight as evidence for the absolute need to lower the price of corn through a duty-free market. While the initial reaction of the press to shortages in England was individualized and sympathetic, generally reaction to the Irish situation was politicized and dismissive.

Ireland in the 1840s was a peripheral entity which differed from the imperial centre in its landholding system, tenant rights, coercion laws, poor laws, language and religion. Nevertheless, Ireland was part of the Union and was represented in parliament. (The most notable Irish member was Daniel O'Connell, long excoriated by the metropolitan press for his insistence on Catholic Emancipation and Repeal of the Union.) Differences between the metropolitan and marginal cultures became an excuse for disparagement and seemingly punitive treatment of the famine-stricken Irish. The predominance of the negative image of Ireland and its leadership in the popular press may have contributed to the failure of the Union to provide for its most severely stricken member.

Well before the potato blight struck, the Devon Commission had concluded in its report to parliament in February 1845, that over half a million Irish men were without work for thirty weeks of the year. This suggested that nearly two million Irish lived in near starvation circumstances on an annual seasonal basis. John Delane, the anti-Catholic and anti-O'Connell editor of *The Times,* sent Thomas Campbell Foster to Ireland in August, dubbing him '*The Times*' Commissioner' as though, as an investigative writer, Foster had near parliamentary powers. His series on 'The Condition of the People of Ireland' appeared in the paper from 21 August 1845 (a few weeks before the blight struck in the autumn), through to 6 January 1846. The letters were published as a book in 1847, with a second edition appearing in 1848.

Foster's situation was very much that described by Claude Levi-Strauss in *Tristes Tropiques*: 'A young man who lives outside his social group for a few weeks or months, so as to expose himself... to an extreme situation, comes back endowed with a power which finds expression in the writing of newspaper articles and best sellers' (Spurr 1993, 149). In Levi-Strauss's analysis, the very extremity of the situation is the source of power for the writer, and exaggeration of differences or contrasts between the familiar and the strange add to the sense of power. Foster's writing follows Levi-Strauss's pattern while at the same time superimposing a colonizing rhetoric on his experience.

Framing the extremity of Ireland, Foster began the series by reassuring his readers that he would be perfectly objective:

> I enter upon the duty which you have committed to me, a stranger to Ireland, and wholly devoid of Irish prejudices, with no motive whatever save an earnest desire to ascertain the truth, and to state it with strict impartiality.
>
> (Foster 1845i, 4)

He believed that in Ireland

> the real origin of every disturbance, and of almost every crime, is the *want of employment*... Ejected from his land, without other means of living, the Irish tenant is rendered desperate by the prospect of starvation.... Bitter sectarian hatred, rebellion and assassinations are the result. But would the foolish and wicked talk about Protestantism, or Popery, or Saxon rule, or harsh landlords... produce such results without the pre-existing, all-exciting cause of mischief – desperation founded on hopeless starvation?
>
> (Foster 1845h, 5)

Foster here makes a purely economic argument for the cause of Irish violence. This pose as the impartial exterior judge of causes is commented on by Edward Said in *Orientalism*. Foster deliberately maintains an outsider's position as he attempts, in Said's phrase, to render Ireland's 'mysteries plain for and to' England (Said 1978, 21). Said suggests that the outsider or Western writer 'is never concerned with [the subject] except as the first cause of what he says. What he says and writes, by virtue of the fact that it is said or written, is meant to indicate that [the writer] is outside [his subject], both as an existential and as a moral fact' (Said 1978, 21). In Said's terms, Foster viewed himself as 'a hero rescuing [Ireland] from the obscurity, alienation, and strangeness which he himself had properly distinguished' (Said 1978, 121).

Foster's impartiality rapidly dissipates. He disparages the Irish, for example, for having dungheaps near their cottage doors (then a common European practice). He is annoyed and suspicious when Daniel O'Connell's tenants insist on speaking Irish, requiring the writer to use interpreters. He is surprised at meeting intelligent, respected Roman Catholic priests who supported Repeal of the Union. Foster's letters are, in fact, far from impartial but decidely Anglocentric. He praises the good landlords, but is outraged at the tenants who rebelled against the bad ones. Most important of all, Foster travels through Ireland at a time when the last season's store of potatoes is almost exhausted, a time of such extremity as Lévi-Strauss suggests, and he gives an outsider's picture of the absolute wretchedness of the poorest peasantry in Europe. Before the first indications of the potato failure have appeared, he describes

> a village called Labgarroo, containing 24 cottages, and almost the whole of its shockingly destitute and half-naked shoeless population immediately swarmed out and surrounded me.... such an assemblage of wretched beggar-like human beings I never saw. Picture to yourself

> the beggars...in the streets of London, dressed up to excite commiseration, and who write with a piece of chalk on the flags 'I'm starving', and then lay themselves down beside this scrawl crouched up in a violent shivering fit as the people pass them from church, and you have an exact *fac simile* [*sic*] of the kind of looking people around me – the *tenants* of the Marquis of Conyngham!.
>
> (Foster 1845g, 5)

Foster's exterior cultural construct of begging insists that the practice rises always and solely from pretence rather than from real need. The idea that some part of the population in Ireland, or even England, might be genuinely destitute eludes him.

Foster does note an explanation for the poverty of southern Ireland from one of his informants, a small farmer who explained that 'The people do what they can to improve, but the landlord does nothing, and they have not the ability to improve. They are tenants at will; and if they improve, their rent is raised accordingly at the next valuation' (Foster 1845f, 6). Still, Foster insisted that the smallest effort could improve the quality of Irish life:

> An hour's labour of the wife with a needle and thread would prevent the man's torn coat hanging in shreds about him. An hour's labour of the man would fill-up the puddle-hole at his door, through which his children paddle every day in and out of his cottage, rendering it and themselves filthy.
>
> (Foster 1845f, 6)

Here, Foster echoes an 1836 report by one of the Poor Law Commissioners, a Mr Nicholls, who first mentioned the husband who is 'too indolent to make a dry approach to his dwelling, though fit materials are close at hand' (Foster 1845e, 6). Foster frequently used other travellers' reports and books as sources. He quoted Henry Inglis's *Journey through Ireland* (1835) on the contrast between Protestant Enniskillen and Catholic Ballinamore. He cited Nicholls again when describing conditions in Donegal and referred to 'Arthur Young, who wrote on Ireland some 20 years before the union, or upwards of 60 years ago [in the 1780s]' and to 'Wakefield, who wrote in 1812' (Foster 1845d, 5). In fact, some of the letters on the condition of the people of Ireland are so close to previous travel writings or commissioners' reports, that one wonders whether Foster took the trouble to be an eye witness at every site he mentioned.

Rather than reporting his own investigation of contemporary events, he seemed satisfied with repeating previous views (and prejudices) regarding the Irish.

The travellers' reports, earlier examples of Said's concept of existential and moral exteriority, emphasized the differences between England and Ireland. Ireland was seen by travellers as a distant country with distinctly different customs which were represented as obscure, alien and strange. The differences were then used by English writers to justify a cultural superiority implicit in the concept of the exotic (Williams 1998, *passim*). Foster's mid-century recycling of earlier travellers' writings emphasized the otherness of Ireland largely in terms of negatives. The differences, for his 1845 audience, were magnified by his reliance on books pre-dating Catholic Emancipation or even the Union itself. Ireland was thus prejudged by Foster as backward, while both observer and observed were mired in a partisan history.

A typical example of Foster's cultural bias may be seen in his criticism of the Irish celebration of All Saints Day at a time when the Irish realization of the crop failure was becoming widespread. On 7 November, Foster reported that the Irish were not as busy as he thought they should be in saving what they might from the rotting crop:

> with famine before them, it will hardly be conceived in England that Saturday last [All Saints Day], a fine bright, sunshiny day, when the potatoes might have been exposed to the wind and sun and the worst grated into flour, was proclaimed a holiday (no doubt for some good purpose), and not a man was to be seen at work in the whole county whilst their potatoes were rotting. Some will not even take the trouble to dig them, because they are diseased, though there is no question about it but much of them, though rotten may be saved. The town of Limerick was like a fair. There were thousands of people idling about in the streets.
>
> (Foster 1845c, 5)

Had Foster been a Catholic or even acquainted with Catholics, he might have understood the nature of All Saints Day as a day of holy obligation. Two weeks after Foster's attack, *The Times* published the Englishman Robert Parker's lament that 'If they [the rotting potatoes] are infected, nothing can save them' (Parker 1845, 6). But Foster published no retraction of his accusation of Irish idleness. Anti-Irish sentiment was unapologetically an intrinsic editorial viewpoint for *The Times* as well as the *Morning Chronicle* (Kinealy 1995, 105), though the *Economist*, eager to use the Irish example as further proof of the need for free trade, appeared more sympathetic at the outset of the crisis.

While Foster continued to criticize the Irish peasant, he reserved his most vitriolic prose for his reports on Daniel O'Connell. From mid-November until the series ended in January, *The Times* published Foster's attacks on O'Connell. His letter of 18 November charged 'that amongst the most neglectful landlords who are a curse to Ireland Daniel O'Connell ranks first, that on the estates of Daniel O'Connell are to be found the most wretched tenants that are to be seen in all of Ireland'. Foster frothed in a partisan frenzy:

> the land under lease by him is in ... the most frightful state of over-population ... the poor creatures are left to subdivide their land and to multiply over population. ... Unaided, and unguided, the poor creatures are in the lowest degree of squalid poverty I have yet seen, and this within sight of Derrynane-house [O'Connell's residence].
>
> (Foster 1845b, 9)

Maurice O'Connell, as his father's estate manager, wrote to *The Times* in a letter published 20 December refuting Foster's accusations, and referring to the emotional ties between the O'Connells and their tenants:

> Is it for not evicting these people ... that he is censured? .... Or is it for not applying some Malthusian anti-population check ...? Most of these people formed the seventh generation of their families who had held those farms under ours. Was no feeling of pity to be entertained toward them?
>
> (O'Connell 1845, 3)

Daniel O'Connell's refusal to evict pauperized tenants was thus valorized on the basis of long-term ties between the landlord and his tenantry, and an acceptance of population growth. Maurice O'Connell invited Foster actually to visit Derrynane rather than invent descriptions of it. Foster revisited the area in the company of W. H. Russell, later noted as *The Times*'s chief correspondent for the Crimean War. Maurice O'Connell described Foster as 'sulky as a bear with a sore head and as thorough a specimen of a cockney as ever clipped the Queen's English' (O'Connell 1846, 3). Foster countered on 6 January with an accusation that Daniel O'Connell's 'disgrace ... which will stick to him as long as he continues to be the curse of Ireland, and to mar her prosperity with his sordid agitation' (Foster 1846, 5).

The timing of Foster's reports could not have been worse. Just as the bad harvest of 1845 threatened to deepen into famine for 1846, '*The Times* Commissioner' constructed a view of the Irish for his metropolitan readers which was calculated to engender disgust and disbelief. While the effect of Foster's reports on British public or parliamentary opinion is difficult to determine, their effect on Irish journalists was one of almost unanimous editorial outrage. Foster reported on 13 December the reaction of the Irish press and was indignant at their accusation that he might be a liar: 'I have been held up to abuse by all the press of all the parties in Ireland' (Foster 1845a, 8). Foster thus cast himself as martyr.

Irish protests that Foster's reports were exaggerated were ridiculed in *The Times*. The newspaper reprinted a satirical report from *Punch* whose editors had invented a fictional correspondent with the elegant *nom de plume* of Narcissus Pink (Anon. 1845a 176). This personage was tremendously handsome and attractive to women, a slap at Daniel O'Connell as womanizer. The exquisite Pink finds that all the worst in Ireland is quite the best. *Punch*'s inversion is meant, of course, to ridicule O'Connell and the protests of the Irish press. Where Foster says the Irish are filthy, Narcissus Pink insists they are clean; where Foster says the dunghill steams, Narcissus Pink says it is the dunghill's effluvia which is responsible for the beauty of the Irish children:

> Now, a dunghill in England is an offensive, foetid thing – at least to my nose. How very different are the dunghills in Ireland! They positively steam with sweetest odours; to which circumstance may, I think, be attributed the lovely complexions and seraphic looks of the swarms of children that abound in the village. They are all, too, so scrupulously clean – and so comfortably clothed.
>
> (Anon. 1845a, 176)

The anti-Irish stance of *Punch* here is entirely consistent with its editor Mark Lemon's prejudice against the Irish which was also expressed in his defamation of a fellow editor, the Irishman Joseph Sterling Coyne, as a 'Paddy' and in his recommendation that the emblem of Ireland be the hyena whom 'kindness cannot conciliate, nor hunger tame' (Foster 1993, 172, 178).

The furious exchanges of late November and December in 1845 and early January 1846 between Foster and Maurice O'Connell in *The Times* led to the *Illustrated London News* and *The Pictorial Times* taking up the cry and republishing Foster's criticisms with 'eye witness drawings' of Derrynane cottages.

The *Illustrated London News*, usually more sympathetic to O'Connell and the Irish since the editor 'Alphabet' Bayley was himself Irish, joined in the attack on O'Connell. On 10 January 1846, 'Views of the O'Connell Property in Ireland' included a series of illustrations of cottages purported to be from the estate of Daniel O'Connell. The text used extensive quotes from '*The Times* Commissioner' which ridiculed the poorly kept, under-capitalized nature of O'Connell's own holdings as proof of his inability to administer his own land, much less govern an independent Ireland. The 'Interior of Culvane's Hut' at Derrynane showed the meagre furnishings of an Irish cabin with a *lit clos*, or box bed, to the right of the fire, basketry, four people, two cows and a pig inside the dwelling. Accompanying the drawing was a quote from Foster, giving his degrading message additional circulation:

> Unaided and unguided, the poor creatures are in the lowest degree of squalid poverty.... the cabins are thatched with potato tops... the doorways are narrow and about four feet and a half high... many are without any hole for a window at all; a cow, or a pig, was usually inside, and a half a dozen children; the cottages inside were almost invariably quite dark and filled with smoke... through the thick smoke [one saw]... half-naked children, pigs, cows, filth, and mud.
>
> (Anon. 1846, 25)

Foster had mentioned that 'the condition of the fields, seems very bad'. The article waggishly added that Mr O'Connell was not 'lord of the soil, but of the rocks and boulders' (Anon. 1846, 25).

Keeping its readers focused on O'Connell's holdings rather than on the developing famine, *The Times* editors published parts of Foster's tirade against O'Connell again on Christmas Day 1845. Similar anti-O'Connell sentiment prompted *Punch* to feature O'Connell as 'The Real Potato Blight of Ireland' (Figure 23. Newman 1845a, 255). The artist, William Newman, drew O'Connell as a lumper, a tasteless bloated potato, seated for his portrait like a smug French aristocrat on a settee. The couch itself refers to the term 'lazy bed', the raised rows of folded sod and dirt in which potatoes were planted. O'Connell-as-potato wears the Repeal cap, whose crown-like brim signalled disloyalty if not treason for chauvinist *Punch*. Critical of Tory pampering of the Irish, *Punch* had already published an earlier Newman cartoon showing Sir Robert Peel

Figure 23 William Newman, 'The Real Potato Blight of Ireland', cartoon. *Punch*, 9, 1845, p. 255.

attempting to grow a shamrock tree with 'Treasury guano' (Figure 24. Newman 1845b, 166). Thus manured, the tree blossomed with repeal caps and had shamrocks sprouting daggers for leaves.

Punch saw the Irish in purely chauvinist terms and was dismissive of both the Irish parliamentary leadership and of the growing reality of

Figure 24 William Newman, 'Peel's Irish Crop', cartoon. *Punch,* 9, 1845, p.166.

starvation in the wake of agricultural disaster. Thus, John Leech's late summer cartoon of O'Connell as a fat beggar, a con man with his hands in his pockets, begging 'rint' for Repeal from a starving, ragged Irish mother and her children made O'Connell seem even more atrocious (Figure 25. Leech 1845, 213). *Punch* ridiculed those who did pay Repeal rent in a cartoon entitled 'Dan's Tribute. – Tim Doolan Paying his Subscription' (Figure 26. Anon. 1845b, 218). The small boy who contributes his sixpence is a ragged street urchin, a *Punch* disparagement of the social level of those committed to Repeal and an independent parliament for Ireland. With a colonizing assumption that poverty and ignorance were manifest in Irish pronunciation, the caption spelling indicates an Irish brogue: 'Here's sixpence an English gentleman gav'd me for holding his horse in Phaynix. Say it's from an Inimy of the Saxon'. The implication is that whatever the English give the Irish would be used against the imperial centre.

The impending famine was, of course, marginalized in all this political comment. The immediate need for food was distanced and veiled by the banter and politics of the metropolitan press. Poverty and mendacity had been communicated as norms for Ireland, and the more extreme the reporting of these, the more powerful the reporter was made to seem. Nothing in Foster's writing points to the blight as a particular cause of

Figure 25 John Leech, '"Rint" *v.* Potatoes – the Irish Jeremy Diddler', cartoon. *Punch*, 9 1845, p. 213.

distress. The cause of distress, *The Times* asserted repeatedly in its lead articles or editorials, was simply bad landlords, O'Connell being the premier example.

The Whiggish quarterly, the *Edinburgh Review*, took another view. Writing as an anonymous author on 'The Irish Crisis' in the January 1848 issue, Charles Trevelyan, then permanent secretary of the

Figure 26 Anon., 'Dan's Tribute', cartoon. *Punch*, 9, 1845, p. 218.

Exchequer, concluded that a switch from farming to industry was the answer:

> A large population subsisting on potatoes which they raised for themselves, has been deprived of that resource, and how are they to be

> supported? The obvious answer is, by growing something else. But...corn cultivation requires capital and skill, and combined labour, which the cotter and conacre tenants do not possess. The position occupied by these classes is no longer tenable, and it is necessary for them to become substantial farmers or to live by the wages of their labour....The people will henceforth principally live upon grain...which they will purchase out of their wages....the pauperism of Ireland...must be supported by adequate industrial efforts.
>
> (Trevelyan 1848, 303–4)

This laudable shift from subsistence farming of potatoes to wage-earning at wheat harvests and in (non-existent) factories may have seemed excellent advice, but it also represented the colonializing imposition of a set of the metropolitan values and desiderata of London upon a completely different culture. As Michael de Nie says, these are 'proposals for Ireland's re-creation in the image of Britain' (de Nie 1997, 64). There was very little industrial work in Ireland and certainly not enough farm labour in grain production sufficient to feed the millions of Irish who were dependent on subsistence potato farming for their nutritional needs. Grain, as Trevelyan acknowledged, was not as efficient for feeding people as potatoes were in terms of nutritional value or calories per acre planted. Nor did the famine disappear promptly on Trevelyan's schedule at the end of outdoor relief on 1 October 1847. It proved tiresomely persistent.

In 1848, as the spectre of famine rose again for the fourth harvest in a row, the *Illustrated London News*'s pro-Irish policy changed drastically. The paper acquired a new Whig editor, Charles MacKay from Glasgow. The *ILN* editorial approach now upbraided the 'unmanageable' Irish using the prose of colonializing superiority. In one scathing attack, MacKay's lead writer, placing his readers at the centre of correct values in England, made the most negative comparisons of the Irish to other exotic and marginalized peoples:

> the people of this country [England] begin to loathe the very name of Irish misery. They would relieve it if they knew how.... They sometimes believe the Irish peasant to be unteachable....He lives in a wigwam, and shares it with a pig. He speaks a barbarous language, and is in arrear with the intelligence of the world...the condition of the Esquimaux or Kaffirs is preferable to theirs.... The Laplander can get rein-deer flesh or blubber to supply his need; but there is nothing

> but the potato, and not enough of that for the Celt in Ireland.... Rent absorbs everything but the potato. All the other produce is exported to England to pay it. When the potato fails, the peasant has but to resign himself to starvation, to fever, and to death. When it is ordinarily abundant, a perilous fecundity among the people lays up a stock of embarrassment for future years...We can, in fact, see no hope for Ireland until the people are raised into the condition of bread-eaters.
>
> (Anon. 1848, 335–36)

So the solution for the *Illustrated London News* editor was now identical to that of the *Edinburgh Review*: let them eat bread. In the process of making the eviction-enforced cultural shift to bread (and thereby profiting the British corn merchants and the more progressive grain-growing Irish landlords), the west of Ireland was depopulated: one million who could afford a passage out emigrated, and another million died. The death toll is very likely larger since there are several sources indicating a dependency of three million people on either public works or soup kitchens in the years 1846 and 1847. When the Whigs stopped all relief from the central government in 1847, those destitute had neither passage money nor food.

The longstanding English view of the Irish as Other, as exotic in custom and unreformed in religion, was re-enforced by Foster's denigrating prose in *The Times*. But *The Times*, in 1843, had supported the old Tory Poor Law solution of outdoor relief, a more humanitarian and more effective solution than workhouses for feeding the poor during the famine. When Charles Trevelyan wrote for the *Edinburgh Review*, was he, in fact, documenting the Whig disengagement from an avoidable holocaust? Had it been politically possible for O'Connell to support Sir Robert Peel's Tory government in 1846 instead of Lord John Russell and Whigs, more humanitarian policies might have prevailed. In a careful reading of contemporary British periodical prose and in examining the cartoons of *Punch*, we see the rigid roles of political leaders and Irish famine victims reinforced daily by a politicized press, in which there is no compromise and no alternative to partisan, sectarian and stereotyped behaviours. Thus, the colonializing and overdetermining prose of the metropolitan press supported a lethal dominance over the marginalized and distanced Irish, and contributed to the inadequacy of supportive political response in regard to the Irish famine.

Works cited

Anon. *Illustrated London News* (27 May 1848), 335–6.
Anon. 'Views of the O'Connell Property in Ireland', *Illustrated London News* (10 Jan. 1846), 24–5.
Anon. *Punch* 9 (1845a), 176.
Anon. *Dan's Trbute, Punch* 9 (1845b), 218.
de Nie, Michael. 'Curing "The Irish Moral Plague"', *Eire–Ireland*, 32. 1 (Spring 1997), 63–85.
Foster, R. F. *Paddy and Mr. Punch*. London, 1993.
Foster, Thomas Campbell. 'The Condition of the People of Ireland':
The Times (6 Jan. 1846), 5;
The Times (13 Dec. 1845a), 8;
The Times (18 Nov. 1845b), 9–10;
The Times (21 Nov. 1845c), 5;
The Times (7 Nov. 1845d), 5;
The Times (12 Sept. 1845e), 6;
The Times (23 Sept. 1845f), 6;
The Times (9 Sept. 1845g), 5;
The Times (25 Aug. 1845h), 5;
The Times (21 Aug. 1845i), 4.
Kinealy, Christine. *This Great Calamity: The Irish Famine 1845–1852*. Boulder, CO, 1995.
Leech, John. '"Rint" *v.* Potatoes – The Irish Jeremy Diddler', *Punch* 9 (1845), 213.
Newman, William. 'The Real Potato Blight of Ireland', *Punch* 9 (1845a), 255.
Newman, William. 'Peel's Irish Crop', *Punch* 9 (1845b), 166.
O'Connell, Maurice. 'Letter to the Editor', *The Times* (2 Jan. 1846), 3.
O'Connell, Maurice. 'Letter to the Editor', *The Times* (20 Dec. 1845), 3.
Parker, Robert. 'Letter to the Editor', *The Times* (21 Nov. 1845), 6.
Said, Edward. *Orientalism*. New York, 1978.
Spurr, David. *The Rhetoric of Empire*. Durham, NC, 1993.
Trevelyan, Charles. 'The Irish Crisis', *The Edinburgh Review* 87 (1848), 229–320.
Williams, W. H. A. 'Into the West', *New Hibernia Review* 2.1 (Spring 1998), 69–90.

19
The Nineteenth-Century Media and Welsh Identity

Aled Jones

The idea of Welsh nationality has long been characterized by ambiguity. The devolution referendum in 1979 exposed the indeterminacy of the sense of nationhood in Wales and led to a prolonged crisis of identity which the narrow majority in favour of a devolved Welsh Assembly in 1997 only partially resolved. To be sure, both tests of the public mood distinguished Wales from the far more self-confident and consensual national political culture of Scotland. But, at the same time, this elusive, often fragmented and sometimes contradictory sense of nationality in Wales offers some distinctive insights into the differing processes of identity-making and the problem of media construction and representation of identities which have occurred elsewhere in the United Kingdom and beyond. This chapter will focus on one element of that process, namely the role played by the print media in generating and elaborating concepts of 'Wales' in the years between the Crimean and the First World wars.

It was during the second half of the nineteenth century that a weak but recognizably modern, post-Enlightenment idea of nation emerged in Wales, along with a renewed drive for the establishment of self-consciously national Welsh institutions. It was also the moment when a reconfigured political culture emerged alongside an expanding industrial society and an urbanizing population.[1] During this time of intense cultural, as well as of unparalleled economic, activity, a domestic periodical press, highly diverse in its formats, its doctrinal commitments and the locations of its places of production, appeared in both English and Welsh languages.[2] The argument that will be pursued here is that the expansion of print helped to generate and, in turn, to be further stimulated by, a cultural mobilization which enabled new forms of national identity to gain currency.[3] In effect, journalism made this shift in perspective possible in three ways: by underpinning an emerging public sphere, by providing tangible evidence of Wales's relative cultural autonomy, and by mapping its territory in new ways. Together, these functions articulated and shaped the desires of an important minority who, eager for self-expression, bequeathed to Wales its modern but fractured array of self-images. The chapter will consider each of them in turn.

First, however, it is necessary briefly to outline the process of press expansion in Wales, which differs in significant ways from that experienced in England, Scotland or Ireland. For one thing, the emergence and growth of a popular press occurred much later in Wales than it did in the rest of the United Kingdom, despite the fact that the defining moment of Welsh-language literacy had come with the translation of the Bible into Welsh from Hebrew, Latin and Greek texts in 1588. A flood of biblical commentaries and theological polemics followed. Almanacks began to appear in the late seventeenth century, in both English and Welsh languages, and monthly periodicals, mainly in the Welsh language, were issued from the 1730s, increasing in number in the 1790s partly in response to the democratic energies unleashed by the French Revolution. '*Dyma ni yn awr ar daith ein gobaith*', wrote the Welsh rationalist and Painite, Morgan John Rhys, in his *Cylchgrawn Cymraeg* (The Welsh Periodical), in 1792. 'Here we are now on the journey of our hope'. It is perhaps significant that Raymond Williams chose this as the epigraph for his last major monograph, *Towards 2000* in 1983. But improvements in print and distribution technology, the development of a skills-base in compositing and the increased availability of advertising revenue from small businesses, did not enable a newspaper press to grow in Wales until the early nineteenth century. The first newspaper printed in Wales, the *Cambrian*, was launched only in 1804, followed by other English-language newspapers in Bangor in 1808 and Carmarthen in 1810. The first Welsh-language news weekly, *Seren Gomer*, was launched in Swansea in 1814, specifically to provide the language and its speakers with a 'modern' form of communication. There followed a series of mainly English-language county newspapers, and titles associated with religious denominations, which were printed almost exclusively in Welsh. Where the former relied on local readerships and advertisers, the latter, especially those associated with Wales's largest Nonconformist denomination, the Calvinistic Methodists, aimed at a more widely distributed readership and the financial support of adherents. Baptist journalists, for instance, complained of the difficulty of sustaining a Baptist perspective on the news when virtually all potential small business revenue had been captured either by the English-language press or by the more upwardly mobile Calvinistic Methodists.[4]

The expansion, as one would expect, was further accelerated following the Repeal of the Stamp Act and Paper Duties respectively in 1855 and 1861. It was then that such key titles as Thomas Gee's *Baner ac Amserau Cymru* (Banner and Times of Wales) and *The Cambrian Daily Leader*, Wales's first daily newspaper, were launched. Other dailies followed in Cardiff in 1869 and 1872, respectively the Conservative-supporting *Western Mail* and the Liberal *South Wales Daily News* (the latter was bought by the *Western Mail* during the circulation wars of 1928). Today, the *Western Mail* in the south and the Welsh edition of the *Liverpool Daily Post* in the north pass for the equivalents of a national press, even though their circulations remain strongly regional. Fiscal deregulation in the 1850s and early 1860s provides only a partial explanation for this expansion, which Wales shared with all other parts of the United Kingdom, and indeed with most Western industrial societies. War in the Crimea and the United States stimulated demand, dramatic increases in industrialization put money into buyers' pockets, urbanization led to new forms of popular leisure, including reading rooms and libraries, and new forms of transport, particularly railways, enabled newspapers and periodicals to reach new markets. Developments in New Journalism in the 1890s, especially halfpenny evening papers in the English language, prepared the way for the great restructuring of the Welsh print media during and after the First World War, when weeklies, especially those printed in the Welsh language, were critically weakened and the dailies emerged strengthened as the main form of print communication, owned and controlled by ever fewer proprietors.

Press expansion in Wales, therefore, came relatively late, but occurred at a very rapid pace. In the eighteenth century, print, in the form of books, almanacks and monthly and quarterly periodicals, had begun to seep into a society which, in 1700, had virtually been print-free.[5] The saturation by print of that society, however, occurred only as a consequence of the coming of newspapers in the early nineteenth century. In the same period, Wales also industrialized, vastly increased its population and had towards the end of the nineteenth century formed quasi-state institutions, including a national university. The very vibrant, domestically-produced periodical and news journalism created during the nineteenth century, however, was to be overshadowed by the London-produced press after the First World War. The Welsh press never fully recovered from the pressures and shortages of that war in the way that the Scottish press did so emphatically succeed in doing. Despite the claims of pretenders, there has never been a national press in Wales comparable to that of Scotland.[6] To find evidence of a convincing degree of media autonomy in Wales after 1918 one needs to look to broadcasting: first to radio, then to television.[7]

The sudden expansion, in particular of newspaper journalism, during the nineteenth century affected the ways in which Wales was perceived, within and without. The consequent shift towards a definition of Wales as a place which had developed its own public sphere, in which leaders acquired celebrity status beyond their locality, where social movements emerged and clashed, a place where things happened and where those events were reported in a way that emphasized both Wales's territorial autonomy from, and its political and other connections with, England, was, in part at least, the consequence of an early negative reaction against the importation into Wales of newspapers edited and printed in England. Whether it was regarded as the source of the seemingly inexorable expansion of the English language, or as the progenitor of news values which neglected or undervalued Welsh affairs and achievements, London, in particular, as both the centre of the British state and the location of the media powerhouse of Fleet Street, was viewed with a combination of awe, anxiety and suspicion. The media relationship with London also raised the larger problem of Wales's status within the United Kingdom. Confusingly designated as both a region of England and a constitutionally distinct Principality of the British realm, Wales was neither a separate nor a fully integrated territory. But the elision of 'historic nation' status with the often reiterated claim that Wales was merely a western extension of England was an important one which shaped the ways in which Wales was defined, and its people represented, in nineteenth-and early twentieth-century Britain.[8] Seen in this light, the printing of a Welsh newspaper, in either language, might be understood as a means of defining an alternative centre to London and the other great corporations of England which were taking an increasingly keen interest in Wales and its abundant supplies of labour and natural resources. Every new title, by its very nature, acknowledged and identified both the presence and the difference of Wales.

Two overwhelmingly important questions are posed by the history of the print media in nineteenth-century Wales. One is why so much money and physical and intellectual energy were expended on the publication of newspapers and journals. The other is why this activity failed to create a distinctively national Welsh press which might have represented or embodied a more cohesive sense of national identity. The number of titles launched between the 1830s and the 1890s was out of all proportion to the size either of the advertising base or the consumer market.[9] Furthermore, the desire to publish a domestic press was sustained at a time when cheap but better-resourced English titles were becoming more readily available, principally as a consequence of the expansion of the railways and the growth of the postal service and such lending and retail outlets as the W. H. Smith chain of bookstores. In certain respects, the vigour of the domestic press signalled the extraordinary vitality of Welsh cultural life in the nineteenth century, nurtured as it was by religious controversy, political conflict, new forms of industrial employment, rapid population growth and inter-regional demographic movements. But there was also a more deliberate edge to the hyperactivity of Welsh journalism in this formative period. Their producers were evidently persuaded of the view that newspapers and journals were agents as well as symptoms of social change, and the press was overwhelmingly perceived as a medium which performed a structuring, rather than a merely reflecting role in the institutions and activities of public life. Emerging initially as private enterprises from small-scale printing firms in towns such as Swansea, Carmarthen, Brecon, Bangor and Denbigh, monthly and quarterly journals and weekly newspapers were soon closely identified by editors and readers alike with specific religious and political movements. These publications nourished communities of belief across a wide area, enabled networks to communicate with each other, and provided a pool of activists from which a social leadership could emerge.

It has often been said that nineteenth-century Wales lacked an entrepreneurial middle class, but this is only partly true. While it lacked a politically cohesive resident commercial bourgeoisie, Wales did however produce a significant middle-class stratum of 'social leaders'. Nonconformist ministers and Anglican clergy, sustained financially by their churches and congregations, played a particularly prominent role, but party activists, small businessmen, industrialists, merchants, trade unionists and, crucially, printers, the controllers of the new communications technology, were also visibly participating in public life. Many were also part-time periodical correspondents, or were the patrons of journalists and editors. Seldom did their social activity and the 'news' that they themselves made in Wales find an echo in the columns of newspapers published in London. For these activists and public figures, the domestic press of both languages not only indicated their presence but also acted as an echo-chamber which amplified their voices. Journalism in Wales thus adapted the news values of London, with its concern for the activities of governments and political leaders, by ordering its priorities around the actions and utterances of its own leading figures and the concerns of its own, in the main voluntary, institutions, notably the religious denominations and political groupings. Members and supporters of the numerous religious and political organizations of nineteenth-century Wales developed a language which, while being predominantly centred on their particular theological or political identities, also took on the semblance of a national identity.

The things that distinguished the Welsh as a people, primarily from the English, such as custom, history, language, literature or moral values, were recurrent themes in its journalism. These explorations of difference were rarely explicitly nationalist in tone, even in the years after 1847 when a government report on the failures of Welsh education generated a heated public debate which led to a heightened form of national consciousness, or in the 1880s, when land agitation threatened violent and highly politicized conflict on the Denbighshire moors.[10] But the efforts that were made to publicize and legitimise ideas of Welsh national difference, and of exemplary individuals who personified those ideas, were unmistakable. The problem for this journalism, however, was that there was no political centre on which to direct its attention. The annual conventions of the main religious groups, trade union conferences, and meetings of voluntary societies and pressure groups acted as valuable proxies. Local government reform in the 1880s, elections and speeches delivered by visiting politicians provided further events around which reporters could generate the excitement necessary to sell newspapers, but even then there was no unifying political centre around which to construct a 'national' political narrative in the fashion of Fleet Street. Instead, separating its reporting of 'the London Parliament' from the activities of the Welsh public sphere, much Welsh journalism at this time placed its emphasis not so much on politics as on culture.

This brings us to the second function of journalism in relation to the emergence of ideas of nationhood, namely the articulation of national identities founded on notions of cultural difference. However, the role performed by the press in the production of identifiably 'national' cultural forms raises a number of difficulties. If, following Ernest Gellner (1983), it is argued that nations are not *a priori* models of human association but are instead the products of industrialization and of its need to centralize power, it follows that an homogenizing culture needs to be socially constructed.[11] As Philip Schlesinger has recently observed (1991, 158–62), the nation and its institutions is for Gellner a political roof for culture, which is in turn the cement which holds together the framework of industrial society. The difficulty with this analysis for the

Welsh case is that Gellner equates nation with nation state, with a national community which expresses itself, or at least part of itself, through the concept if not always the practice of sovereignty. Wales, like Scotland and Ireland at this time, shared a British state roof over its cultural life, and while cultural production in Wales on the whole supported the political roof of the centralized British state, it also laid the basis for the later challenge to its legitimacy. In any case, the centralizing tendencies of nineteenth-century capitalism did not lead to the construction of a cohesive national British culture. Not only were state endeavours to build such an edifice weak and ambivalent, particularly when compared to the efforts made by more thoroughgoing centralizing European states such as France, but the tendency was in any case resisted. By and large this resistance was not effected as much by a self-conscious mobilization of nationalist sentiment, or a nationalist party, but rather through the activating of cultural forms and practices: eisteddfodau, religious organizations, education, all linked and embodied in cheap and accessible forms of print communication, particularly monthly periodicals and weekly newspapers. It was within these formations that the idea, and the language, of nationhood was sustained and developed. Consequently, the Welsh case suggests that we invert Gellner's thesis. It was not so much that the nation provided a roof over the head of culture as that culture provided a roof over the concept of nation, a nation which did not exist other than in cultural practice, in a sense of the past, in the consciousness of difference and, increasingly, in the rhetorical web cast by the print media.

A further complicating feature of culture-as-national-identity in Wales was the use of two languages, English and Welsh, which were related in a clearly understood but also openly challenged hieararchy. In a population which had more than doubled from 700,000 in 1811 to 2 million in 1901, more than half (54.4 per cent) were found by the language census of 1891 to speak Welsh. In some counties, such as Meirionnydd, 93.7 per cent of the population were Welsh-speaking, while in Glamorgan, the most populous and industrial of the Welsh counties, the proportion of Welsh-speakers was as high as 43 per cent.[12] For much of the preceding century, however, it had been assumed (with few exceptions) that Welsh was, in the longer term, a dying language and an obstacle to modernity. The Anglican propagandist, David Owen (1795–1866), writing and editing under the pseudonym Brutus, was for most of his life a fierce critic of the Nonconformist attachment to the Welsh language. For thirty years between the mid-1830s and mid-1860s, in his own brand of Welsh-language 'muscular journalism', he embraced the English future. He was not alone. Prevailing concepts of Liberal progress also demanded and predicted the triumph of English as the paradigm of modernity, and strong pressure was exerted, internally as well as externally, to be rid of Welsh as an obstacle to national integration. *The Times* in the 1860s, for example, continued to excoriate the language as the 'barbaric' residue of a primitive past. There is here, however, an historical paradox. Those who sought to undermine the Welsh language from within, by celebrating the power of English, expanded the public sphere in which Welsh was read and written, and thus helped to lay the foundations for its emergence as a modern, codified language which acted as a form of communication for an increasingly literate society. By the 1860s, Welsh was not only the language of Nonconformity, but also of radical politics, and as the historical demographer, Professor Brinley Thomas, has demonstrated, far from being weakened by the massive in-migration of workers mainly from south-west England and Ireland into the iron and coal industries of the south from mid century, Welsh, largely through its utility in the language domain of the workplace, retained its vigour and strengthened it public presence (Thomas 1954 and 1962).

The prevalence of two languages complicated the making of what Benedict Anderson (1983) has identified in his study of the origins and spread of nationalism as the unity-conferring phenomenon of print-language. Exploring the internal agencies that actively and consciously constructed nationhood as an historical option, he regards print as a technology which demanded not only a language hierarchy, where 'privileged' forms of language emerged by virtue of being used by the press, but also the acceleration of the processes of standardization within a dominant language, the 'language-of-power' (1983, 43–5). Print-language thus provided a collective synchronic structure to the lives of newspaper readers, fixing the variation of knowledge and helping to define perceptions of reality. In so doing, as Anderson has suggested, it enabled readers to form 'in their secular, particular, visible invisibility, the embryo of the nationally-imagined community' (1983, 45). In nineteenth-century Wales, however, this was a two-way, and potentially a three-way, process. The expansion of English by means of the circulation of print in Wales powerfully reinforced the tendencies towards an English-speaking British national identity. But, rejecting the subordination imposed by the technological power of English-language print, printers and journalists, from 1814 onwards, deliberately fashioned newspapers, magazines and books in Welsh, whether or not they were financially viable.

By breaking into print in this way, Wales was provided with a second 'language-of-power', one which also underwent the same process of standardization. From the 1860s, for example, regular columns in the Welsh-language weeklies on orthography, grammar and spelling explicitly sought to improve standards, and by the 1880s a new self-confidence may be discerned in the public use of the language. The journalist, novelist, playwright and Liberal activist, Beriah Gwynfe Evans, declared in 1888 that Welsh was no longer a 'national embarrassment' (Evans 13 Jan. 1888), and J. E. Vincent in *The Times* saw fit to invoke the Indian Vernacular Press Act as a means of suppressing the activities of political radicals in whose hands he regarded the Welsh language to have become a vibrant but seditious instrument (Vincent 27 Sept. 1887). Certainly, for many of the Nonconformist ministers who had controlled so much of Welsh-language newspaper journalism between the 1850s and the 1880s, the language was regarded as a *cordon sanitaire* which prevented the transmission of alien 'English' political and cultural values. Hardly any sport reporting, for example, may be found in Welsh-language newspapers until the 1900s. The use of Welsh was without doubt a defining centre, a badge of difference.

But a third process was also under way, namely the emergence of a Welsh writing in English. While the use of the Welsh language was the most visible, and for that reason sometimes the most exaggerated, sign of difference, English-language journalism in Wales, without doubt the majority press in terms of the size of its readership, also resonated with a distinctively Welsh timbre. While in its style it owed much to English county journalism, at least until the coming of the New Journalism in the 1890s, and while it failed to produce a credible 'national' title, in its content, its editorial concerns, and its mode of address the English-language press provided its readers with new and increasingly 'national' points of reference. As one activist in the Young Wales movement in 1888 observed: 'The English Liberal press in Wales has not made Wales less national; it has fortified our nationality, and at the time enlarged Welsh vision' (*Cymru Fydd* March 1888, 126). Newspapers and periodicals of both languages, therefore, contributed substantially towards the creation of a civil society which could begin to be defined as a national community.

Finally, by means of mapping territorial space through the reporting of contemporaneous events, journalism enabled readers (then as now) to see a place called Wales virtually on every page. It did so in two ways, first by linking, through juxtaposition on the same pages, places, individuals and events from the same geographical area – the counties of Wales and

Monmouthshire. By iteratively associating places within a specified territory, a subjective cultural role previously undertaken in the poems, stories, and ballads of the oral tradition, and simultaneously objectified in forms of English travel writing,[13] news journalism generated a powerful sense of place. Second, through the formal separation of those reports within the format of the paper from 'other' news, including Parliamentary, City and foreign news, journalism defined the borderline between what was 'Welsh' and what was not. The semiotics of the national may be seen most clearly in the contents of those titles, in both languages, which deliberately sought to reach a readership throughout Wales by reporting events and printing letters from all its regions, but to some extent it may also be seen in those which were more explicitly local or regional in scope. By representing Wales as a place where things happened, as a territory 'covered' and thus defined by its own news media, another layer of meaning was added to 'Wales' as a national community, imagined in one sense, but all too tangible in another. Where, as Edward Soja observes, 'national patriotism and citizenship were usually couched more in a cultural than a geographical identity and ideology, another example of the inherently spatial defined as something else', such assertions of territorial 'ownership', at least in a journalistic sense, framed the idea of the nation in even more concrete and 'common-sense' terms (Soja 1989, 5). While this territory appropriately extended its reach into the neighbouring city of Liverpool, whose growing Welsh population and relatively prosperous Welsh middle class provided an important impetus for Welsh cultural activity, particularly in commerce and publishing, until well into the twentieth century, the nineteenth century witnessed the sharp decline of the cultural significance of the once dominant Welsh diaspora, especially in London. By the end of the nineteenth century, as O. M. Edwards and other reformers concerned by the 'condition of Wales' realized, Wales, not London or Oxford, was the place to be. Thus, an indigenous print media had by the 1890s given this Wales, in two languages, a renewed visibility as a place which was subjectively defined, and separated from the rest of the world, by cultural and territorial boundaries. The symbolism of the dragon, the harp, women in 'traditional' dress, the leek and other emblems of Welsh societies and educational institutions adorned the mastheads of such journals as O. M. Edwards's *Cymru* (Wales) from the 1890s, the ritual repetition of which were intended to signify to the reader an unmistakable identification with a concept of nationality (*Cymru* 15 Aug 1891, 1). Later, for example in Keidrych Rhys's journal *Wales* (Winter 1946),[14] this manifestation of place would take the form of a hand-drawn map of

Wales as an island, disconnected from the rest of the United Kingdom, resembling, and intended to carry the same iconic charge, as similarly drawn maps of the Holy Land.

The three general tendencies which have been outlined above are well illustrated in the newspaper *Baner ac Amserau Cymru*. A study of the paper also indicates why, ultimately, the press failed to sustain an inclusive and consensual sense of Welsh national identity. Launched in 1857, shortly after the repeal of the Stamp Act, the Denbigh-based weekly became a commercially sound, news-packed paper which enabled its owner and editor, Thomas Gee, not only to project his own political personality but also to embody a clear image of the 'Wales' to which it belonged and whose events it reported. Gee, born the son of a printer in Denbigh in 1815, lived through the social and political transformation of Wales alluded to earlier. He was 16 at the time of the Merthyr Riots, 17 when the Reform Act was passed, 21 at the end of the War of the Unstamped and the reduction of the Stamp Duty from 4d. to 1d., and 24 at the time of the Newport Rising. He was nearly 30 when the Rebecca Riots ended and was in his early 50s when the Liberal Party spectacularly broke the Tory domination of Wales in the General Election of 1868. He was still an active journalist when Gladstone formulated his Irish policy in the 1880s, and lived to see, to comment on, and in important respects to influence a new political class that emerged in Wales in the 1890s. Not surprisingly, he was referred to by David Lloyd George at the time of his death in 1898 as the Welsh Grand Old Man, a term normally reserved for Gladstone himself. His newspaper was initially established as a self-conscious bid for modernity: its role, he wrote in 1857, was to supply 'something more of a railway and steam character than the monthlies' (the railway, incidentally, had reached Denbigh in 1852). The paper campaigned against what it termed 'landlordism', supported and made celebrities of politically victimized and evicted tenants during the Welsh Tithe War, and was an agent in the Liberal conquest of local government in north Wales (Gee was himself the first chair of the Denbighshire County Council)[15] But Gee's vision extended well beyond the town where his paper was printed. He reported news from every part of Wales, and, equally importantly, covered national British politics from a distinctively Welsh perspective. The analysis of content in Table 1, taken from the 21 July 1880 issue, ideally typifies the format and content of the paper in its late nineteenth-century heyday.

British reports (Parliamentary, London and crime stories), appear on pages 4, 6 and 8, while local and north Wales items are grouped on page 13. News stories originating in other parts of Wales, the largest category

Table 1 Format and content of *Baner ac Amserau Cymru*

Page	*Content*
1	Welsh property and commercial advertising
2	Local advertising
3	Main stories: redistribution of seats, south Wales colliery disaster
4	Parliamentary news, and items on the Welsh in Manchester
5	Serialized fiction
6	Parliamentary and political. Crime in Wales and UK
7	Colliery disaster, Risca
8	British and foreign news
9	Political comment. Welsh news in brief
10	Fuller items of Welsh news, by region and district
11	ditto
12	Farming and markets
13	Readers' letters. Local Denbighshire news
14	Welsh religious items. 'Latest Intelligence'. Poetry
15	Advertisements
16	ditto

in terms of column inches, appear on pages 7, 9, 10, 11 and 14. Led by a full report of the Risca colliery disaster, a large number of other items are grouped by place, combining news from north and south Wales on the same pages. In terms of the quality of its format and range of content, *Baner ac Amserau Cymru* was not noticeably different from a host of other good English county and city newspapers at the time. But the world which it reproduced on its pages was a self-consciously Welsh one, both in itself and in terms of its relations with the British state: its parliament, monarchy, and judiciary. By the careful commissioning of items, and their skilful editing, Gee, and others engaged in similar enterprises, constructed with ink and paper a new, inclusive and self-consciously national landscape. It could give the impression that Wales was no longer an imagined community, but a 'natural' and given entity in which related events occurred and which could establish relations with other centres of power. In the absence of another national centre of any significance, the press in Wales in the period before the First World War appropriated that role for itself.

But it did so under false pretences. For one thing, Gee's newspaper never acquired a national readership base. For another, its content, like that of many others, receded increasingly into regional news and comment, and were sustained by local advertising. Lastly, the demographics of Wales were such by 1918 that it was clear that a truly national press would need to appeal to readers of both languages, and the surest means of doing so was by selling the same newspaper to the bilingual Welsh-speaker and the mono-lingual English-speaker; in other words, to switch to English as the sole effective print-language. Newspapers such as *Baner ac Amserau Cymru* might have been the prototypes for national newspapers, but they were not. Without other administrative and political institutions to anchor a national community, the fragile structures of journalism were simply incapable of bearing the strain. Commercial imperatives demanded a turn to the local or, at best, the regional, and, having tantalizingly promised a working definition of Wales as a nation, no newspaper with a truly 'national' coverage emerged.

With the decline of a national territorial coverage in the weeklies, editors were increasingly preoccupied by what they perceived to be the demands of more localized readers and advertisers. By the end of the nineteenth century, the better-resourced and more dynamic English-language newspapers were concerned almost exclusively with satisfying the needs of expanding regional markets, a process which was further strengthened by the creation of chains and local monopolies in the early twentieth century. Where editors sought to speak to or for their readers, they did so with a strong local or regional accent: at best, titles associated themselves with north, or with an even more emphatically capitalized south, Wales. The even greater social and cultural divisions between west and east were similarly exacerbated. The same forces which had sought to build an inclusive national culture led to the intensification of regional difference within that territory. However, it is at the same time important to bear in mind that whereas the idea of Wales as a unified, or as a potentially unifiable, political and cultural entity may eventually have been de-centred by the Welsh print media, distinctive but regionally specific notions of Welshness were, to the contrary, strengthened.

Given the relative weakness of other nationwide Welsh institutions during the second half of the nineteenth century, individual newspapers and periodicals ceased to be mere vehicles for political organizations, religious movements and commercial advertisers, and became, as a collective social entity over which no single body had control, the structure that produced and sustained a new language of nationality. Whether instigated by political ambition, religious difference, a commitment to preserving the Welsh language or the promise of ready profit, the cumulative effect of this prodigious activity was to restructure the public sphere and to redefine the imaginative and spatial boundaries of Wales. This process, crucially, enabled some at least of its people to see the ground on which they stood from the inside, rather than from the outside, as a proactive centre rather than a reactive periphery. But, in spite of a number of significant instances of enthusiasm for the idea of the nation that was represented in and by important sections of the Welsh press, Wales, unlike Scotland, did not succeed in re-inventing itself as a media centre that could challenge London's dominance over the Welsh newspaper market.

It failed to do so broadly for three reasons. First, the production quality, the reporting and commercial resources, the speed of distribution and the advertising power of the Fleet Street daily and Sunday papers ensured their long-term commercial viability throughout Wales. Second, the shift towards the ability to read English amongst growing sections of the Welsh readership corresponded with a changing pattern of newspaper reading. Increasingly, readers in Wales looked to the London dailies and Sundays for general news, and to their local or regional papers for more immediate information. The third, intermediary option of a newspaper with an all-Wales focus was not sustainable within that restructured market. Finally, and perhaps more importantly, the indigenous press came to define not one but several notions of Welsh identity, which remained divided, often bitterly so, by religion, region, and language. If journalism, particularly in the form of daily and weekly newspapers, provided a cultural 'roof' under which a Welsh national identity might have been, and to an extent was, nurtured, it was also evident that the concept of nationhood which it employed was increasingly complex and problematic, and that its identity was fractured into a plurality of forms. The search for an alternative centre to London had led to a further de-centring within Wales itself, a problem which continues to affect the broadcasting media and contributes to the tensions that exist between English-and Welsh-speakers in the present day.

Yet it is precisely in this historical failure to establish a centre and a strong sense of national identity that the case of Wales acquires a more universal significance. As national identities are redefined in increasingly fractured and contested terms worldwide in the postcolonial era, the idea of a national centre is itself being undermined. The condition of the contemporary world may be better described by the absence of centres than by the continuing operation of the centre–periphery power relations of the modern imperial, industrial and communication age. As a consequence, we have witnessed the eruption of a reinvigorated rhetoric of nationality, and an escalation rather than a dimunition of the politics of identity. As in Wales in the nineteenth century, the emergence of new forms of communications media in the present century have signalled the beginning, not the end, of a struggle over the representation of national identities.

Notes

1. For fuller accounts, see K. O. Morgan 1970 and 1981, and I. G. Jones 1987. For a recent perspective on the economic expansion of south Wales, see L. J. Williams 1995, and for the growing power of Cardiff see John Davies 1981 and M. Daunton 1977.
2. See E. Morgan Humphreys 1944, R. Tudur Jones 1981, L. Brake, A. Jones and L. Madden 1990, and Aled Jones 1993.
3. For a further exploration of the concept of 'cultural mobilization' in relation to Welsh media, see D. Bevan 1984. See also essays by Dai Smith, Peter Stead, Michelle Ryan, John Hartley and Trevor Wright in Curtis (ed.) 1986.
4. See Aled Jones 1993, 122.
5. See Geraint H. Jenkins 1993 and 1978.
6. See Penrose in Jackson 1993, 27–49 for the implications of this process.
7. For a fuller examination of this claim, see John Davies 1994.
8. Anthony Smith puts this succinctly: 'the English (Anglo-Norman) state expanded into Wales, destroying the Welsh kingdom and bringing most Welshmen into the realm as a peripheral cultural community under the domination of the English state', which was followed by a process of 'ethnic bureaucratic incorporation' (Smith 1991, 39, 57).
9. For a statistical analysis of title production and distribution, see Aled Jones, 'The Welsh Newspaper Press', in Hywel Teifi Edwards (ed.), *Companion to the Literature of Wales* (Cardiff, forthcoming).
10. For an evaluation of the role of 'social leadership' in the tithe agitation, see Howell 1977.
11. See also Tom Nairn (1994, 72), who argues that the origins of nationalism 'are located not in the folk, nor in the individual's repressed passion for some sort of wholeness or identity, but in the machinery of world political economy'.
12. For the most comprehensive history of the language, consult Janet Davies 1993.
13. See, for example, Pennant 1784, Borrow 1862, and Morton 1934.
14. See also Wilkins 1882, 3.
15. T. Gwynn Jones, *Cofiant Thomas Gee* (Denbigh, 1913) remains the most comprehensive, if deeply partial, biography.

Works cited

Anderson, Benedict. *Imagined Communities: reflections on the origin and spread of nationalism*. London, 1983.

Bevan, D. 'The mobilisation of cultural minorities: the case of Sianel Pedwar Cymru', *Media, Culture and Society* 1 (1984).

Borrow, George. *Wild Wales*. London, 1862.

Brake, L., Jones, A., Madden L., (eds). *Investigating Victorian Journalism*. London, 1990.

Curtis, Tony (ed.). *Wales: the imagined nation. Essays in Cultural and National Identity*. Bridgend, 1986.
Daunton, M. J. *Coal Metropolis, Cardiff 1870–1914*. Leicester, 1977.
Davies, Janet. *The Welsh Language*. Cardiff, 1993.
Davies, John. *Broadcasting and the BBC in Wales*. Cardiff, 1994.
Edwards, Hywel Teifi, ed. *Companion to the Literature of Wales*. Cardiff, 2000.
Evans, Beriah Gwynfe. *North Wales Observer and Express* (13 Jan. 1888).
Gellner, Ernest. *Nations and Nationalism*. Oxford, 1983.
Howell, David W. *Land and People in Nineteenth-century Wales*. Cardiff, 1977.
Humphreys, E. Morgan. *Y Wasg Gymraeg*. Caernarfon, 1944.
Jackson P., and Penrose, J. (eds). *Constructions of race, place and nation*. London, 1993.
Jenkins, Geraint H. *Literature, religion and society in Wales, 1660–1730*. Cardiff, 1978.
Jones, Aled. *Press, Politics and Society. A history of journalism in Wales*. Cardiff, 1993.
Jones, I.G. *Communities: Essays in the social history of Victorian Wales*. Llandysul, 1987.
Jones, R. Tudur. *Ffydd ac Argyfwng Cenedl*. Vol. i, Swansea, 1981.
Jones, T. Gwynn. *Cofiant Thomas Gee*. Denbigh, 1913.
Llafur. Journal of Welsh Labour History, 7.2 (1997): special issue on media and history.
Morgan, K. O. *Rebirth of a Nation: Wales 1880–1980*. Oxford, 1981.
Morton, Henry V. *In search of Wales*. London, 1934.
Nairn, Tom. 'The Maladies of Development'. Repr. in John Hutchinson and Anthony D. Smith, *Nationalism*. Oxford, 1994.
Pennant, Thomas. *A Tour in Wales*. 1784.
Schlesinger, Philip. *Media, State and Nation. Political Violence and Collective Identities*. London, 1991,
Smith, Anthony D. *National Identity*. Harmondsworth, 1991.
Soja,Edward. *Postmodern Geographies: the reassertion of space in critical social theory*. London, 1989.
Thomas, Brinley. *Migration and Economic Growth: a study of Great Britain and the Atlantic Economy*. Cardiff; 1954.
Vincent, J. E. *The Times* (27 Sept. 1887).
Wilkins, Charles. *The Red Dragon*: *A National Magazine of Wales* (Feb. 1882).
Williams, L. J. *Was Wales Industralized*? Llandysul, 1995.

20
'Long and Intimate Connections': Constructing a Scottish Identity for *Blackwood's Magazine*

David Finkelstein

Any investigation of Scottish cultural and national identity in the nineteenth-century periodical press must invariably confront the underlying myths and narrative strategies used in such textual constructions. Yet surprisingly few attempts have been made to initiate such investigations. In part this is due to the politically orientated and sociological nature of most recent critical work on the subject.[1] These texts tend to give short shrift to nineteenth-century constructions of Scottish identity because of their perceived lack of nationalist sentiment in the wake of an industrialized and ideologically conservative period in Scottish history. By the mid to late nineteenth century, according to most surveys of Scottish cultural and economic history, the political yoking of Scotland to England, which had begun with the union of the Scottish and English parliaments in 1707, and the subsequent economic and industrial development of Scotland in tandem with the spreading of the British Empire, had resulted in a complacently depoliticized Scotland, one in which there existed what one critic has neatly summarized as 'an ideology of noisy inaction' (Womack 1989, 147). In this reading of Scottish cultural history, the general consensus is that cultural and literary stagnation followed the Scottish Enlightenment of the late eighteenth and early nineteenth century, culminating in, among other things, the *fin-de-siècle* Kailyard literary movement, with its sentimental, romantic and idealized depiction of a pre-industrial rural Scotland.

However, such a reading fails to consider the paradox within which much of the cultural analysis and literary production of this period operated, the paradox of constructing a separate Scottish identity within the constraints of a larger British unit. The solution was a concentric construction, built around an uneasy but safe romanticization and celebration of past symbols of Scottish greatness, whereby Jacobite aspirations, Highland culture, Mary Queen of Scots and the Battles of Bannockburn and Culloden could be reflected upon within the safe constraints of an Anglicized present. Nineteenth-century constructions of Scottish cultural heritage generally represented an establishment view that the Union between Scotland and England had in the end been nothing if not beneficial to the Scots nation. As a recent survey trenchantly summarized, the view was that it had 'transformed Scottish manners, and rescued Scotland from barbarism, anarchy and unruliness and introduced in their place economic development, socio-political order and cultural enlightenment' (Donnachie and Whatley 1992, 52). I wish to illustrate how such nineteenth-century paradigms of identity find their way into the non-fiction periodical prose work of Margaret Oliphant, one of the most underrated Scottish authors of the period. More specifically, I will focus on the rhetoric and narrative strategies through which she constructed an official identity for one of Scotland's best known and most influential literary journals of the time, *Blackwood's Magazine*. In highlighting Oliphant's role as approved architect of an image for *Blackwood's Magazine*, I shall touch upon such issues as editorial and authorial roles, and the effects of these often conflicting interests in the final published text, the two volume *Annals of a Publishing House*.

In June 1860 an anonymously written article appeared in *Blackwood's Magazine* entitled 'Scottish National Character'. The veiled author was Margaret Oliphant, a frequent contributor to 'Maga' since 1852, the year her novel *Katie Stewart* made its serialized appearance in the journal's monthly pages. The 1860 article was ostensibly a review of Dean Ramsay's memoirs and anecdotes of Scottish culture and customs, *Reminiscences of Scottish Life and Character*, but Oliphant took the opportunity to discourse at length and in sweeping terms on national identity, and in particular on Scotland's past cultural heritage. 'Of all our national possessions', she announced dramatically:

> there is none so precious and so dear to us as that Past of ours, with all its noble and touching lessons, the chart of our great voyage – a chart with annotations how sad, tragical, and passionate; its breakers signalled by what blood-lines; its sandbanks distinguished by what cost of lives; its dangers all marked and chronicled by the last heroical effort of some courageous, victorious, drowning hands, snatching from its own destruction a security against its brother's! That wonderful record of a nation's experience.
>
> (Oliphant 1860, 716)

The destruction referred to here, which contradictorily results in security is, of course, the uniting of the parliaments of Scotland and England in 1707 following, among other things, the disastrous and economically costly attempt to establish the Scottish colony of Darien on the Panamanian coast in the late 1690s. As Scotland's imperial ambitions foundered, so too did its economic independence. But what fascinates Oliphant is not the economic realities of Scotland's heritage, but the more personal, individual characteristics making up Scottish identity. Scottish character is presented as having been formed and forged primarily through opposition to England:

> Few countries, perhaps, have been placed in a position so well adapted for the development of *character*... as this our kingdom of Scotland, ancient, hardy, pugnacious, and poor; – always dwelling next door to the rich brother, who vexed her soul with ostentatious display of his greater wealth – often hardly set to it holding her own against him – driven now and then to her hills and rocks – tenacious, indomitable, not to be conquered or silenced – bearing herself in more peaceable times with that most excusable fashion of arrogance, the pride and brag of a poor gentlewoman wrapping herself close in her pinched cloak in self-defending bitterness, while her plump neighbour laughs beside her, full and lavish, mocking the pomp of poverty. We do not suppose that any just estimate of the Scotch character of former generations can be formed, without keeping in perpetual recollection this juxtaposition with the richer nations, which the poorer one never can forget her perennial opposition and antagonism to.
>
> (Oliphant 1860, 717)

What is interesting is that Oliphant presents supposedly positive aspects of Scottish character in negative terms. Independent spirit, pugnacity and hardiness are categorized as contrary aspects of a poverty-stricken nation that looks bitterly across at its wealthier neighbour. Indeed, in this text, independent Scottish patriotic sentiment is denied and demoted for possessing within it the seeds of such past grievances. While England, the more benevolent power, benefits from having moved beyond such past 'quarrels' as Bannockburn, Scotland 'has never been able to forget it' (Oliphant 1860, 717). The result is a Scotland which 'in her heart of hearts hugged the remembrance still of having once, if but once, thoroughly thrashed and beaten that big brother, whose bigness, and lavishness, and plentitude have afflicted her life' (Oliphant 1860, 717).

Equally interesting is the gendered context of these passages. Scotland as the weaker, female partner in this union of nations is a trope Oliphant returns to frequently throughout her writing career, often representing it in the context of an enforced, matrimonial state. Thus in an 1861 review, the union between Scotland and England is compared to that of bride and groom, with often antagonistic results: 'Wedded, not without boastfulness like many a bride, she [Scotland] has had her struggles in the new family relationship, which has bound the hands but not the tongues of the two spouses' (Oliphant 1861, 267). I shall return to this particular trope of meaning in discussing Oliphant's role in constructing an identity for *Blackwood's Magazine*.

Lest anyone should begin to question the sentiment underlying these statements, Oliphant hastily suggests that such antagonism has now been usefully absorbed and modulated into a different if somewhat undeclared national identity:

> This strong national sentiment has given a certain edge to the national character. That instinct of opposition, contradicting resistance, which kept Scotland so long independent, seems to have acted upon her offspring, as a mother's passion would do, all unawares and unconsciously, upon her child. There is nothing so clearly marked in the national character as this thread of contradictoriness.
>
> (Oliphant 1860, 718)

While Oliphant's analysis, simultaneously laudatory, nostalgic and condemnatory, points to a contradictoriness in Scottish character, her text struggles within the paradox already noted regarding contemporary analyses of Scotland's cultural position. Oliphant wishes to celebrate qualities which differentiate Scots from the neighbouring English, but at the same time operate within a conservative ideological framework emphasizing British unity and a single national identity. The result is a text which identifies and revels in a marked Scottish heritage that is safely labelled as part of an independent, pre-union past, but is commented on from the perspective of a current, united British national identity. It is a curious mixture of admiration for the past combined with a present smugness, a smugness emanating from a vision of a larger and safer identity into which the Scots have been subsumed. Thus while it is safe to praise past Jacobites for their staunch patriotic and religious fervour (and their opposition to King George in favour of Bonnie Prince Charlie and his heirs), such praise is qualified by a hasty reassurance that any such anti-English, anti-monarchical sentiments are now safely buried: 'Jacobitism, to be sure, is dead and gone by this time; and if the old sentiment of loyalty exists anywhere, it is in this island – and in this island finds no warmer or stronger adherents than among the descendants of those who poured out the utmost torrent of their disdain, broadly, bitterly, with stinging sneer and sarcasm, upon the "wee, wee Germain lairdie" who had displaced their ancient line' (Oliphant 1860, 723).

This is not to suggest that Oliphant always felt comfortable with this construction in her writing. As past studies have made clear, Oliphant's best work, both fictional and non-fictional, derives its power from her involved engagement and contrasting of Scottish themes, landscapes, language and personalities against English conceptions (Jay 1995; Colby 1979). But in analysing the tropes of meaning within the essays on national identity written for *Blackwood's Magazine,* and in illustrating their general conformity to contemporary establishment views of Scottish cultural positions, I wish to suggest that the pattern, language and tone used in Oliphant's discussions prove of even greater importance when we consider her development of a public image for *Blackwood's Magazine* in her two-volume *Annals of a Publishing House.*

In August 1894 William Blackwood III, head of the Blackwood publishing house, wrote to Margaret Oliphant asking her to undertake a three-volume celebration of the firm's history and the magazine's relations with the great authors of the nineteenth century. 'I have long wished to have a history of the House of Blackwood and the Magazine written to do full justice to those who have gone before me in the management of its affairs', he stated, 'and to let the present and rising generation realize the class of men they were, and how close and friendly their relations were with the leading authors of their time, and the loyalty that subsisted between them and their publisher' (Blackwood 24 Aug. 1894). He offered Oliphant a yearly stipend of £500 until the work was complete, and concluded, 'You have been so long and intimately connected with the House and the Magazine, and moreover you are so thoroughly in touch with Maga's past, that you would I think and hope find the task a congenial one' (Blackwood 24 Aug. 1894). Oliphant accepted the opportunity, telling William:

> I began my married life by my first story in Maga – the proof of which (Katie Stewart) I received on my wedding day. I should like to wind up the long laborious record (which seems to me now to have been so vain, so vain, my life all coming to nothing) with this.
>
> (Oliphant 3 Nov. 1894)

Oliphant completed two volumes of the work shortly before her death in 1897. The third was written by Mary Porter, daughter of the previous director John Blackwood.

Oliphant's work utilizes a specific narrative structure, constructed from overlapping but sometimes conflicting agendas. Oliphant sought a method of presenting the past greatness of *Blackwood's Magazine*, and of emphasizing its role both as cultural product of Edinburgh and as a direct inheritor of the more general cultural heritage of the Scottish renaissance movement. At the same time, the difficulty lay in developing these themes within the straitjacket of present-day contexts, in having also to suggest that *Blackwood's Magazine* had moved from a purely Scottish identity to appeal to and embrace a wider national and colonial audience. The result is a text which constructs national and cultural identities by employing strategies similar to those seen in Oliphant's articles on Scottish national character and history.

Oliphant begins the first volume with a discussion of eighteenth-century Scottish identity and literary achievements. Noting the gathering of talent in Edinburgh during the Scottish enlightenment, Oliphant states:

> There were thus brought together many of the men who swayed and were born to sway the conquering race of the world, the united but various peoples whom it sometimes vexes our little Scotland, which has contributed so much to its force, to consent to hear identified among the heathen as 'English', – a whimsical yet by no means unreal, though we fear inevitable, grievance. She has always had plenty of revenges upon the more abundant neighbour who, for general purposes, has swallowed up in his, like a husband with his wife, an equally dignified and considerable, if not so wealthy, name. She has never been without her large share in actuating the policy of co-partnership; and in those days she moulded the minds of almost all the budding statesmen of the time, English as well as Scottish.
>
> (Oliphant 1897, 4)

Yet once more the shadow of a more powerful, male figure looms over the fate of a female Scotland. This image harks back to Oliphant's articles written thirty-odd years before, developing a similar identification of Scotland and England in gendered terms. The metaphor used here is that of husband and wife, and the effect is the same – the smaller, female figure is subsumed into the larger, male, 'English' entity. Even at this late stage in Oliphant's career, there remains a note of piquancy, even bitterness, in the description, for we are no longer discussing political realities, but identifying the literary and intellectual forces which once made Scotland, and Edinburgh, such a dominant cultural force. One may ask whether this reflects a certain resentment that Scottish voices are given short shrift in an 'English' identity? Or perhaps even a resentment at the need to establish a presence south of the Scottish border in order to be heard and recognized?

At the same time, there is the difficulty of discussing an independent identity within the context of a unified Britain. Oliphant gets round this by suggesting that past cultural differences which existed in Scotland, differences which fed into and so greatly influenced *British* culture at large, were due to Scotland's relative isolation from London-centred influences, and its strong intellectual links with Europe.

> Scotland then was much more widely separated than now from the other, larger, but not more distinct capital and metropolis of the south. The University of Edinburgh flourished greatly... by the fame, which was European, of many of its teachers, and the large invasion of pupils... to partake the instructions which gave an intellectual stimulus beyond their immediate sphere of action to half the world... Even now, when everything tends towards London, Edinburgh preserves a very distinct stamp of her own; but in those days she was as individual and distinct as Paris or Vienna.
>
> (Oliphant 1897, 3–4)

It is notable that here, literary identity is defined in context of the rivalry between Edinburgh and London. This acts as a microcosmic reflection of the larger debates over cultural and literary identity which surface throughout the rest of the work. The inevitable exodus to London from Scotland in general, and Edinburgh in particular, of publishers, editors, and writers from the 1820s onwards, a move resisted by the Blackwood firm until 1840, when it opened a London office, reflected the shifting balance of literary and cultural power between the two rival capitals.

This pattern of movement southwards also finds its way into Oliphant's construction of an identity for *Blackwood's Magazine*. It is a linear construction, suggesting a rise from humble beginnings as an Edinburgh-based monthly journal, to the heights of dominance in London, and so by implication British, literary circles and culture. This narrative of success weaves a series of intersecting metaphors and images, including romance – the romance of writing during the early days of Maga, writing perceived as a leisure activity, not as a profession; growth, from childhood through adolescence into sober maturity; and paternal guardianship, the benevolent, restraining hand of the editor guiding the literary production of his sometimes errant literary children.

The romance of writing and the cult of the author are motifs which permeate most publishing memoirs and literary memorials of the late nineteenth century. Indeed, central to any construction of editorial and periodical identity is the appropriate association with important literary figures. In official histories, the financial details and mundane tasks that characterize much of daily editorial and literary duties (how many copies were sold, what sort of articles were commissioned), are inevitably subsumed into narratives emphasizing personalities and connections to perceived literary and cultural touchstones. This holds true in the *Annals*. Chapters are grouped around individual contributors and authors who played their parts in the pages of Maga. And Maga's beginnings are perceived in romantic terms, its early issues brought forth due to a committed and loyal core of individuals. More importantly, these individuals are represented as writing with a view to rousing their Edinburgh compatriots, not with an eye cast towards their southern neighbours:

> It is clear, for one thing, that the opinion of London and the world – almost convertible phrases nowadays, and the chief, almost the only, aim of literature and literary ambition – did not occur at all to these young men. It was for Edinburgh they thought, which is a most singular thing to think of among all the changes which time has brought about. No doubt Edinburgh is quite enough to make a reputation still; but there is perhaps no one nowadays, certainly no number of men, who would venture to leave the rest of the world out of their calculations.
>
> (Oliphant 1897, 116)

In keeping with this construct, Oliphants also suggests, for example, that the securing of tangible economic benefits such as advertisements were initially ignored because of the founder's interest in producing art and literature, not in generating business profits:

> The Magazine had been very careless of all these aids, being in its beginning a romantic adventure altogether, and not founded upon the principles which the smaller fry of trade literary enterprises in London were laboriously working out.
>
> (Oliphant 1897, 500)

Thus Oliphant establishes and asserts that aesthetic principles, not economic considerations, are the underlying *raison d'être* in the early, romantic days of Maga, while simultaneously implying that in the case of southern rivals, and specifically London-based ones, the reverse holds true.

Crucial to this romantic enterprise is the cautious figure of the editor and publisher. Oliphant conjures up the figure of William Blackwood, founder of the firm and of the magazine, not as a grasping businessman, but as a daring adventurer, seeking nuggets of gold in the mines of Literature. 'It is a common belief', she notes at one stage:

> in the literary world that publishers are the most grasping of middlemen, eager only to have the lion's share of the profits. But in those days there was a certain spirit of daring and romance in 'the Trade'. The Revival of Literature was like the opening of a new mine: it was more than that, a sort of manufactory out of nothing, to which there seemed no limit. You had but to set a man of genius spinning at that shining thread which came from nowhere, which required no purchase of materials or 'plant' of machinery, and your fortune was made.
>
> (Oliphant 1897, 25)

The founder William Blackwood (1776–1834) is represented as the canny Scot who is able to set the men of genius at their task of spinning golden threads of literature, befriending, encouraging and gathering together a coterie of authors to form an engaged and directed unit. Throughout the heady, early days of Maga's development, from the publication of the infamous Chaldee Manuscript in 1817, to its attacks on Coleridge, Wordsworth, and the Cockney School of Poetry, William Blackwood is cast as a bedrock of levelheadedness in the face of a whirlwind of literary production and intellectual high jinks. He is the guardian in rough times, behind whom shelter contributors such as John Gibson Lockhart, William Maginn, Thomas de Quincey and John Wilson ('Christopher North'). He is also the benevolent editor, patiently restraining his headstrong coterie:

> Anxious but indomitable, holding in his Pegasus as well as he could, sometimes permitting himself to be run away with, sometimes pulling up hard with a great effort, but always steady as a rock to his engagements, his opinions, and his friends, William Blackwood stood alone to take all the risks and fight all the battles.
>
> (Oliphant 1897, 173)

Thus are authors and editors linked together, with the editor firmly in control of the final product. Throughout the *Annals*, Oliphant makes frequent reference to the firm's use of personal contacts with authors to restrain and channel their energies and activity. These relationships are explicated utilizing images similar to those of dominance and compliance present in Oliphant's constructions of Anglo-Scottish relations. The influence and control exercised by the editor, much like that benevolently exercised by England over Scotland, is portrayed by Oliphant as beneficial to and gratefully accepted by the compliant author. Of John Gibson Lockhart, for example, Oliphant concludes, 'The young man who, when he had become a literary personage by the agency of the Magazine, wrote the above, had the best of reasons for appreciating the generous publisher who began to influence his life from his very first appearance in Edinburgh' (Oliphant 1897, 184). Commenting on the Irishman William Maginn, who initially wrote to Blackwood under a pseudonym, Oliphant uses the opportunity to stress the protective nature of the editor during a threatened libel case:

> R.T.S. was at the very outset of his career, and known to nobody; but he too sheltered behind the steady personality of the publisher, without even a word of reproach from that much-tried man. So early an alarm might well have broken the newly formed bond, but there is nothing but the warmest cordiality in Mr. Blackwood's first letter on the subject to the veiled prophet of Cork.
>
> (Oliphant 1897, 368)

The larger-than-life figure of the editor thus shields his impetuous authors under the larger, unified protection of anonymity, and envelops them under a unified, periodical identity, much in the same way as Scotland is shielded and enveloped by and under England in Oliphant's texts.

Youthfulness and impetuous energy also form part of Oliphant's construction of the magazine. Just as the journal is perceived to move from the purely local in material to the national and even global, treating of Irish politics and colonial possessions, so too it is seen to move from youthful initiative to sober maturity. Of the magazine's early days, Oliphant concludes that its activities were those of a headstrong youth, now irrevocably changed but nevertheless much missed:

> This is now the mythic period, the heroic age of its history. The rights and wrongs once so fiercely contending, have died away in silence.

> Youth plays the same or very similar pranks around us at the present time in many papers and magazines; but we miss the boyish laughter, the redeeming element of fun, which was so much of it. The young critics of 'Blackwood', in the exuberance of their mischievous fancy, had not the portentous gravity which is so general now.
>
> (Oliphant 1897, 179)

The tone with which Oliphant elucidates this change in tone is seemingly steeped in regret and nostalgia for a romantic past:

> These scraps of hasty letters take us behind the scenes, and let us see how hard it was to keep all in working order: and how doubly hard to drive a winged steed in the vehicle which is to carry your eggs to market, over all the rough roads and harsh macadam of the half-made ways. It is much steadier driving nowadays, when the teams are so much tamer, and the roads crushed smooth by endless merchandise. And yet perhaps it was a different rate of going, with all its risks and continual danger of upsetting, in the old heroic days.
>
> (Oliphant 1897, 315)

But if Maga is now seen to lack the bite of earlier times, short shrift is given to those unable to recognize and adapt to these changes. James Hogg, for example, is soon cast out into the literary wilderness once his style of writing begins deviating from editorial policy:

> Hogg had not outgrown the age of glorious sport, when to bait an unfortunate victim and pursue him about the world for the laughter of the reader was the inspiration of the moment; but the Magazine, not any longer a dashing and reckless adventurer, but a very important undertaking, meaning both fame and fortune, had outgrown it.
>
> (Oliphant 1897, 348)

Oliphant notes, quoting a letter on the issue from William Blackwood, that it 'pointed out very decidedly the particular points of reference to which we have referred – the advancement of the Magazine in seriousness and sobriety, and the stationary character of the belated contributor, to whom there was no triumph higher than that of the Chaldee Manuscript' (Oliphant 1897, 349). The magazine had changed with time, moving beyond local satire to weightier, international subjects. Oliphant's text modulates itself to accommodate this shift in identity.

In the introduction to the first volume of the *Annals*, William Blackwood restates what he had told Oliphant was the underlying reason for instigating this construction of literary and cultural identity:

> to pay to the memory of his [John Blackwood's] father and elder brothers the tribute of putting on record the not uneventful annals of the publishing house and the 'Magazine' which William Blackwood founded, together with some account of the brilliant band of authors and contributors whom his energy and his very genuine love of literature succeeded in rallying to his support.
>
> (Oliphant 1897, vii)

The overall result of this celebration of literary relations is a complex and sometimes uneasy textual construction, weaving together a jostling variety of editorial, authorial and cultural identities. Chronicled within the text are shifting centres of literary power, neatly paralleling contemporary cultural accommodation to the political realities of the relationship between Scotland and England. Thus the paradox of containing, and at the same time emphasizing a unique Scottish national identity within a larger British context, can also be seen at work in Oliphant's textual yoking of authorial roles to a more powerful periodical and editorial identity.

Note

1. See, for example, Ian Donnachie and Chris Whatley, *The Manufacture of Scottish History*, Edinburgh, 1992; David McCrone, Stephen Kendrick and Pat Straw, eds., *The Making of Scotland: Nation, Culture and Social Change*, Edinburgh, 1989; T. C. Smout, *A Century of the Scottish People, 1830–1950*, London, 1987; Peter Womack, *Improvement and Romance: Constructing the Myth of the Highlands*, Basingstoke, 1989. In literary studies, little attention has been paid to the discourse of the periodical press, with most discussions centring on investigating Scottish identity in the context of fiction and the novel. Exceptions exist, including William Donaldson, *Popular Literature in Victorian Scotland: language, fiction and the press*, Aberdeen, 1986, and various chapters in Douglas Gifford, ed., *The History of Scottish Literature, Volume 3: Nineteenth Century*, Aberdeen, 1989.

Works cited

Blackwood III, William. 24 August 1894, Blackwood Papers, National Library of Scotland, MS 30380, 581–2.

Colby, Robert and Vineta. 'Mrs. Oliphant's Scotland: The Romance of Reality', in Ian Campbell, ed., *Nineteenth Century Scottish Fiction*. Manchester, 1979, 89–104.
Donnachie, Ian and Chris Whatley. *The Manufacture of Scottish History*. Edinburgh, 1992.
Jay, Elisabeth. *Mrs. Oliphant: A Fiction to Herself*. Oxford, 1995.
Oliphant, M. O. W. *Annals of a Publishing House: William Blackwood and his Sons*. vol. 1, Edinburgh, 1897.
Oliphant, M. O. W. 3 November 1894, Blackwood Papers, National Library of Scotland, MS 4621, 211–12.
Oliphant, M. O. W. 'Scotland and Her Accusers', *Blackwood's Magazine* 90 (Sept. 1861), 267–83.
Oliphant, M. O. W. 'Scottish National Character', *Blackwood's Magazine* 87 (June 1860), 715–31.
Womack, Peter. *Improvement and Romance: Constructing the Myth of the Highlands*. Basingstoke, 1989.

21

Making News, Making Readers: The Creation of the Modern Newspaper Public in Nineteenth-Century France

Dean de la Motte

Rarely, if ever, is the mass press associated in the popular imagination with utopian discourse or political idealism of any kind. It is a truism of our over-mediated society that the news has long been moving away from politics and toward entertainment. Yet it does not take a thinker as eminent as Neil Postman, whose *Amusing Ourselves to Death* (1985) explores the effects of television on public discourse in America, to realize that questions of social justice and political change (unless the latter involves violence or scandal) are of little interest to – and perhaps are incompatible with – the business of making the news.

Of course the mass media are inscribed in a larger ideological and inherently political context, but often make claims of objectivity that are, it should be remembered, relatively recent in the history of the press. In France, these claims reach back little more than 150 years, and coincide exactly with that country's early efforts at industrialization – almost a half-century after the British and Americans, retarded by a particularly unstable period of revolution, empire and reactionary restoration, from 1789 to 1830.[1] Paradoxically – or so it may seem from our perspective – apolitical, 'objective' or mass journalism was ostensibly developed to serve the interests of the people by rising above partisanship in a utopian gesture of enlightenment. The development of the new penny press – *la presse à bon marché* – is accompanied by the melioristic rhetoric of utopian socialism, which reached the height of its influence under the July Monarchy (1830–48).[2] In the pages that follow I trace the transformation of certain elements of a potentially subversive 'counter-discourse' (Terdiman 1985) – utopian socialism – into a tool for the business of selling newspapers.

From well before, but especially during and after the French Revolution, the periodical press throve upon a clash of opinion generated by a vast array of partisan publications of every conceivable political shade. Not until Emile de Girardin articulated the conception of a *colourless* information-based daily, which simultaneously seeks to make itself invisible and foreground its own materiality as commodity, do we see a fundamental shift that in turn allows us to examine the actual formation of the dominant discourse we still know today as the mass press. The Revolution of 1830 – itself arguably a product of the press – created the conditions necessary to the development of the first truly mass medium, the inexpensive daily newspaper: a hitherto-unknown freedom of the press and educational reforms culminating in the Guizot Law of 1833, which guaranteed French males access to elementary education, ultimately raising literacy rates to a considerable extent (Allen 1991, 8–9). These changes coincided with innovations in printing and marketing to allow the production and consumption which, in turn, would enable the introduction of the first mass dailies in 1836 (Bellanger, *et al.* 1969, 11–26; 114–24).

The facts of the 'revolution in the periodical press' in France are well known and often cited: Girardin and his erstwhile collaborator Armand Dutacq introduced, respectively, *La Presse* and *Le Siècle* on 1 July 1836 for an annual subscription of 40 francs, half the amount then charged by the established papers. Profits were to come largely from advertising revenues rather than from subscriptions and, as Richard Terdiman points out, 'the effect of commercializing information was to reduce the influence of other social and ideological determinants upon the formation of newspaper discourse…then, like money, it began to seem colorless'. (Terdiman 1985, 126)

While Girardin's role in the history of the French press has taken on almost mythical dimensions, and significant study has been devoted to his life and writings,[3] little attention has been paid to the influence of utopian socialism on the rhetoric which frames his project for *La Presse*. If anything, he is portrayed as a Tallyrand figure, an opportunist who was, as early as 1830, persuaded that 'only strict political neutrality would allow him to reach a wide readership' (Pellissier 1985, 99).[4] The pages that follow focus on the extent to which Girardin took from the early socialists the notion that *production* and *consumption*, liberated from the partisan shackles that hampered the free circulation of ideas and inventions in the marketplace, would *necessarily* have a democratizing effect on French society.

Until 1836, daily newspapers were typically the organs of opposing political parties; each paper had its *colour*, which was in fact the current image in use at the time. The metaphor is particularly apt, for it allows us to see that the objectivity Girardin claimed for his paper was to be an undistorted, utilitarian instrument, a mirror-image of the world. Positioning *La Presse* as a philanthropic enterprise whose sole aim was to disseminate information for the good of the public, Girardin dons the damning colourlessness as a badge of honour in an editorial published 13 August 1836. Quoting *Le Journal du Commerce*, in which it was written that '*La Presse* cannot easily be classified, due to the vague character of its writing', he responds:

> No, certainly, we don't have a single one of the colours that make up the rainbow of contemporary politics, and heaven forbid!...
>
> And so – let us declare once and for all how delighted we are to have the newspapers of the old school call us *colourless*.
>
> (Girardin 1836b, 1)

And yet, Girardin claims to belong to a party of sorts: 'Our party is the one that has undertaken the organisation of the new France made possible by the social revolution of 1789, acting while others only talk' (Girardin 1836b, 1). The new, free France will be based on the full exploitation of its natural and human resources. If the editor's language is reminiscent of Saint-Simon, Fourier, and their disciples, it is not by chance, for it could be argued that Girardin almost single-handedly adopted and popularized the broadest of the utopian socialists' common concerns and, in the context of growing literacy and rapid industrialization, ultimately transformed them into a marketing tool.

After all, Girardin's credentials as a *vulgarisateur* were unrivalled: in 1828 he had co-founded *Le Voleur (The Thief)*, his first great success; it was a weekly digest later described in the *Physiologie de la presse* (1841) as 'one of those digest rags whose editor-in-chief was a pair of scissors' (Anon. 1841, 81). As early as 1831, Girardin seems to have intuited just which aspects of utopian socialism could be adapted to the development of the mass press: the desire to find a unity of purpose and to create a new social organization that would benefit the greatest number of people. He seizes upon the pragmatic and rejects the visionary or mystical; if we compare his statement of purpose in *Le Journal des connaissances utiles* to early socialist formulations, the similarities are striking. In *L'Organisateur* of 1819 Saint-Simon writes that 'the question of social organisation cannot be too promptly or completely illuminated, for men only quarrel through a lack of mutual understanding' (Saint-Simon 1819, 1). In the 'Prospectus' for *Le Producteur* of 1825, we learn from his disciples that the paper's goal will be 'the greatest possible development of production in its broadest sense' (Anon. [1825], 3–4). For *Le Producteur,* journalism is 'that new means of communication created by modern societies, [which] has never been more necessary that it is today, and never have those whose task it is to wield this powerful tool had a more noble mission to accomplish' (Anon. [1825], 5). Before it has even come into existence, the mass press is assigned a social task – one for the masses – to be *useful* by spreading understanding, disseminating knowledge in its multiplicity of forms, and enabling and promoting *production* in every sense of the word.

No one was better placed to take up this civilizing mission that Girardin, whose popular *Journal des connaissances utiles* [*Journal of Useful Information*] was ostensibly published by a 'National Society for Intellectual Emancipation', which received considerable praise from various individuals and newspapers, all of which was duly reprinted in the *Journal* itself, recycled for the further edification of its readership: 'The society... publishes a newspaper, whose low price should solve the problem of the moral, political, agricultural and industrial education of the masses in both city and country' (Girardin 1832a, 2).[5] Indeed, the Society would have its members join in the crusade, asking that 'the noble citizens who would accept the title of "correspondent" do their best to sell at least five subscriptions' (Girardin 1832a, 2). And if one's social conscience were not sufficient motivation, 'A SILVER MEDAL is to be awarded to those "correspondent members" who can obtain twenty-five subscriptions in their local area' (Girardin 1832a, 2).

Thus, with the liberalization of the press ushered in by the July Revolution and educational reform underway, the *Journal des connaissances utiles* solemnly assigns itself the following goals:

> To maintain and improve social order by simultaneously teaching each class: ITS RIGHTS, ITS DUTIES, ITS INTERESTS. To reinvigorate national industry by publicising discoveries in the arts and sciences, new inventions, and economic developments – in short all USEFUL things – so that they will be USED.
>
> (Girardin 1832b, 241)

The *new* is to be moved as quickly as possible from curiosity to commodity, and the newspaper is to serve as an educational intermediary between producer and consumer. Like today's *Consumer Reports,* the

Journal des connaissances utiles both encouraged consumption and sought to educate the consumer. For Girardin this incipient consumerism is inseparable from the social question:

> since the masses are the greatest [potential] consumers, all political economy, all questions of social well-being could well be summed up in these shopkeeper's words: SELL CHEAP, SELL A LOT; – SELL A LOT, SELL CHEAP.
>
> (Girardin 1832b, 238)

This slogan, the very soul of mass marketing, will remain Girardin's throughout his long career, and under its aegis *La Presse* will be founded four years later. Social wellbeing is tied to production, production depends upon consumption, and consumption can spread among the masses only if their own productivity rises and the cost of the products they wish to consume falls. In Girardin's epigrammatic remarks is inscribed the circular logic (some would call it the myopia) of the utopian socialists' emphasis on *production*, on 'trickle-down' economics *avant la lettre*. Internal, political strife and partisan bickering have stifled the industrial revolution in France, and only the harmonization of competing ideologies will foster a productive society.

The success of the *Journal des connaissances utiles* is due to the application of these 'economic doctrines', as Girardin calls them. Already he is laying the groundwork for *La Presse; publicité* (advertising) will replace *polémique*, and the newspaper of the future will transcend party differences (or *couleurs*). In this first article for *La Presse*, he will echo Saint-Simon, making plain how closely linked his paper is – indeed how entirely it depends upon – a leveling of ideological differences:

> In creating this paper our thought was that the numerous differences of opinion had to do with the diversity of interests and points of view, and that these differences were more apparent than real, superficially pervasive but in depth almost non-existent. A close analysis and a serious study quickly told us that the most eminent minds of our time have built upon a foundation of almost identical ideas. By probing, we have uncovered the same layers – as we will later reveal – under the edifices of their speeches and writings.
>
> Socially there is little more than a single idea in men's minds: *The most happiness for the greatest number possible*. Our intention is not to reform the press as it exists today – which would be futile – but to gather together the brightest lights shining from every direction, and

> which together will illuminate why we have been stagnating for the last six years [since the Revolution of 1830], and why we must abandon our current position for a normal, progressive life. We do not intend to redo what the old press has done, but to do different things and in a different way.
>
> (Girardin 1836a, 1; emphasis mine)

I quote Girardin at length to give a sense of his talent for the rhetorical flourish and his tendency to speak in generalities of little detectable substance. His contemporary, the novelist, sometime collaborator and would-be newspaper magnate himself, Honoré de Balzac, may well have had Girardin in mind when, in his 'Monographie de la presse parisienne' of 1843, he describes the 'Rienologue' ['Nothingologist'] or 'Vulgarisateur' ['Popularizer'] as follows:

> The *vulgarisateur* takes a derivative notion, sloshes it around in a wash-bucket of clichés and then mechanically rattles off a horrid philosophico-literary hodge-podge that goes on for several pages. The page *seems* to be full, it *seems* to contain some ideas, but when a truly educated man takes a whiff, all he can smell is the odor of an empty cellar. It's deep but there's nothing there: the mind has been extinguished like a candle carried into a tunnel deprived of oxygen. Yet the *Rienologue* is the god of today's bourgeoisie.[6]
>
> (Balzac 1991, 60)

Still, we must surely view Girardin's 'invention' of the penny press in France as part of the larger intellectual moment of the July Monarchy, which witnessed an extraordinary (if temporary) development of free expression in the press, industrialization and, above all, a preoccupation among intellectuals with the social question so pithily stated by Girardin in his first issue of *La Presse*. His paper is the first which, instead of confirming or forming the specific opinions of a given political party or tendency, will *publicize*, and in two senses: *La Presse* will convey to the public all that is new(s), and will in turn help to create a newspaper public dependent not upon the political contexualization of current events (the role of the 'old' press), but rather upon the apolitical decontextualization and commodification of those same events.

At this time the French word *publicité* meant not just advertising but publicity, information. While it is suggestive to reflect upon the tendency to blur such distinctions, Girardin left no doubt what *he* intended. Later in life, and with a characteristic lack of modesty, the great journalist – perhaps hoping to rival Pascal and La Rochefoucauld in one fell swoop – allowed his *Pensées et maximes* to be excerpted from his journalistic writings. Under the rubric 'Publicité', he bluntly states: 'Without commercial advertising, there can be no true progress, no serious competition, no progress of use to the masses', and 'Where advertising does not flourish, freedom cannot reign' (Girardin 1867, 612). Only in this context can we understand what Girardin meant when, under the July Monarchy, he labelled himself a 'progressive conservative'. Sainte-Beuve was more direct:

> He is one of the leaders of this modern society... that is out for itself alone, for its own growth, for its own progress, for its own expansion in every direction and for its own well-being. This new society is for peace and all that it brings, and lobbies neither for a reversal of the 1815 treaties [imposed upon France following Napoleon's final defeat at Waterloo] nor for a change in the border of the Rhine, but rather, and above all, for the development of the railroads.
>
> (quoted in Morienval 1934, 85–6)

Elsewhere I have discussed the surprising parallels between *La Presse* and that strain of French Romanticism that will evolve into 'art for art's sake'.[7] The aesthete's apolitical devotion to form and craftmanship is strangely analogous – at once mirror and negative image – to Girardin's vision of a colourless newspaper serving no master but an *objective truth* in the service of social progress. Just as certain optical illusions seem to change shape with time, however, Girardin's self-effacing, colourless newspaper undergoes a curious transformation; in its effort to avoid becoming – as were the traditional papers – an extension of something beyond itself (of *representing* someone or something else), the text of the paper is unexpectedly, splendidly autonomous. Its purely referential project – presenting the news – is ultimately self-referential, for in disposing of a known political *colour* that would filter or mediate information in order to promote a political agenda, it is presented *for its own sake*, as a decontextualized commodity to be consumed along with the other 'articles' listed in the advertising section on the paper's final page or pages.

It is all too easy for us to forget that in Girardin's (and many of the early socialists') view there existed no fundamental contradiction between increased industrialization and productivity on the one hand, and social harmony and even equality on the other. Indeed, it was widely held that the first would of necessity engender the second; the genius of Girardin is revealed in his relentlessly practical application of socialist ideals to the business of making and selling the news.

In the end, however, *La Presse* necessarily backed away from the social question and the utopian underpinnings of its own utilitarianism. Long on an amicable footing with utopian thinkers and journals, Girardin was accused by the Fourierist *Phalange* of 1837 of offering vague, impracticable solutions to social problems. On 27 July, Girardin allowed publication of a *feuilleton* by Eugène Pelletan that ridiculed the utopian Charles Fourier, and *La Phalange* responded in its next issue with an appeal to public opinion and the government to put an end to 'les crimes de la presse'; the pun was doubtless intended (Aug. 1837). Within a few days Charles Fourier himself was dead, and while it would certainly be hyperbole to say that Girardin had killed him – as he had rival editor, Armand Carrel, in a pistol duel a year earlier, often cited as a defining moment in the transition from the old political to the new information-based press – the event is just as heavily-laden with symbolism; it might be more accurate to say that Girardin had slain the Fourier within, for the social question would never again regain the urgency, real or feigned, that it had in the early years of the *Journal des connaissances utiles*.

While he was clearly not alone, Girardin's master coup of popularization was to take the potentially subversive utopian discourse of Saint-Simon, Fourier, and others – for it was certainly 'in the air' – and domesticate and commodify it by harnessing it at the most opportune moment to advertising, industry, and a growing desire for information on the part of a class of newly-literate citizens (all of which, of course, were then further promoted in Girardin's papers), including the entertainment offered by the important new literary genre he helped to introduce and promote in that same year of 1836: the serial novel (*roman-feuilleton*). The social question and cult of objectivity – part of the Enlightenment legacy of the *philosophes* – fuse in a project to create the first mass newspaper in France, a press for the masses that will at once respond to public opinion and help to create (by enlightening) it. As Terdiman has pointed out, Girardin (and later, Moïse Millaud)

> turned the daily into a commodity in terms of both its means of circulation and its content...the newspaper was adapted to the implicit needs of commerce. It *became* the institutional incarnation of the dominant discourse...for which...it could be taken – and indeed was taken – as a characteristic figure.
>
> (Terdiman 1985, 129)

Of more interest to us here, however, is the extent to which a potentially subversive political discourse was domesticated and *made to serve* the interests of a stratified bourgeois capitalism devoid of – and we might say hostile to – any efforts to resolve the 'social question'. *La Presse* owed allegiance to no one but itself; how splendidly appropriate, then, was its very name! Unlike *Le Constitutionnel* or *Le Bon sens* [*Common Sense*], it referred not to a political tendency or an abstraction, but to itself, to the materiality of its own (re)*production*. The paper that would represent the world of reality suggests by its very title a self-enclosed, self-referential text whose objectification (in both senses of the word) will recommend it as an alternative to the highly contextualized, subjective discourse of the traditional political press.

In conclusion it is important to place the introduction of the apolitical, mass-market newspaper in perspective. Although Girardin's experiment generated enormous animosity from the traditional press, it did not attract as many subscribers as he had hoped. It may be, as Terdiman suggests, that such discursive innovation could not yet be absorbed by the culture: in 1863 Moïse Millaud launched *Le Petit Journal* along similar lines, and it became the first wildly successful mass daily, rising from 38,000 subscribers in that year to 600,000 in 1880 (Terdiman 1985, 132–3).

What Terdiman does not mention is the extent to which Girardin's ostensibly neutral or colourless *La Presse* was indebted to many of the utilitarian tenets of utopian socialism, which in turn he 'neutralized' with great skill. The very notion of an 'objective' or 'colourless' press is obviously grounded in a specific politics, one that identifies the economic expansion of a nation with the socio-economic empowerment of the masses. While historians of the French press, like Thomas Ferenczi in his recent *L'invention du journalisme en France* (1993), continue to view its evolution within a largely political and literary framework, it is clear that Girardin's conception of the daily as an impartial purveyor of information *in the interest of the common good* – a 'good' generated by consumer capitalism – is the very model of all mass media to this day. And so began the process – viewed favourably by Girardin but far more often with distaste since then – of the *américanisation* of the French press and its public.[8] At the founding moment of the mass press he grasped what attentive students like Neil Postman have associated primarily with the 'peek-a-boo world' of electronic media,[9] where visual 'objects' appear briefly in the viewer's perceptual field, and just as quickly drop out of sight – and mind: 'The true name of the press', Girardin always maintained, 'is oblivion' (quoted in Morienval 1934, 86).

Notes

1. The authoritative text on the history of the French press remains Claude Bellanger, *et al.* Of particular interest in this context is *Tome II*, 11–26 (on the evolution of printing techniques) and 114–24 (on the *presse à bon marché*). For a schematic overview of these same questions in the wider context of Europe and North America, see Pierre Albert, esp. 32–54. For a brief, highly accessible and richly illustrated history of the press, see Jacques Wolgensinger, esp. 49–85. Finally, for an account of the press in the context of French literary history, see Max Milner and Claude Pichois, 32–77.
2. For an exhaustive account, see H. J. Hunt.
3. In addition to venerable biographies by Jean Morienval and Maurice Reclus, see Pierre Pellissier.
4. All translations from the French are my own.
5. As with many nineteenth-century periodicals, *Le Journal des connaissances utiles* was collected and published yearly in bound volumes. Often such volumes are more easily accessible to the scholar, and in the case of the *Jdcu* as well as the Fourierist *Phalange* I am referring to the page numbers of these volumes rather than to microfilms of the original publications.
6. For a fascinating account of Balzac's journalism and his relationship with Girardin, see Roland Chollet.
7. 'Utopia Commodified: Utilitarianism, Aestheticism, and the *presse à bon marché*, in Dean de la Motte and Jeannene Przyblyski, *Making the News: Modernity and the Mass Press in Nineteenth-Century France* (Amherst, MA, 1999) 141–59.
8. Although de Tocqueville's *De la démocratie en Amérique* (1835; 1840) comments significantly on the role of the press in American political life, and Girardin frequently holds up the anglo-saxon 'commercial' press as a progressive ideal, in the United States the mass press dates from the same period as Girardin's *La Presse*, with the establishment of the New York *Sun* (1833), *Herald* (1835), and *Tribune* (1841). In England, the penny press did not develop for almost another twenty years, with the introduction of the *Daily Telegraph* in 1855 (Albert 1970, 50–2).
9. Postman's brilliant analysis does, in fact, briefly discuss the early role of the American penny press (Postman 1985, 66) in the systematic decontextualization of news that will be accelerated by telegraphy, railroads, and then of course photography, cinema, radio and television. I would only venture that the 'age of exposition' – where the printing press was sovereign – was far less impervious to distortion, irrelevance and even sheer silliness that Postman's somewhat nostalgic interpretation will concede. It is interesting to note that *exposition* – from exhibiting or showing – has a further meaning in French, as in the *Expositions Universelles* (World Fairs) which in the late nineteenth century showcased the very inventions that would shape daily life and perception in the early twentieth century. A *rapprochement* of the two meanings was recently

attempted by Philippe Hamon, in his *Expositions: Literature and Architecture in Nineteenth-Century France*, Berkeley, CA 1992.

Works cited

Albert, Pierre. *Histoire de la presse*. Paris, 1970 [corrected 1996 edition].
Allen, James Smith. *In the Public Eye: A History of Reading in Modern France, 1800–1940*. Princeton, NJ, 1991.
Anon. *Physiologie de la presse*. Paris, 1841.
——.'Question adressée au journal LA PRESSE', *La Phalange* 1837 (annual volume format), 441–2.
——. *Le Producteur* 1 (1825).
Balzac, Honoré de. 'Monographie de la presse parisienne' (1843), *Les Journalistes*. Paris: 1991, 13–144.
Bellanger, Claude, *et al. Histoire générale de la presse française: tome II: de 1815 à 1871*. Paris, 1969.
Chollet, Roland. *Balzac Journaliste: le tournant de 1830*. Paris, 1983.
Ferenczi, Thomas. *L'invention du journalisme en France*. Paris, 1993.
Girardin, Emile de *Pensées et maximes extraites des oeuvres de M. Emile de Girardin* par Albert Hetrel. Paris, 1867.
——. No title, *La Presse* 1 (1, July 1836a), 1–2.
——. 'France', *La Presse* 38 (13, Aug. 1836b), 1.
——. *Le Journal des connaissances utiles* II, 1 (1832a).
——. *Le Journal des connaissances utiles* II, 9 (1832b).
Hamon, Philippe. *Expositions: Literature and Architecture in Nineteenth-Century France*. Berkeley, CA, 1992.
Hunt, H. J. *Le Socialisme et le romantisme en France: Etude de la presse socialiste de 1830 à 1848*. Oxford, 1935.
Milner, Max, et Claude Pichois. *Histoire de la littérature française: de Chateaubriand à Baudelaire*. Paris, 1985.
Morienval, Jean. *Les Créateurs de la grande presse*. Paris, 1934.
Pellissier, Pierre. *Emile de Girardin: Prince de la presse*. Paris, 1985.
Postman, Neil. *Amusing Ourselves to Death: Public Discourse in the Age of Show Business*. New York, 1985.
Reclus, Maurice. *Emile de Girardin: le Créateur de la presse moderne*. Paris, 1934.
Saint-Simon, Claude Henri de. *L'Organisateur* 1 (1819).
Terdiman, Richard. *Discourse/Counter-Discourse: The Theory and Practice of Symbolic Resistance in Nineteenth-Century France*. Ithaca, NY, 1985.
Wolgensinger, Jacques. *La Grande aventure de la presse*. Paris, 1989.

22
The Virtual Reading Communities of the *London Journal*, the *New York Ledger* and the *Australian Journal*

Toni Johnson-Woods

Just prior to Queen Victoria's ascension to the throne, the print industry was preparing for its greatest revolution: the creation of the first mass media. A variety of political, technological and cultural forces combined to foster a print medium that was both affordable and accessible to more people than ever in history. The new penny weekly journals sold millions of copies and were powerful cultural products, yet they have been relatively ignored. The public who embraced the cheap weekly journals of 1832 demonstrated that readers were willing to invest pennies in weekly reading material, but these publications aimed at enlightenment, education, 'relatively objective facts' (Anderson 1991, 54). Some readers sought entertainment not improvement. The *feuilletons* in the daily French papers demonstrated that leisure reading material boosted circulations. Soon proprietors worldwide learned the economic imperative that there was money to be made from printing publications that contained entertaining literature. Three journals, all similar in format and content were to capitalise on the trend for popular miscellany reading – they outsold all other periodicals in their own countries. In the London semi-detached, the New York brownstone, and the Ballarat canvas millions of readers shared the same reading culture and thus formed a 'virtual community' (Gellner 1987, 7) as they perused the *London Journal and Weekly Record of Literature, Science, and Art* (1845–1912), the *New York Ledger* (1855–1903), the *Australian Journal: A Weekly Record of Amusing and Instructive Literature, Science, and the Arts* (1865–1962). The periodical world was truly much smaller than the geographical one.

In 1845 an unemployed engraver, George Stiff, launched a periodical 'devoted to amusement and instruction of the people', the *London*

Journal. After a faltering start the *London Journal* secured a niche between the genteel *Family Herald* and the radical *Reynolds's Miscellany* and sold nearly half a million copies per week. At one stage it 'boasted the largest circulation in the English-speaking world' (Noel 1954, 62). Indeed, it provided much of the material for the other phenomenally successful periodical of the century: the *New York Ledger*. At the end of 1850, an enterprising Scottish-Irishman Robert Bonner, paid $900 for a flagging trade periodical that he had been printing in his small office. He changed the name, enlarged it, and slowly, in order not to alienate current subscribers, increased the literary content, largely by pirating serials from the *London Journal*. The *Ledger* at $2.00 per year or 4 cents per copy outsold its competitors by between 300,000 and 400,000 copies each week (*Harper's* Jan. 1860, 7) and made Bonner one of the 'richest and most famous men in the United States' (Noel 1954, 56). Ten years later in the colony of Victoria, Clarson Massina and Company launched another *London Journal* lookalike: the *Australian Journal* which was '[b]ased on the plan of the English Weeklies, such as the *Family Herald, London Journal, Cassell's Papers* &c' (*Walch's Literary Intelligencer* 2 Sept. 1865, 1). During its first years, the *Australian Journal* positioned itself as a local publication, its motto was 'colonial literature for colonial readers', but imported serials from the *Ledger*, generally the same serials which also appeared in the *London Journal*, often usurped the local material. They deserved their appellation 'universal' because the *London Journal*'s success spawned imitators throughout the world (Collins 1858, 217).

In the mid nineteenth century, the new genre of literature had infiltrated in such enormous quantities that both Wilkie Collins and Margaret Oliphant devoted essays to them in August 1858:

> Day after day, and week after week, the mysterious publications haunted my walks, go where I might... I left London and travelled about England... There they were in every town, large or small. I saw them in fruit-shops, in oyster-shops, in lollypop-shops. Villages even – picturesque, strong-smelling villages – were not free from them... I made acquaintance with one of them among the deserts of West Cornwall, with another in a populous thoroughfare of Whitechapel, with a third in a dreary little lost town at the north of Scotland. I went into a lovely county of South Wales; the modest railway had not penetrated to it, but the audacious picture quarto had found it out. Who could resist this perpetual, this inevitable, this magnificently unlimited appeal to notice and patronage...

> [T]he mysterious, the unfathomable, the universal...penny-novel Journals.
>
> (Collins 1858, 217)

Collins' tongue-in-cheek description of their ubiquity encapsulates the extent of the penetration of the penny weeklies. These publications were outstanding not only because of their profusion but because of the similarity. They all looked the same: 16 unbound quarto pages, most had a picture on the upper half of the front page and two or three columns of small print underneath (Collins 1858, 217). Another critic noted 'a strong family likeness pervades them all... You detect at a glance the tribe to which they belong...the temperament of some may be more lively of others somewhat more serious; but it is impossible to mistake the fact, that they all come from the same lineage' (A.A. 1859, 329). The lineage of course is cheap literature hawked in the streets and contrasts to the book-like shilling monthlies. Thus they developed a penny periodical 'look' which was not only adopted by many English publications but also the *Ledger* and the *Australian Journal*. As they hung in shop windows, appearances played an important part in their marketing strategy, 'marking' them as a certain type of periodical, a signifier for shop-gazing consumers. The external appearance was a simulacrum for that which was inside, again varying little among the three.

As the appearances were uniform, so too were the contents. Almost without exception penny publications contained two serials, a few short stories, miscellaneous items of interest (such as recipes, curious newsy items etc), snippets of scientific information, literary gossip and the ubiquitous 'Answers to Correspondents'. The main attraction was the continuing serial that started on the first page, under the illustration (which may or may not be related to the story). A popular serial realized immediate circulation increase:

> J. F. Smith's '"Minnigrey"...raised the circulation of the *London Journal* to between three and four hundred thousand... "Woman and Her Master"...was so successful...that it raised the circulation of the *London Journal* to the greatest number it has ever reached – namely 500,000...Sir Walter Scott...failed to excite the enthusiasm of its readers, and it was found that the circulation had gradually fallen to about 250,000.
>
> (Masson 1866, 102–3)

thus proving that readers were discriminating in their reading matter, even if the vendors found little to differentiate among them: '[s]ometimes I sells more of one, and sometimes I sells more of another. Take 'em all the year round, and there ain't a pin, as I knows of, to choose between 'em' (Collins 1858, 218). This was not peculiar to English periodicals only; apparently the cultural currency was within the genre and crossed national boundaries. The same serials appeared in New York and Australia as Mary Noel wrote, 'the serial contents of the *London Journal* could scarcely be separated from [a] United States story-paper' (Noel 1954, 25). The *Australian Journal*, as the latecomer to the field, had a wide variety of serials from which to chose, yet it remained true to its penny weekly forebears and preferred those which had appeared in the *Ledger/London Journal*.

The universality of contents reflects the penny reader's expectations. First, material had to be moral. Despite the undeserved reputation for being rather scurrilous, '[the *London Journal*] exhibit[s] a morality not quite diabolical' (Dixon 1847). Such claims were later refuted by critics who actually read them, and one recalled:

> About a dozen or fifteen years ago, when people began to hear of their [penny journals] enormous sales, some enquiries were made into their character. It was then discovered that our lower classes were being entertained with tales of seduction, adultery, forgery, and murder. Since then the subject has been allowed to drop, and an impression still remains that the penny journals run riot in pictures of debauchery and crime, many of them showing what infamous lives the higher classes of English society lead . . . But these twelve or fifteen years have wrought a change, and anyone who will look at such a periodical as the *London Journal* will be much struck, perhaps a little abashed, by the strong moral tone that pervades the writing.
>
> (Masson. 1866, 97)

Even cautious Margaret Oliphant agreed (Oliphant 1858, 209). The controlling agents were the editors who seemed to adhere to the following principle as described by Robert Bonner, editor of the *New York Ledger*:

> I pictured to myself an old lady in Westchester with three daughters, aged about twenty, sixteen and twelve, respectively. Of an evening they come home from a prayer meeting, and not being sleepy, the mother takes up the *Ledger* and reads aloud to the girls. From the day I got the *Ledger* to the present time there has never appeared one line

> which the old lady in Westchester County would not like to read to her daughters.
>
> (Letter, Bonner Papers 1 April 1884)

Bonner castigated his writers who wrote inappropriate material. Editors of the *Australian Journal* also shied away from controversial material. Would-be contributors were told that there was 'an objectionable Frenchiness' to his/her submission, and that the journal would not accept 'sectarian or bigoted' material (*Australian Journal* April 1895 463). As one writer noted:

> Virtue was always rewarded and vice always punished; the lower you descended in search of the dramatic element, the more triumphant virtue was, and the worse punishment awaited the vicious. Strange as it might appear, in the regular right down bad neighbourhoods the morals were more strict, and villain more deeply execrated, the virtue of the heroine above price, more intently admired, and her welfare watched over with a more anxious solicitude.
>
> (Robinson 1858, 230)

The tone of the serials might be bland, but the genre was sensational.

The sensational story was the perfect form for the serial writer. The event-heavy narratives, incredible coincidences, and large casts of characters maintained reader interest by imposing many smaller climaxes, all of which kept readers eager for the next instalment. The fact that the most recurrent stories were aristocratic romances set in Europe says that worldwide readers sought 'escape' in their leisure reading. Such formulaic fiction is associated 'with times of relaxation, entertainment and escape' (Cawelti 1976, 1) because it places few demands on the reader. As many of the stories were written piecemeal, readers could also contribute to the development of a story by writing to the editors expressing their dissatisfaction, and sometimes this affected the outcome of a story; on occasion editors appealed to the public to send in their solutions to a hero/heroine's dilemma or to predict the outcome of a story. Given that the preferred genre was the sensational story, one would anticipate that Wilkie Collins and Mary Elizabeth Braddon would be the doyens of the penny press. But Braddon was 'by no means a star of the first magnitude in the world of the penny journals,' because, as one critic postulated, 'Miss Braddon does not moralize enough to suit the taste of the penny people' (Anon. 1866, 97). Those who were moral mavens then ran the risk of having their work appear anywhere, without their permission.

Publishers pirated or purchased material from any source – stories from New York, fashion from France, information from Australia filled the democratic pages of the mid nineteenth-century miscellany. The Americans stole from the French and the English; the English stole from the French and the Americans; and the French, well the French had no need to steal from anyone (to paraphrase Noel 1954, 6). Reynolds, has the dubious fame of being the first English writer to be pirated in America: in 1844 the *Mysteries of London* (unattributed and printed under another title) appeared in the *Yankee*. In 1855 the *London Journal* was running E.D.E.N. Southworth's 'The True and False Heiress' when Bonner pirated one instalment for the *New York Ledger*; soon he received an irate letter from T. B. Peterson, Southworth's publishers in Philadelphia, stating that Bonner had stolen a writer currently being published in America. This was not the first time she had been pirated by the *London Journal*; 'The Curse of Clifton', which first appeared in the *Saturday Evening Post*, was reworked with 'substituted English places and names' and appeared as 'Brandon of Brandon' (Mott 1947, 140). Needless to say, Bonner could not let such a worthy writer pass by and opened his chequebook. Southworth joined the *Ledger* stable of writers. Pirating was a cheap alternative to paying writers.

It is no wonder that proprietors were keen to procure tried and true writers for their publications, legally or otherwise. The potential audience was in the millions:

> I have no estimate of the circulation of similar papers in America, Australasia, and South Asia, but seeing that these countries possess twice as many inhabitants as the 'old country,' we may safely assume that their newspapers are at least twice as numerous and have more than twice as many subscribers. Assuming, further, that every copy is ready by two individuals, we can reckon a total of eighteen million readers of serial-running newspapers, of whom six millions belong to the United Kingdom and twelve millions to the United States and the Colonies.
>
> (Westall 1890, 79)

One way to sustain circulation figures was to publish a successful serial because it attracted new readers. *Le Constitutionnel* paid Eugene Sue 100,000 francs for *Le Juif Errant* (*The Wandering Jew*); circulation increased from 3500 to 25,000 realizing a profit of *circa* 600,000 francs. Even after the serial was finished, the numbers remained constant, demonstrating that reader loyalty outlasted the life of a serial. As one American newspaper found to its delight:

> A considerable sum of money was offered to the reader for the most successful forecast of the termination [of 'The Postmaster of Market Deignton']. In the case of the 'Chicago Record' 41,186 guesses or forecasts were received. The circulation of the paper increased 23 per cent; or 31,000 copies a day, and this increase was held permanently, after the publication of the story had ceased.
>
> (repr. in *Australian Town and Country* 10 Jan. 1897, 38)

The phenomenal sales of the popular press created an economic cycle of immense repercussions for the literary scene.

As can be seen by the preference for pirating the same authors, the same moralizing mavens appeared again and again. The following table demonstrates the incestuous nature of the three periodicals:

Table 2 Ranking of authors appearing regularly in the popular press

Ranking	*Australian Journal*	*London Journal*	*New York Ledger*
1	Harriet Lewis	Pierce Egan	Sylvanus Cobb Jnr
2	Leon Lewis Sylvanus Cobb Jnr	Harriet Lewis	Leon Lewis
3	E. Dupuy A. Rochefort	J. F. Smith	E.D.E.N. Southworth
4	E. D. E. N. Southworth	E. D. E. N. Southworth	E. Dupuy
5		Percy B. St John	A. Rochefort Harriet Lewis J. F. Smith

Across the board, writers for penny periodicals remained within that genre; Braddon is the only author I have found who moved up the literary feeding chain to the shilling monthlies. Attempts to offer 'better' reading material resulted in decreased circulation figures; for example, when the *London Journal* replaced Pierce Egan with Sir Walter Scott (*Ivanhoe* and *Kenilworth*) it was found that the circulation had gradually fallen to about 250,000 (Masson 1866). Egan was reinstalled and stayed until he died. The editor of the *Northern Daily Telegraph* summarized it best: tales which have been very popular and highly spoken of when published in volume form fall the flattest as newspaper serials (Westall 1890, 81). Cheap periodicals' weekly writers produced stories which conformed to readers' expectations, no matter what their country of origin.

The most consistently serialized author in the *London Journal*, the *New York Ledger* and the *Australian Journal* was the American Mrs Harriet Lewis. Twenty-three serials appeared in the *New York Ledger*, twenty-two in the *London Journal* between 1869–1878, and thirteen in the *Australian Journal*. Though the same stories, subtle changes were made to the titles to underscore the personality of the journal; first there was the feminized *London Journal* versions and then the de-imperialized *Australian Journal* versions.

Table 3 Feminized titles in the *London Journal*

New York Ledger	*London Journal*
A Dazzling Fraud	Hunted for Her Money
Buried Secrets	Two Husbands
A Daring Game	Neva's Three Lovers

Table 4 De-imperialized titles in the Australian Journal

New York Ledger	*London Journal*	*Australian Journal*
The Bailiff's Scheme	The Buried Legacy	The Buried Legacy
The Old Life's Shadows	Lord Thornhurst's Daughter	Old Life's Shadows
A Daring Game	Neva's Three Lovers	A Daring Game
The Lady of Kildare	The Lady of Kildare	The Rival Claimants
The Lord of Strathmore	Found Guilty	Found Guilty

Lewis's trans-Atlantic publications gesture to some systematic cooperation between the two periodicals: all of her serials appeared in London exactly three weeks after their New York debut. Most probably Lewis or Bonner had contracted an advance sheet arrangement such as the one E. D. E. N. Southworth had negotiated with the *London Journal* (Mott 1947, 141). Why Harriet Lewis was more published in Australia than E. D. E. N. Southworth, it is impossible to say, for they both wrote similar types of story; perhaps it was the fact the Harriet was already dead or perhaps the Australian readers still preferred stories about 'home'.

Harriet Lewis might have dominated the *London Journal*, but her husband, Leon Lewis, reached Australian shores before she did. His tales did not tempt the *London Journal* or any other English periodicals; indeed none of the popular male writers of the *New York Ledger*, such as Leon Lewis, Colonel Alfred Rochefort, or Sylvanus Cobb Jnr, found favour in the *London Journal*. Nevertheless they enjoyed huge success in the USA: Cobb was the 'most consistently read novelist of his time' (Hart 1950, 99), whose 'Gunmaker of Moscow' was, according to one contemporary source, second only to *Uncle Tom's Cabin* in popularity:

> Nine out of every ten (nay we may say 99 out of every 100) persons prefer the stories of Sylvanus Cobb Jr., to the novel productions of Miss Mulock [*sic*] and George Elliott [*sic*]. *The Hidden Hand* and 'The Gunmaker of Moscow' are far more universally read than *John Halifax* and *The Mill on the Floss*. Even the great Bulwer is voted a bore.
>
> (*National Quarterly Review* Dec. 1860, 33)

That male adventures crossed the Pacific first suggests that frontier fiction appealed during the early settlement of the colonies as if to document, fictionally, the settling of a new country. Domestic stories and urban tales came later. Also the United States' early history of penal servitude, expansion and exploration more closely mirrored Australian ideals of wronged convicts, heroic bushmen and earnest settlers.

Not all male writers were popular in Australia. John Frederick Smith, according to a contemporary report in *Cassell's*, 'met with the most unequivocal success, not only in England but the Colonies and United States of America', and therefore was not 'cribbed, cabined, and confined by national predilections' (*Cassell's Illustrated Family Paper*, 22 May 1858, 386). Yet Australia ignored Smith and Pierce Egan and Percy B. St John, much as Britain ignored US frontier writers. Most imported works by male novelists in the *Australian Journal* were adventure tales. Real men wrote tales such as 'The Russian Spy; or Vladimeer the Nihilist' (Alfred Rochefort) and 'The Trapper's Last Trial' (Leon Lewis), not 'The Lost Wife' (J. F. Smith) or 'The Pet Heiress' (Percy B. St John), let alone 'Love Me; Leave Me Not' (Pierce Egan).

As the *Australian Journal* entered the field last, it had a distinct advantage over its predecessors: it could see what had worked before and thus could capitalize on other's mistakes by selecting only the most popular authors. Considering the cultural heritage, it is surprising to find that the United States and not England provided most of the imported material, but if one considers the setting of the story, then the balance is tipped in England's favour. This may have been for economic reasons; purchasing/pirating from the United States was cheaper or easier because of its closer proximity and faster mails. How much of the imported material was appropriated is uncertain. The *Australian Journal* had to defend itself against charges of plagiarism over Henry Stoddard's 'The Two Secrets' in June 1873 and wrote that the story had been

> specially furnished to the *AUSTRALIAN JOURNAL* by Mr. Stoddard.... We deem this statement necessary, in consequence of a most malignant allusion recently made, by the editor of a Melbourne evening paper, to the United States stories printed in these pages.
> (June 1873, 461)

The incestuous nature of publishing is highlighted in some of the above mentioned examples of piracy – Southworth taken from the *Saturday Evening Post* by the *London Journal* and then from the *London Journal* by the *New York Ledger.* There were also appropriations of themes, such as the 'Mysteries of...'; hot on the heels of the *Mysteries of Paris* (1842) came the *Mysteries of Berlin* (1843) and the *Mysteries of New York* (1843), and the *Mysteries of London* (weekly parts Oct. 1844 – Sept. 1846), culminating in that perennial favourite, the *Mysteries of Fitchburg, Massachusetts* (1844) (mentioned in Noel 1954, 21).

Another of the most distinguishing features of the penny weeklies was their almost simultaneous change in format and their consequent demise. The demise of the six-month serial in the 1880s coincides with the disappearance of the three-decker novel and marks the beginning of the end for the penny miscellany. Both the *New York Ledger* and the *London Journal* tried to regain readers by reissuing their most popular serials from earlier years, decreasing scientific, newsy items, and increasing the number of illustrations. Thus the miscellanies developed into 'novel-papers', to use Mary Noel's term. In 1880 the *London Journal* announced 'a complete novel every week' with special novel supplements at Summer, Christmas and other special occasions. The *Australian Journal* fared better than its overseas counterparts, no doubt because it adapted best. It returned to publishing local authors and became a primary vehicle for new talent. But the age of George Newnes's *Tit-Bits* had arrived, and with it the demise of the *New York Ledger,* closed in 1898, and the *London Journal,* 1912. Only the *Australian Journal* survived until 1963, probably because it had evolved into a literary magazine.

Why did the nature of journals change? Like many cultural phenomena, a multitude of factors no doubt combined to change the market. Previously where the miscellany had sufficed for the family, a number of specialized periodicals targeted specific audiences: girls' and boys' periodicals, women's journals, Answers to Correspondents. Robert Bonner blamed the Sunday supplement, but these did not contain the amount of fiction readers absorbed in these weekly magazines and certainly they had been in the United Kingdom for many years. Classics and reprint novels could be purchased for sixpence, one quarter of their previous price; dime novels and colonial libraries jostled for the literary pennies. George Manville Fenn considered that the journals themselves were ultimately to blame:

> [the disappearance of the three-volume novel] had to come, and the introduction and success of cheap... good periodical literature was mainly responsible for its advent... The *LONDON JOURNAL*, and others like it, seriously competed with the novel. A new class of readers was arising, thanks to the spread of education and the increasing pressure of the times, a class which wanted its fiction short and sharp, and as time went on the one-volume novel was called into being (Fenn, 1990, 124).

Perhaps the journals themselves had educated the public into reading and demanding shorter and shorter pieces of fiction. After all, the weekly instalments had maintained them; now their public wanted their weekly fiction brought to a satisfactory conclusion in a shorter period of time. The economics of the circulating libraries had kept the three-decker afloat and had pressured writers into bloating their narratives in order to gain three times the value from one piece of fiction. Whatever the cause/s, the serialization of long novels and the day of the novel newspaper has passed for the moment. Today the penny miscellany remains a nineteenth-century aberration, but one which had an international textual community to validate it.

Works cited

A. A. 'Cheap Literature', *British Quarterly Review* 1 (April 1859), 313–45.

Altick, Richard. *The English Common Reader: a Social History of the Mass Reading Public*. Chicago, 1957.

Anderson, Patricia. *The Printed Image and the Transformation of Popular Culture 1790–1860*. Oxford, 1991.

Australian Journal. London.

Australian Town and Country. Sydney.

Barton, G. B. *Literature in New South Wales*. Sydney, 1866.

Bonner Papers. New York Public Library.

The British Quarterly Review. London.

Cassell's Magazine. London.

Cawelti, John G. *Adventure, Mystery and Romance: Formula Stories as Art and Popular Culture*. Chicago and London, 1976.
Collins, Wilkie. 'The Unknown Public', *Household Words* (21 August 1858), 217–22.
Dalziel, Margaret. *Popular Fiction 100 Years Ago: An Unexplored Tract of Literary History*. London, 1957.
Dixon, Hepworth. 'Literature of the Lower Orders', *Daily News* (9 November 1847).
Fenn, Manville. 'Changes I Have Seen in the Literary World', *Pearson's Weekly. A Checklist of Fiction 1890–39*, ed. George Locke. London, 1990, 123–5.
Gellner, Ernest. *Culture, Identity, and Politics*. Cambridge, 1987.
Haining, Peter (ed.). *The Penny Dreadful, or Strange, Horrid and Sensational Tales*. London, 1975.
Harper's Magazine. New York.
Hart, James D. *The Popular Book: A History of America's Literary Taste*. New York, 1950.
[Masson, David]. 'Penny Novels', *Macmillan's Magazine* (June 1866), 96–105.
James, Louis. *Print and the People 1819–1851*. London, 1976.
Mott, Frank Luther. *Golden Multitudes: The Story of Bestsellers in the United States*. New York, 1947.
National Quarterly Review. New York.
Noel, Mary. *Villains Galore: The Heyday of the Popular Story Weekly*. New York, 1954.
Oliphant, Margaret. 'The Byways of Literature. Reading for the Million', *Blackwood's Magazine* (August 1858) 200–16.
Robinson, Emma. *Mauleverer's Divorce*. London, 1858.
Walch's Literary Intelligencier. Hobart.
Westall, William. 'Newspaper Fiction', *Lippincotts* 45 (May 1890), 77–88.

Index